Statistical Physics
of Macromolecules

BOOKS IN SERIES

Statistical Physics
of Macromolecules

ALEXANDER YU. GROSBERG

Institute of Chemical Physics
Russian Academy of Sciences
Moscow, Russia

ALEXEI R. KHOKHLOV

Physics Department
Moscow State University
Moscow, Russia

Translated by: Yuri A. Atanov

American Institute of Physics   **New York**

AIP Press
American Institute of Physics
500 Sunnyside Boulevard
Woodbury, NY 11797-2999

Library of Congress Cataloging-in-Publication Data
Grosberg, A. IU.
    [Statisticheskaia fizika makromolekul. English]
    Statistical physics of macromolecules / Alexander Yu. Grosberg,
Alexei R. Khokhlov; translated by Yuri, A. Atanov.
            p. cm. -- (AIP series in polymers and complex materials)
    Includes bibliographical references and index.
    ISBN 1-56396-071-0
    1. Polymers. I. Khokhlov, A. R. II. Title. III. Series.
QD381.8.G7613 1994
530.4' 13--dc20   94-6347
                                                              CIP

10   9   8   7   6   5   4   3   2   1

Contents

Editors' Foreword

Complex materials are multi-component systems such as polymers, colloidal particles, micelles, membranes, foams, etc.; they are distinguished from simple crystalline solids and simple liquids in that they generally possess molecular or structural length scales much greater than atomic. They often exhibit unusual properties or combinations of properties, such as light weight combined with strength, toughness combined with rigidity, fluidity combined with solid-like structure, or the ability to dissolve in both oil and water. Complex materials turn up as surfactants in soaps and facial creams, as lubricants on magnetic disks, as adhesives in medical supplies, as encapsulants in pharmaceuticals, etc. Polymers are ubiquitous as fibers, films, processing aids, rheology modifiers, and as structural or packaging elements.

Some recent additions to the list of complex fluids include liquid crystalline polymers, sheet-like polymers, polymers with non-linear optical properties, bicontinuous block copolymers and interpenetrating networks, advanced zeolites and molecular sieves made from mixtures of liquid crystals and silicates, and new and exotic chiral liquid crystalline phases.

The study of complex materials is highly interdisciplinary, and new findings are published in a bewildering range of journals and periodicals and in many scientific and engineering societies, including ones devoted to physics, chemistry, ceramics, plastics, material science, chemical engineering, and mechanical engineering. Experimental techniques used to study polymers and complex fluids include scattering of neutrons, x-rays, and light, surface-force measurements, cryomicroscopy, atomic force microscopy, magnetic resonance imaging, and many others. The aim of this series is to bring together the work of experts from different disciplines who are contributing to the growing area of polymers and complex materials.

It is anticipated that these volumes will help those who concentrate on one type of complex material to find insight from systematic expositions of other materials with analogous microstructural complexities, and to keep up with experimental and theoretical advances occurring in disciplines other than their own.

We believe that the AIP Series on Polymers and Complex Materials can help spread the knowledge of recent advances, and that experimentalists and theoreticians from many different disciplines will find much to learn from the upcoming volumes.

Ronald Larson
AT&T Bell Laboratories
Philip A. Pincus
University of California, Santa Barbara

Editors' Preface

It is a distinct pleasure to introduce the English version of the Grosberg and Khokhlov volume on the **Statistical Physics of Macromolecules** to the AIP Series on Polymers and Complex Fluids. Polymer physics has been experiencing an accelerated expansion of activity since the late 1960's. This development has been engendered by several factors including: the technical feasibility of neutron and photon scattering techniques to provide detailed microscopic probes of macromolecular structure, organization, and dynamics; the recognition that polymer statistical mechanics has strong analogies with the explosion of theoretical ideas in critical phenomena; the realization that polymers are essential ingredients in biological machinery. The theoretical aspects of modern macromolecular science have been greatly influenced by three "schools" inspired by P. G. de Gennes (Paris), S. F. Edwards (Cambridge), and I. M. Lifshitz (Moscow). There are already noteworthy books which have appeared from the first two groups. Thus it is appropriate that this text, authored by two of Lifshitz' most reknowned students, becomes readily available to English readers.

Some unique features of this volume which set it apart from other polymer physics texts include: significant treatment of subjects such as polyelectrolytes, coil-globule transitions, etc. Particularly noteworthy and timely is the extensive chapter on biopolymers. This editor hopes that this will provide some impetus for the two rather distinct communities, treating synthetic and naturally occurring macromolecules, to achieve a great commonality of viewpoint. This book will be a welcome addition to the macromolecular physics literature.

Ronald Larson
AT&T Bell Laboratories

Philip A. Pincus
University of California, Santa Barbara

Preface to English Edition

This book is devoted to polymers, which are both excellent materials for many technological purposes and the key element of biological machinery. The primary objective of writing this book, which appeared originally in Russian in 1989, was to cover the whole field of the physical theory of polymers. Beyond the importance for a broad range of applications, this theory is both interesting and beautiful from a purely physical point of view. The first aim of this book was thus to provide a basic introduction to the main concepts, methods and results of statistical physics of macromolecules for the beginners who plan to be active in this field. In this spirit, only firmly established methods and results are included. Our second aim was to provide sufficient knowledge for qualified experts in one of the related fields to understand modern scientific papers on statistical physics of macromolecules, and to give an idea of how to apply the methods of this field of science. In this sense the book is half-way between a textbook for graduate students and a scientific monograph. We hope that this approach, which appeals to the Russian speaking scientific community in polymer physics, physical chemistry and biophysics, both in Russia and abroad, appeals to the English speaking community too.

Certainly, "the characteristic time" of scientific development is very small in our century, and, since this book was originally written, many important scientific results have been achieved in statistical physics of macromolecules. Some of these results should at least be mentioned here, in particular:*

(i) Superstructures, domains, and microphase segregations of various architectures and physical natures become available experimentally and well understood theoretically, providing fascinating technical applications.

(ii) A new level of deeper understanding of heteropolymers and other disordered polymer systems is reached and the corresponding concepts are applied fruitfully for various biophysical issues, such as protein folding and others.

(iii) New achievements in mathematical theory of knots and links, and in experimental methods of molecular biology, brought new concepts and ideas to the field of polymer topology.

(iv) Many objects of very different non-polymer natures were fruitfully analyzed by means of polymer or polymer-like approach; in particular, flux lines

---

*We apologize that we do not mention names everywhere is this book.

xiii

in superconductors, dislocations, membranes and fluctuating crumpled surfaces ("2D polymers"), fluctuating manifolds of general dimensionality, including disordered ones, strings, rays of light in the media with multiple scattering, etc.

However, "if you write a book at the time moment T and try to review all the results known up to the time $T - t$, your book will become obsolete not later than at the time $T + t$: don't try to make t too small!" Remembering this principle, we made only several minor improvements and additions in the book for this English edition. We are indebted to A. M. Gutin and I. A. Nyrkova for suggesting to us some of the improvements and corrections. In particular, we have included the "ray optics" approximation for stretched chain, some details concerning the reptation model (like the reptation of star-branched macromolecules), some basic equations of the classical matrix theory of rotational isomers, etc. We have improved the discussion of basic mathematical methods in Sec. 6 and of viscosity of polymer solutions in Sec. 32, 33, etc. This comparatively small number of changes and alterations seems justified also since the material originally included proved to provide a reasonable basis for understanding new developments in the statistical physics of macromolecules, including the ones mentioned above.

Last, but not least, in this introduction for the English speaking reader we feel compelled to mention some historical or personal aspects. The work in the field of theoretical physics of polymers in the USSR in the fifties and sixties is associated with the name of Professor Mikhail Vladimirovich Volkenstein (1912–1992), and for the subsequent two decades this field developed in the atmosphere of the person of Professor Ilya Mikhailovich Lifshitz (1917–1982).

Certainly, this is not a good place for memoirs, and we will not describe the fascinating scientific atmosphere around I. M. Lifshitz, although we enjoyed being his students. But in this atmosphere I. M. Lifshitz's school of polymers was formed, and we would like to write a few words here about this.

When I. M. Lifshitz began to work on polymers in the sixties (being mainly interested in some biological applications) he was already one of the leading theoretical physicists in the USSR, and his results were very well recognized in solid state physics, electron theory of metals, and in the theory of disordered systems. I. M. Lifshitz's own way of doing theoretical physics was based on the absolute priority of physical understanding while still using advanced mathematics. Beyond the mathematical formalism and particular results, Lifshitz's works contained a very important general physical insight. As for statistical physics of macromolecules, he started with thinking about biology, and probably this is why the dramatic difference between expanded coil with giant flucutations and condensed globule with broken correlations was the cornerstone of his physical insight on polymers.

Many particular results were discovered independently by I. M. Lifshitz. For example, the quantum-mechanical analogy of polymer statistics, in which, compared to S. F. Edwards presentation, accents were shifted from path integral toward Schrödinger's equation with imaginary time. There were also many differences on a more technical level. For example, I. M. Lifshitz used integral operator $\hat{g}$ to describe what he called "linear memory' (clear reminiscence of disordered systems), i.e., chain-like connectivity of monomers. While very often $\hat{g} = 1 + (a^2/6)\Delta$, there are also a lot of other possibilities (for example, in the theory of liquid crystals etc.) and, moreover, the use of $\hat{g}$ is often helpful for physical understanding.

In writing this book we tried to preserve the "original flavor" of the Lifshitz approaches which turn out to be very successful in solving many problems of the physics of polymers and biopolymers.

Of course, this does not mean that this book deals only with the original results obtained by Lifshitz's school. We tried to cover the whole field. In particular, we have benefited from many ideas reviewed in the excellent monographs by P. G. de Gennes "Scaling Concepts in Polymer Physics" (Cornel University Press, NY, 1979) and by M. Doi and S. F. Edwards "Theory of Polymer Dynamics" (Academic Press, NY, 1986). However, the qualified reader will find that even these topics are presented from a slightly different point of view.

We thus hope that our work will be helpful to readers of different backgrounds, interests and areas—both in the scientific and the geographical sense of this word.

Preface

The statistical physics of macromolecules rouses much interest among specialists doing research in solid-state and condensed-matter physics, chemistry of high-molecular-weight compounds and polymer technology, biophysics, and molecular biology. Despite the major recent scientific advances, the statistical physics of macromolecules has hardly been touched on in the latest academic literature. This book is intended to make up this deficiency.

We do not assume that the reader is versed in polymer systems. At the same time, the amount of knowledge of physics and mathematics required to read this book does not generally transcend a typical university course. Thus, we expect that this book may be useful not only to future specialists in statistical physics of macromolecules but also to physicists, chemists, and biologists dealing with macromolecules and polymers.

As the book explores different polymer systems and can be of interest to diverse groups of readers, we have tried to facilitate efficient reading of its individual chapters. For example, to comprehend Sec. 29 on polymer networks or Secs. 37–44 on biopolymers, one needs only very limited data from the foregoing sections, which can be found easily by the references given in the text.

To make reading easy, each subsection title is patterned as an abstract or a synopsis of basic ideas and results. A more qualified reader thus can skip familiar stretches of text. A few subsections marked with an asterisk call for more advanced knowledge of physics, mathematics, or both.

Typically, in polymers as well as other complex physical systems, the specific values of some definite quantities are less interesting to an investigator compared with their dependence on some large or small parameters characterizing the system. In such cases, we use the following widely adopted notation. If N denotes a large parameter and ε a small one, then $\cong$ indicates an asymptotic equality (i.e., an approximate equality whose accuracy improves with N growing or ε diminishing). Also, $\approx$ denotes an approximate equality whose numeric uncertainty is independent of large or small parameters; $\sim$ indicates that the orders of magnitude are equal [i.e., the dependence on large (N) and small (ε) parameters is the same]. "Much greater than" ($\gg$) and "much less than" ($\ll$) signs designate strong inequalities growing still stronger as N increases or ε decreases. In addition, $\gtrsim$ means $>$ or $\sim$, and $\lesssim$ corresponds to $<$ or $\sim$.

Because this book is not a monograph but a textbook, we provide only a limited list of references, comprising mostly books and large review papers. In particular, we recommend the monographs[1-15,52] for further studies of the subject, while the popular books[16-20] are suggested for a first acquaintance.

In this book, such polymer systems as glasses, crystals, sophisticated biopolymer complexes, and so on are hardly mentioned. At present, we cannot describe these with an acceptable degree of generality and clarity.

Our colleagues and co-workers regarded our work on the manuscript with great kindness. No doubt, the book has improved owing to the comments by the reviewers Profs. A. V. Vologodskii, A. M. Elyashevich, Yu. S. Lazurkin, and M. D. Frank-Kamenetskii. Having read various parts of the manuscript, Drs. A. A. Darinskii, A. V. Lukashin, E. I. Shakhnovich, and especially I. A. Nyrkova, helped to correct some errors and inexact expressions. We wish to express our sincere thanks to all of them.

A system of concepts underlying the physics of polymers that we try to present here has been built under the influence of Professor Ilya Mikhailovich Lifshitz (1917–1982). We were lucky to have been his students and would like to express our everlasting gratitude for everything he taught us.

Introduction

I.1. *The subject of the statistical physics of macromolecules is the conformations and conformational motion of polymer chains on large space–time scales that determine the most significant properties of polymers and biopolymers.*

Macromolecules (i.e., molecules of polymer to be studied here) are of great interest to many people. A technologist may see in natural and synthetic polymers materials with a wide assortment of practically useful properties (mechanical, thermal, electric, optical, and so on), and this attitude is indeed confirmed by our everyday practice. A chemist specializing in high-molecular-weight compounds investigates the chemical reactions taking place in polymers and develops ways to synthesize new polymers. Finally, and perhaps most significantly, a contemporary molecular biologist or biophysicist knows that the properties of biopolymers define the structure and function of all biologic systems at the molecular level and provide a basis for their evolution. Hence, the physics of macromolecules is a principal theoretic basis for molecular biophysics as well as for the chemistry and technology of polymers.

The physical properties of macromolecules can be conditionally subdivided into two large classes; electronic, and conformational. The former, such as electric conductivity, superconductivity, and optical properties, are determined by the state of electron shells and are not discussed in this book. The latter are associated with the spatial arrangement and motion of atoms and atomic groups in a macromolecule.[a] The conformational properties determine the basic behavior of both biologic and synthetic polymers. Moreover, the most distinguished and significant conformational phenomena in macromolecules and polymers proceed on space–time scales substantially larger than the atomic scales. Obviously, classical statistical physics is capable of providing the principal methods for the description of such large-scale conformational effects. The corresponding branch of science is appropriately called the *statistical physics of macromolecules*.

This book deals with specific characteristics of polymer substance. Its specificity consists in a chain structure of molecules and their great, in a certain sense macroscopic, length. Although the manifestations of this specific feature are

---

[a]A spatial arrangement of atoms in a molecule is termed a *conformation*; a *configuration* is a state of electronic shells (i.e., an arrangement of electrons)

numerous, it is useful to describe briefly the principal ones:

1. Molecular chains exercise long-range correlations that reveal themselves in polymer systems in strong fluctuation effects, anomalously small entropies, and accordingly, anomalously high susceptibilities to external forces.

2. Macromolecular systems possess a long-term, or even practically unlimited, memory for the formation conditions and previous history of motion. This is conditioned by fixing of the sequence and arrangement of chain links (linear memory) and by topologic exclusion of the mutual crossing of macromolecular filaments (topologic memory).

I.2. The history of the statistical physics of macromolecules is divided into several periods; the current period is characterized by the active ingress of new ideas and methods of modern theoretical physics.

The statistical physics of macromolecules as a branch of science appeared in the 1930s, after H. Staudinger demonstrated experimentally in 1922 a chain structure of polymer molecules. The word *macromolecule* itself was introduced by Staudinger.

The next cardinal step was taken by W. Kuhn, E. Guth, and G. Mark. They examined the so-called phenomenon of high elasticity, that is, the elasticity of polymer substances similar to rubbers. High elasticity was found to have an entropic nature: during the stretching of a polymer sample, the constituent chain molecules straighten out (i.e., pass from a randomly coiled conformation to a more extended one). The entropy obviously diminishes in the process, because there is only one straight conformation, many coiled ones, and the free energy of the sample thus grows. The development of such simple physical ideas has finally brought about the realization that the conformational statistical properties of macromolecules determine the whole complex of physical properties of polymer materials.

In the following years, the range of problems in the statistical physics of macromolecules has grown considerably. Many significant ideas and results in this field are associated with the name of the eminent American physical chemist P. J. Flory. Since the early 1950s, the problems of the conformational statistics of polymers have been successfully tackled by a group of Soviet physicists headed by M. V. Volkenstein.

After the discovery of the double-helix structure of DNA by J. Watson and F. Crick in 1953, the progress in molecular biology gave additional impetus to the studies of macromolecular conformations. The first thirty years of statistical physics of macromolecules produced many important results and formulated several fundamental concepts. Still, this area of research remained rather detached and isolated, being sooner regarded as belonging to physical chemistry than to condensed-matter physics.

The situation dramatically changed during the late 1960s. First, it had been made clear by then that some fundamental problems of molecular biology could be formulated in terms of the physics of macromolecules. Second, the problems of the statistical physics of macromolecules proved to be closely related to the most urgent and imperative problems of general physics. As a result, polymer theory attracted the attention of some leading theoretic physicists, I. M. Lifshitz in the USSR, S. F. Edwards in the UK, and P. G. de Gennes in France. From that moment on, the methods of modern theoretic physics began to permeate polymer science. Quite soon, this ended the isolation of statistical physics of macromolecules, the formation of a harmonious system of simple models and qualitative concepts about the fundamental physical properties of polymers on the molecular level, and the successful utilization of these concepts in both the physical chemistry of polymers and molecular biophysics.

I.3. *Macromolecules are multilinked chains; they may differ in composition (i.e., the links may be either different or identical), degree of elasticity, number of branches, charged groups, and in the case of ring macromolecules, topology.*

The simplest polymer macromolecule consists of a sequence of a large number of atomic groups joined into a chain by covalent bonds. For example, a molecule of the well-known polyethylene can be depicted by the structural chemical formula

$$
\begin{array}{cccc}
\text{H} & \text{H} & \text{H} & \text{H} \\
\diagdown & \diagup & \diagdown & \diagup \\
\text{C} & & \text{C} & \\
\diagup & \diagdown & \diagup & \diagdown & \diagup \\
\cdots & \text{C} & & \text{C} & \\
\diagup & \diagdown & \diagup & \diagdown \\
\text{H} & \text{H} & \text{H} & \text{H}
\end{array}
$$

or $(—CH_2—)_N$. Simple chain-forming atomic groups (CH_2 groups in the case of polyethylene) will be called *links*. A basic characteristic of the macromolecule is the number of links N. This quantity is called the *degree of polymerization* or, for short, the *chain length*. It is proportional to the molecular mass of the chain.

Polymer chains are usually very long, $N \gg 1$. As a rule, molecules of synthetic polymer contain hundreds to tens of thousands of links, $N \approx 10^2 - 10^4$. The length of protein molecules has the same order of magnitude. The number of links in a DNA molecule, the longest of all known, reaches 1 billion, $N \approx 10^9$.

Long polymer chains are formed by a synthesis from low-molecular-weight compounds, or monomers. There are two basic methods of synthesis: 1) polymerization, or consecutive addition of monomers to a growing polymer chain according to the scheme $A_N + A \rightarrow A_{N+1}$; and 2) polycondensation, or gradual merging of chain sections having free valencies at the ends according to the scheme $A_N + A_M \rightarrow A_{N+M}$. The chain stops growing after the addition of a univalent compound or, in the case of polymerization, on monomer exhaustion.

Clearly, the polymer chains formed by random chemical reactions of polymerization or polycondensation from a monomer mixture have a wide length distribution. In this case, the polymer system is called *polydisperse*. While discussing properties of polymer solutions and melts, we do not take into account their inevitable polydispersity; the main qualitative conclusions pertaining to the phenomena to be considered are not affected.

If the synthesis of a polymer chain proceeds in a monomer mixture containing not only monomers with two functional groups, (i.e., with two groups capable of making valence bonds with other monomers) but also compounds with three or more functional groups, then branched macromolecules are due to be formed (Fig. 1.14). The simplest branched macromolecules are shaped like combs or stars. However, in most real cases, various irregular structures, that is, randomly branched macromolecules (Fig. 1.14c), are likely to form under conditions of synthesis in the presence of multifunctional groups. The branched macromolecules also can be obtained by cross-linking sections of linear macromolecules.

A macroscopic polymer network is a kind of ultimate case of a branched macromolecule. This giant molecule emerges in the process of chemical cross-linking of a large number of chain macromolecules. One such molecule can have a length of many centimeters.

Several different methods exist to accomplish the cross-linking of macromolecules. One can use chemically active linking agents to establish covalent bonds between the chain sections, ionizing irradiation of the polymer systems, and so on. The most common example of the cross-linking of macromolecules, well-known in everyday practice, is vulcanization, a process of treating plastic crude rubber to make a high-elastic polymer network.

In the simplest polymer chains, all links are identical. Such macromolecules are called *homopolymers*. The previously discussed macromolecule of polyethylene belongs to this class. On the other hand, chains composed of links of several different types also can be synthesized. These are *heteropolymers* or, as chemists prefer to call them, *copolymers*. The most interesting, although by no means unique, heteropolymers are biopolymers of proteins (20 types of links) and DNA (4 types). Another important class of heteropolymers is represented by block copolymers, consisting of long sections (blocks) of links of different types.

All polymer chains are characterized by a definite degree of flexibility; this notion will be examined in detail in Sec. 2. For now it is sufficient to point out

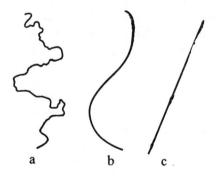

a   b   c

FIGURE I.1. Sketches of flexible- (a), semiflexible- (b), and stiff-chain (c) macromolecules.

that there are both flexible-chain macromolecules, in which a substantial bend can form over the length of several links (Fig. I.1a), and stiff-chain macromolecules, in which a bend becomes appreciable over much greater lengths (Fig. I.1b). In the ultimate case of vanishing flexibility, the macromolecule behaves as a practically stiff rod (Fig. I.1c).

Macromolecules also can differ by the presence (or absence) of charged links. Macromolecules containing charged links are called *polyelectrolytes*. Because a polymer system as a whole is to be electrically neutral, the presence of charges of one sign on some links causes charges of the opposite sign also to be present in the system. As a rule, the latter charges are low-molecular-weight counter ions. There are, however, some heteropolymer macromolecules whose individual links carry the charges of opposite signs; such substances are referred to as *polyampholytes*.

Finally, there are closed-ring polymer chains. Their physical properties are determined to a large extent by topology, that is, by a topologic type of knot formed by polymer rings or by mutual entanglement of the rings.

I.4. *Apart from general macromolecular properties, biopolymers possess a number of specific features.*

The principal distinction of biomacromolecules consists in their heteropolymeric nature, with the link sequence (called a *primary structure*) being strictly fixed in each biopolymer in contrast to the random or block structure of synthetic heteropolymers. From the standpoint of biology, the primary structures of biopolymers are products of biologic evolution.

Next, the biopolymers of DNA and proteins may form helic and folded structures on a small scale, which are called *elements of secondary structure*.

Lastly, the spatial structure of a biopolymer as a whole is formed in a very specific way and is called a *tertiary structure*. The tertiary structure of biomacromolecules is determined by the whole of their properties, including the primary heterogeneous structure, mechanism of flexibility, secondary structure, presence of charged links, and topology.

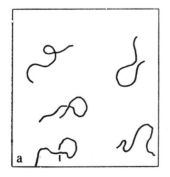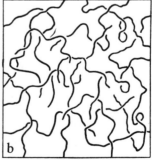

FIGURE I.2. Dilute (a) and semidilute (b) polymer solutions.

I.5. *Polymer systems, both solutions and pure compounds, can exist in many qualitatively different macroscopic states.*

A conventional low-molecular-weight substance can exist in three simple phase states: gas, liquid, and (crystalline) solid. Specificity of the molecules composing a polymer substance manifests itself in the peculiar properties of each of these phase states.

The gas state is not typical for polymer substances. It can be realized only under very low pressures (ultrahigh vacuum). There is however, a polymer system that can (and should) be considered as a kind of gas. Such a system is a polymer solution (Fig. I.2a), where the macromolecules are separated by long distances and are practically non-interacting. Hence, it is clear that the dilute solution is in a certain sense the most significant and fundamental polymer system, because its properties are directly associated with the properties of its individual macromolecules.

As the concentration of the polymer solution grows, the macromolecules become more entangled and begin to interact (Fig. I.2b). It is essential that the mean concentration of monomer links in a polymer coil is very low (*see* subsection 5.2); therefore, the entanglement of macromolecules takes place at low concentrations of the polymer in solution. Consequently, there exists in polymer solutions a wide range of concentrations in which the chains are strongly entangled and yet the volume fraction taken by the polymer in the solution is still very low (Fig. I.2b). The polymer solution in this concentration range is called *semidilute*. The existence of the semidilute region is a distinctive feature of a polymer solution.

On further increase of the concentration of a polymer solution, the interaction between the links of macromolecules becomes stronger. In stiff-chain polymers, this interaction as a rule eventually leads to orientational ordering of the solution, that is, to the emergence of a preferred orientation of the polymer chains (Fig. I.3). In most cases, this happens because an isotropic packing of a sufficiently concentrated system of asymmetric particles is inconceivable. The state of

FIGURE I.3. Orientationally ordered (liquid crystalline) state of a polymer system.

a polymer solution as depicted in Figure I.3 is called *liquid crystalline*.

Polymer solutions with comparable volume fractions of polymer and solvent are called *concentrated*. In the total absence of solvent, a pure polymer substance is obtained. Depending on the character and intensity of interaction, the concentrated solution or pure polymer substance can exist in one of the following four phase states: crystalline, glassy, high elastic, and viscoelastic.

Many properties of the crystalline state of polymers are similar to the corresponding properties of the crystals of low-molecular-weight substances. There is, however, an essential difference in the properties. Because the individual links of a polymer substance are joined in long chains, formation of an ideal, defect-free structure is greatly hampered by kinetic effects. Therefore, the crystal-forming polymers as a rule produce only the partially crystalline phase (i.e., crystalline domains separated by amorphous interlayers).

The three remaining phase states of polymers correspond to the liquid state of low-molecular-weight substances. Polymer glasses are liquids whose viscosity is so high that their flow is not visible during a reasonable time of observation. They are on the whole analogous to low-molecular-weight glasses. Many plastics are actually glassy polymers.

At temperatures above the so-called glass transition temperature, the mobility of individual links increases substantially. However, if the relative motion of macromolecules remains hampered (e.g., because of cross-links present in the polymer network or occasional glassy domains acting as effective cross-links), the polymer substance turns into a high-elastic state. A substance in this state is capable of sustaining extremely large, reversible elastic deformations.

If the macromolecules of a polymer substance are not cross-linked by covalent bonds (i.e., there is no polymer network), a further rise in temperature "defrosts" not only the motions of the individual links and chain sections but also the motion of the macromolecules as a whole relative to one another. The polymer substance begins to flow. This phenomenon indicates that the high-elastic state turns into the viscoelastic state. The viscoelastic state of polymer substance is also called a *polymer melt*.

The relative motion of macromolecules during the flow of a polymer melt proceeds by means of an extraordinary mechanism. Because of the strong entanglement of polymer chains, the surrounding macromolecules impose substantial limitations on possible displacements of a given chain. In fact, each polymer chain in the melt can be visualized as if confined within an effective "tube" created by the surrounding chains, so the only feasible mechanism for large-scale motion of the polymer chain is a diffusional, snake-like motion inside that tube. Such motions have been called *reptations*. It is this uncommon mechanism of macromolecular motion in the melt that is responsible for such special properties of polymer liquids as an anomalously high viscosity, long memory of the flow prehistory, and the dependence of the type of response (whether elastic or viscous) on the frequency of external influence. The latter property is called *viscoelasticity of polymer liquids*.

Needless to say, the polymer chains are miscible not only with a solvent but also with polymers of different kinds. However, the miscibility conditions for two polymers are very rigorous: a very weak mutual repulsion of links is sufficient to cause separation of the polymer mixture. This is because the entropy of mixing of polymer molecules is appreciably less than that of low-molecular-weight substances as the individual links, being joined in a chain, do not possess the entropy of independent translational motion.

Phase separation of a special kind can be also observed in a melt of block copolymers. In this case, a separation into macroscopic phases cannot occur, because the individual blocks in the macromolecule are joined in a common chain. As a result, the so-called microdomain structure containing regions enriched with the links of different components appears. Variations in block lengths or temperature make it possible to change the architecture of the micro-domain structure.

Another specific polymer effect is the realization of intramolecular condensed phases. Indeed, the macromolecular chains are long and flexible, and an intramolecular condensation of the links therefore can take place provided they are attracted to one another. A globular state forms in the process so that the intramolecular condensed phase is analogous in local terms to any known condensed system (liquid, liquid crystal, homogeneous or separated solution, glass, and so on).

I.6. *A polymer system is characterized by the linear memory and volume inter-actions.*

The success of the theoretic description of the system is determined primarily by the adequacy of an idealized model chosen for that system. Clearly, the required model in our case is one that can be called an ideal polymer. As experience shows, any idealization is closely associated with the possibility of characterizing the system in terms of large or small dimensionless parameters.

It is known, for example, that the theory of simple liquids is less advanced than some other branches of condensed-matter theory, because neither large nor small parameters can be identified to describe a liquid. At the same time, the exploration of seemingly more complicated polymer liquids (solutions and melts) is in fact a simpler problem. Their theoretic investigation is made easy, because the specific chain structure of macromolecules permits one to identify two large parameters. One of these has been already mentioned: the number of links in a chain, $N \gg 1$. Next, any system consisting of macromolecules is characterized by a definite hierarchy of interactions. The energies of covalent bonds E_1 (~ 5 eV), such as the bonds that each link forms with nearest neighbors in the polymer chain, are much higher than characteristic energies E_2 (~ 0.1 eV) of all other interactions, such as the interactions of the links with solvent mole-cules, off-neighbor links of the same chain, links of other chains, and so on. Hence, $E_1/E_2 \gg 1$. As a result, the covalent bonds practically cannot break at room temperature, either through thermal fluctuations (because $E_1/T \gg 1$) or various interactions.[b]

This implies that the link sequence in the chain is actually fixed by the high energies of longitudinal valence bonds. Each link "remembers" its position number acquired in the formation process of the macromolecule. This can be briefly summarized by the following statement: a polymer chain possesses fixed linear memory.

All link interactions that are not associated with covalent bonds between the neighbors in the chain are called *volume interactions*. These interactions with intrinsic energy E_2 are much weaker than the forces responsible for linear memory. In the first approximation, such interactions can be totally ignored to obtain what is called an *ideal polymer chain*. In Chapter 1, the systematic presentation of the statistical physics of macromolecules begins with the study of the ideal chain approximation.

Finally, we clarify the terms *link, monomer,* and *monomer link*. In chemistry, monomers are small molecules from which polymer chains are formed. A monomer link is a part of a polymer chain corresponding to one monomer. In physics, all three terms are used as synonyms, and chains are arbitrarily divided into links or monomers (*see* Sec. 18).

---

[b]In this book, the temperature is expressed in energy units so that the Boltzmann constant $k_B = 1$.

CHAPTER 1

Ideal Polymer Chain

The model of an ideal macromolecule plays the same role in polymer physics as the notion of an ideal gas in traditional molecular physics. This model represents a chain of immaterial links, each joined with two nearest neighbors and having no interaction either with solvent molecules or with other links of the same or another macromolecule. There are several models of an ideal chain, just as there are many different ideal gases (monatomic, diatomic, and so on) whose molecules do not interact. The ideal chains differ from one another in link structure and the type of bonding between the nearest neighbors. A common "ideal" feature of all these chains is the absence of volume interactions. The range of actual conditions for which the macromolecules behave as ideal ones is not very wide. In the main, only polymer solutions diluted in so-called θ solvents and polymer melts satisfy the conditions. Nevertheless, the ideal models are quite helpful, because they allow one to form an idea about the character of thermal motion of macromolecules or, in other words, about the entropic properties of a polymer substance.

1. A FREELY JOINTED CHAIN

Let us begin with a very simple polymer model, a freely jointed chain, composed of a sequence of N rigid segments, each of length l and able to point in any direction independently of each other (Fig. 1.1). We shall assume the chain ideal applies by disregarding the interactions between segments that are not chemically bonded.

To characterize the chain conformation, we consider the end-to-end vector R (Fig. 1.1) and calculate the mean square $\langle R^2 \rangle$ by averaging over all possible conformations. This quantity is indeed the simplest characteristic of the average size of the macromolecule, because $\langle R \rangle = 0$ (any value of the vector R can be found as frequently as the opposite value $-R$). We denote the radius vector of the beginning of the i-th segment by x_i and that of its end by x_{i+1}. Additionally, we introduce the "bond vectors" $u_i = x_{i+1} - x_i$. Therefore, the end-to-end vector R can be written as

$$R = \sum_{i=1}^{N} u_i, \tag{1.1}$$

1

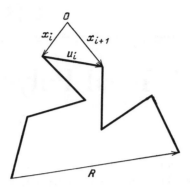

FIGURE 1.1. Freely jointed chain.

and

$$\langle R^2 \rangle = \left\langle \left(\sum_{i=1}^{N} u_i \right)^2 \right\rangle = \sum_{i=1}^{N} \langle u_i^2 \rangle + 2 \sum \sum_{1 \leqslant i < j \leqslant N} \langle u_i u_j \rangle. \qquad (1.2)$$

Because the segment directions in a freely jointed chain are not correlated and the angle between the vectors u_i and $u_j (i \neq j)$ takes, with equal probability, any value from 0 to 2π, $\langle u_i u_j \rangle = l^2 \langle \cos \vartheta_{ij} \rangle = 0$. In addition, $\langle u_i^2 \rangle = l^2$, so that

$$\langle R^2 \rangle = N l^2. \qquad (1.3)$$

Thus, for a multilink chain $(N \gg 1)$, the mean size of a macromolecule $R = \langle R^2 \rangle^{1/2} \sim N^{1/2} l$ is much less than the total length Nl measured along the contour of the polymer chain. This significant conclusion implies that in the set of conformations that the freely jointed chain assumes in the process of thermal motion, stretched (nearly straight) conformations constitute a minor fraction. The absolute majority of chain conformations are depicted by lines that are strongly coiled in space. In other words, the state of a randomly shaped coil corresponds to the thermodynamic equilibrium of the ideal freely jointed chain, that is, to the entropy maximum (because all of the conformations have the same energy).

This conclusion is valid not only for a freely jointed ideal chain but also any adequately long ideal chain. Such behavior is caused by the flexibility of polymer chains.

2. FLEXIBILITY OF A POLYMER CHAIN

2.1. *Any long macromolecule is flexible, but different polymers have different mechanisms of flexibility.*

The flexibility of the freely jointed chain shown in Figure 1.1. is caused by freely rotating connections between rigid segments. It can be said that flexibility is concentrated at the connection points. This so-called freely jointed flexibility mechanism is easy to describe but very difficult to realize in practice; it is observed in very few real substances. All sufficiently long polymer chains are quite flexible, however, the main reason being their great length.

To clarify this remark, consider the most unfavorable limiting case. Suppose that the straight-chain conformation corresponds to the absolute minimum of energy and that all links and bonds are so stiff that the thermal excitation energy $\sim T$ produces only small deformations of their stereochemical structure. For small deformations, the atomic framework of a molecule can be regarded as a classical elastic construction, which in the case of the polymer is approximated by a thin, elastic, homogeneous filament obeying Hooke's law under deformation. Such a model of a polymer chain is called *persistent* or *worm-like* (Fig. 1.2).

Examination of the persistent model shows that even small fluctuation bendings of its sections lead to the total coiling of a sufficiently long chain, because different sections bend to different sides. Flexibility thus is a fundamental property of any long chain structures, caused by their linear shape. The immediate consequence of the flexibility of macromolecules is that any sufficiently long polymer chain looks like an irregular statistical coil.

Although flexibility proper is a generic attribute of all macromolecules, the flexibility mechanism may be different for different polymers. Many stiff-chain polymers and helic macromolecules are characterized by the persistent flexibility mechanism, that is, by uniform flexibility along the whole length. For the double-helix DNA macromolecule, for example, the persistent flexibility mechanism provides a satisfactory approximation. Some macromolecules correspond to a simple isotropic persistent model; others exhibit appreciable anisotropy in

FIGURE 1.2. Persistent chain.

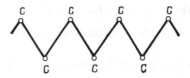

FIGURE 1.3. Planar *trans*-zigzag transformation.

the lateral plane and can be depicted as an elastic ribbon or band.

The persistent flexibility mechanism is in a certain sense the most fundamental, because it occurs whenever thermal oscillations are harmonic. The existence of persistent flexibility can be demonstrated even without any knowledge of the chemical nature of polymers. On the contrary, all other mechanisms of flexibility are caused by various anharmonic effects so that the realization of a specific mechanism depends on the individual chemical structure of the polymer. In particular, the so-called rotational-isomeric flexibility is typical for the most abundant class of macromolecules with a carbon backbone and for other single-filament polymers with single (i.e., σ-) bonds between the links.

Distinct from double π-bonds, σ valences are intrinsically, axially symmetric, and a rotation around the σ-bond only leads to a small increase in energy. This rotation, however, is equivalent to a chain bending if the σ-bond is not parallel to the axis of the macromolecule. For example, in simple carbon chains, the planar *trans*-zigzag conformation (Fig. 1.3) corresponds to a minimum of energy. The angle γ between the neighboring C—C bonds, called the *valence angle*, usually lies within the interval from 50° to 80°. It is essential that the valence angle remains essentially constant during any change of the conformation of the given polymer chain. In other words, the only permitted motion is a rotation around each C—C bond along the conical surface, with the axis directed along the neighboring C—C bond and the vertex angle equal to 2γ (internal rotation, Fig. 1.4a).

Assuming the rotation around single bonds to be free, we obtain the simplest rotational-isomeric model. Just as in a freely jointed model, its flexibility is

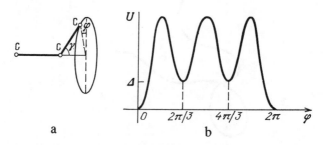

FIGURE 1.4. (a), Rotation around a single bond. γ—valence angle, φ—internal rotation angle. (b), Potential energy as a function of internal rotation angle.

concentrated at individual points. However, it is more realistic, because it is reduced to the free variation of the internal rotation angle φ (Fig. 1.4a), the valence angle being fixed.

As a rule, the internal rotation is not quite free in real chains. The dependence of the potential energy on the internal rotation angle shows several characteristic maxima and minima (Fig. 1.4b). The lowest minimum at $\varphi = 0$ corresponds to the planar *trans*-zigzag. In the process of thermal motion, the system spends most of its time in low-energy conformations (called *rotational isomers*), jumping now and then from one isomeric state to another. Any isomeric state of the link except $\varphi = 0$ signifies that this link is a bending point of the chain. Because the difference ΔU in potential energies of rotational isomers is usually the same order of magnitude as the temperature, these bendings can appear at any chain joint with a probability of order unity. This gives rise to chain flexibility by the rotational isomeric mechanism.

2.2. The directional correlation of two segments of a macromolecule diminishes exponentially with the growth of the chain length separating them.

Refer back to Eq. (1.2). For a chain model differing from the freely jointed one, $\langle u_i u_j \rangle \neq 0$, because the directions of different chain segments are correlated. As $\langle u_i u_j \rangle \sim \langle \cos \vartheta_{ij} \rangle$, this correlation, determining a degree of chain flexibility, can be qualitatively expressed by the mean cosine of the angle between different segments of the polymer.

Thus, we introduce the magnitude $\langle \cos \theta(s) \rangle$, the mean cosine of the angle between the chain segments separated by the length s. This function of s for many polymer chain models possesses the property of so-called multiplicativity: if the chain has two neighboring sections with lengths s and s', then

$$\langle \cos \theta(s+s') \rangle = \langle \cos \theta(s) \rangle \langle \cos \theta(s') \rangle. \tag{2.1}$$

The function having this property is exponential, i.e.,

$$\langle \cos \theta(s) \rangle = \exp(-s/\tilde{l}), \tag{2.2}$$

where the preexponential factor is equal to unity, because $\cos \theta(s=0) = 1$ and $\tilde{l}$ is a constant for each given polymer. This constant is the basic characteristic of polymer flexibility and is called the *persistent length of the polymer*.

Now consider the origin of multiplicativity defined in Eq. (2.1) for a persistent chain isotropic in the lateral plane. Denote the ends of the sections s and s' by a, b, and c (Fig. 1.5) and introduce the unit vectors of chain direction u_a at each point a. Then $\cos \theta(s+s') = (u_a u_c)$. Taking the direction of vector u_b as the z-axis of the Cartesian coordinates, we write the scalar product $(u_a u_c)$ in this coordinate system as

$$\cos \theta_{ac} = \cos \theta_{ab} \cos \theta_{bc} + \sin \theta_{ab} \sin \theta_{bc} \cos \varphi, \tag{2.3}$$

where the first term on the right-hand side is the product of the z components of the vectors u_a and u_c, the second term is the scalar product of the projections of

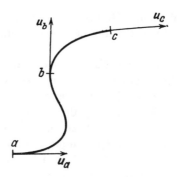

FIGURE 1.5. Explanation of multiplicativity.

the vectors u_a and u_c on the XY plane, φ is the angle between these projections, and $\sin \theta_{ab}$ and $\sin \theta_{ac}$ are their lengths. In an isotropic chain $\langle \cos \varphi \rangle = 0$, because for any conformation of the section bc there exists an equally probable conformation turned through an arbitrary angle around the vector u_b. On the other hand, the bendings of the sections ab and bc are independent, because the chain is ideal. Consequently, by averaging Eq. (2.3), we indeed obtain Eq. (2.1).

Thus, Eqs. (2.1) and (2.2) are exact for the persistent chain isotropic in the lateral plane. The given proof can be applied without any modifications to the chains isotropic in the lateral plane and composed of rigid segments. In so doing, however, it should be understood that in Eqs. (2.1) and (2.2), only the length s comprising the integer number of segments should be considered, that is, the exponential decrease (2.2) is determined for a discrete set of points $s = nb$ where b is the segment length and $n = 0, 1, 2, \ldots$.

For polymer chains anisotropic in the lateral plane, when, for example, the internal rotation potential depends on φ (Fig. 1.4b), the relations (2.1) and (2.2) are not valid for arbitrary values of s. At sufficiently large values of s, however, the correlations diminish exponentially, obeying Eq. (2.2). Hence, the notion of a persistent length also can be applied to these polymer chains.

2.3. The persistent length can roughly be considered as a maximum chain section that remains straight; at greater lengths, bending fluctuations destroy the memory of the chain direction.

Let us clarify the physical sense of Eq. (2.2). First, consider the angle θ between two close sections of the chain, separated by a chain section whose length is much less than the persistent length: $s \ll \tilde{l}$. Under these conditions, $\langle \cos \theta \rangle \approx 1$. This implies that the chain section shorter than the persistent length does not exhibit any flexibility and essentially behaves like a stiff rod ($\theta \approx 0$). In the opposite limiting case $s \gg \tilde{l}$, when the considered sections are separated by a large portion of the chain, Eq. (2.2) yields $\langle \cos \theta(s) \rangle = 0$. Hence, the angle θ takes on any value between 0 and 2π, and the chain flexibility leads to almost total independence of macromolecular sections separated by a distance equal to or exceeding the persistent length. In other words, the memory of the chain direction is lost over a distance comparable to the persistent length.

How long are the persistent lengths of real macromolecules? Although they vary strongly from one polymer to another, two extreme cases can be mentioned:

1. A simple flexible synthetic polymer of polystyrene where $\tilde{l} \approx 1.0$–1.4 nm (depending on the conditions), which corresponds approximately to the length of 4 to 5 longitudinal chain bonds.

2. A double-helix DNA where $\tilde{l} \approx 50$ nm (i.e., approximately 150 base pairs).

2.4. *The persistent length is determined by the valence angle for the chain with free internal rotation and the bending modulus for the persistent chain.*

Consider a simple model consisting of stiff segments of length b connected so that the valence angle γ between any two neighboring segments is fixed, while the rotation of the segment around the axis, whose direction is defined by the neighboring segment (internal rotation), is free, that is, $U(\varphi)=0$ (Fig. 1.4b). The property of multiplicativity expressed by Eq. (2.1) yields the following for this polymer chain model isotropic in the lateral plane:

$$\langle \cos \theta_{i,i+k} \rangle = (\cos \gamma)^k. \tag{2.4}$$

Comparing Eqs. (2.4) and (2.2), we find the persistent length for the given model:

$$\tilde{l} = b/|\ln \cos \gamma|. \tag{2.5}$$

Thus, the smaller the valence angle, the longer the persistent length. This is because at small γ, the close segments have almost identical directions.

To find the persistent length of a persistent macromolecule with isotropic flexibility, we consider its short ($s \ll \tilde{l}$) section, for which according to Eq. (2.2)

$$\langle \cos \theta(s) \rangle \cong 1 - s/\tilde{l},$$

and because the angle $\theta(s)$ is small and $\cos \theta(s) \cong 1 - \theta^2(s)/2$,

$$\langle \theta^2(s) \rangle \cong 2s/\tilde{l}. \tag{2.6}$$

The bending fluctuations of a short elastic rod, however, are easy to describe, because the rod can be assumed to bend with an approximately constant radius of curvature. In this case, the elastic bending energy is proportional to the square of deformation, that is to the square of the curvature:

$$\Delta E = (1/2)s\varkappa(\theta/s)^2 = \varkappa \theta^2/2s, \tag{2.7}$$

where θ/s is the curvature and $\varkappa$ the effective modulus of "bending elasticity" for a unit length of the macromolecule. Subsequently, the mean square of the bending angle is

$$\langle\theta^2(s)\rangle = 2 \int \exp(-\Delta E/T)\theta^2 d\theta \bigg/ \int \exp(-\Delta E/T)d\theta = 2sT/\varkappa. \quad (2.8)$$

The factor 2 allows for the fact that the bendings occur in the two planes independently. Comparing Eq. (2.8) with Eq. (2.6), we obtain

$$\tilde{l} = \varkappa/T. \quad (2.9)$$

Generally, it is quite comprehensible that in the model characterized by a fixed modulus $\varkappa$ of bending elasticity, the persistent length grows as the temperature decreases: the bending fluctuations subside. However, in reality, the value of $\varkappa$ itself varies with temperature; therefore, the relationship between $\tilde{l}$ and T becomes rather complicated. As a rule, however, this relationship is not very essential.

Hence, we have found the persistent length for the simplest macromolecular models using elementary means. To examine more realistic models, however, it is desirable to have a more general method. (Such a method is discussed in Sec. 6.)

3. SIZE OF AN IDEAL POLYMER CHAIN

3.1. *The simplest quantity characterizing the spatial size of a polymer chain is the root-mean-square (rms) end-to-end distance; its comparison with the contour length defines the degree of chain spatial entanglement.*

Because of the flexibility of a polymer chain, macromolecules are never straight. On the contrary, any sufficiently long chain wriggles and twists to form a random coil. The spatial size of this coil is not characterized by the contour length of the chain. The difference in these values results from the random conformation of the chain. This is why the problem of determining the size of coil-shaped chain conformations have to be discussed. This section investigates the spatial end-to-end distance of the chain, just as in Sec. 1, where a freely jointed chain was dealt with. Other parameters of the coil are examined in Sec. 5.

3.2. *Any long macromolecule can roughly be pictured as a freely jointed chain of straight segments; the rms end-to-end distance for any long ideal chain is proportional to the square root of its length.*

The size of a long polymer chain can easily be evaluated from the results of the previous section, namely, from the fundamental fact that the memory of the direction prevails along the chain for only a finite distance, which is on the order of a persistent length $\tilde{l}$. Indeed, chain sections with lengths of approximately $\tilde{l}$ can be regarded as practically stiff, because their end-to-end distance is of order $\tilde{l}$. The number of such sections in the chain is $N_{ef} \sim L/\tilde{l}$, where L is the contour length of the polymer. Because the directions of these sections are essentially independent, the size of the whole chain can be found from Eq. (1.3):

$$\langle R^2 \rangle \sim (L/\tilde{l})\tilde{l}^2 \sim L\tilde{l}. \tag{3.1}$$

Proportionality of $\langle R^2 \rangle$ to the chain length is the most fundamental property of an ideal coil. Alternatively, it can be expressed as

$$R \sim N^{1/2}, \tag{3.2}$$

where N is the number of links in the chain. Numerous consequences of this result are discussed later; however, it should be immediately emphasized that Eq. (3.1) was obtained by rough evaluation and the numeric coefficient remains indefinite. One may expect that this coefficient depends on the macromolecular structure, in particular on the flexibility mechanism.

3.3. *Apart from the persistent length, the flexibility of a macromolecule can be characterized by the effective (Kuhn) segment length.*

Although the size of the coil R and the chain length L can be determined experimentally, the persistent length cannot be found from Eq. (3.1) without knowledge of the numeric factor. It therefore is necessary to introduce another qualitative parameter of macromolecular flexibility that is directly associated with the quantity $\langle R^2 \rangle$. In 1934, such a characteristic was introduced by W. Kuhn and also by E. Guth and G. Mark, and it was later called the Kuhn (effective) segment. The length of the Kuhn segment l for a long ($L \gg \tilde{l}$) ideal macromolecule is defined as

$$\langle R^2 \rangle = Ll. \tag{3.3}$$

From Eq. (3.3), it follows that when a real macromolecule is effectively treated as a freely jointed chain of $N = L/l$ Kuhn segments, the numerically correct value of $\langle R^2 \rangle$ is obtained. Conversely, the model of a sequence of persistent lengths connected together yields only the correct order of magnitude.

Comparing Eqs. (3.1) and (3.3), shows that the parameters l and $\tilde{l}$ are of the same order of magnitude. Therefore, both can be used to describe the degree of polymer chain flexibility. Depending on the specific problem, either quantity can be used. The length of the Kuhn segment is easier to measure experimentally, whereas the persistent length has an unambiguous microscopic meaning.

Thus, in terms of the end-to-end distance, the problem of the microscopic theory is reduced to a calculation of the effective segment for each specific flexibility mechanism. The following subsections delineate methods for the solution of this problem for various macromolecular models.

3.4. *For the simplest macromolecular models (i.e., persistent and with free internal rotation), the size of chains of any length is expressed by simple, exact formulas; the effective segment also can be found for these models.*

Begin with the persistent chain model with isotropic flexibility. Suppose the conformation of a persistent chain of length L is given by the vector $r(s)$. Introduce $u(s) = \partial r/\partial s$, a unit vector of chain direction at the point removed by the distance s from the beginning of the chain measured along the chain contour. Then the end-to-end vector R can be written as

$$R = \int_0^L u(s)\,ds. \tag{3.4}$$

In the case of a freely jointed chain, this general formula turns into its discrete analogue, Eq. (1.1). Correspondingly, $\langle R^2 \rangle$ is calculated in analogy with Eq. (1.2):

$$\langle R^2 \rangle = \int_0^L ds \int_0^L ds' \langle u(s)u(s') \rangle = 2 \int_0^L ds \int_0^{L-s} dt \langle \cos \theta(t) \rangle, \tag{3.5}$$

where $t = s' - s$. Using Eq. (2.2), which is exact for the persistent chain with isotropic flexibility and calculating the integrals therein, we obtain

$$\langle R^2 \rangle = 2\widetilde{l}^{\,2}[(L/\widetilde{l}) - 1 + \exp(-L/\widetilde{l})]. \tag{3.6}$$

Equation (3.6) defines the mean square of the end-to-end distance for an ideal persistent chain of arbitrary length. Analyzing this formula by examining the two opposite limiting cases, that is, short ($L \ll \widetilde{l}$) and very long ($L \gg \widetilde{l}$) chains, we obtain

$$\langle R^2 \rangle \cong \begin{cases} L^2, & L \ll \widetilde{l}; & (3.7a) \\ 2L\widetilde{l}, & L \gg \widetilde{l}. & (3.7b) \end{cases}$$

The first equality means that a short molecule hardly bends, and its end-to-end distance is practically equal to the contour length $R \cong L$. The second equality corresponds to the evaluation (3.1)–(3.3) and shows that the effective segment for the persistent chain model is twice as long as the persistent length:

$$l = 2\widetilde{l}. \tag{3.8}$$

A factor of 2 in this relation can be interpreted as an indication that the memory of the segment orientation "spreads" in the two opposing directions along the chain.

Now consider a model of segments with free internal rotation and fixed valence angle (subsection 2.4). For this model, Eq. (1.2) can be rewritten as

$$\langle R^2 \rangle = Nb^2 + 2b^2 \sum_{i=1}^{N} \sum_{k=1}^{N-i} \langle \cos \theta_{i,i+k} \rangle. \tag{3.9}$$

After substitution of Eq. (2.4) into Eq. (3.9), the calculation of $\langle R^2 \rangle$ is reduced to the summation of a geometric series. The exact result takes the form

$$\langle R^2 \rangle = Nb^2 \left[\frac{1 + \cos \gamma}{1 - \cos \gamma} - \frac{2}{N} \cos \gamma \frac{1 - (\cos \gamma)^N}{(1 - \cos \gamma)^2} \right]. \tag{3.10}$$

As in the previous example, the limiting cases can be considered. For a short ($N = 1$) chain, the result is trivial: $\langle R^2 \rangle = b^2$, and for a long ($N \to \infty$) chain

$$\langle R^2 \rangle \cong Nb^2(1 + \cos \gamma)/(1 - \cos \gamma). \tag{3.11}$$

Comparing Eq. (3.11) with Eq. (1.3) for the freely jointed chain, one can infer that the fixed valence angle $\gamma < \pi/2$ leads to a certain increase in the end-to-end distance, which becomes greater at lower values of γ. At the same time, according to the estimation (3.1)–(3.3), the value of $\langle R^2 \rangle$ remains proportional to N. For this model

$$l = \tilde{l} |\ln \cos \gamma| (1 + \cos \gamma)/(1 - \cos \gamma). \tag{3.12}$$

Note that numerically the ratio $l/\tilde{l}$ weakly depends on the valence angle γ (at least at $\gamma < 80°$); at $\gamma \to 0$, it equals 2 and slowly increases with γ, reaching 2.03 at $\gamma = 50°$, 2.08 at $\gamma = 60°$, and 2.19 at $\gamma = 70°$. In real polymers, and for many other models, the ratio $l/\tilde{l}$ is approximately 2.

The limiting case of the small valence angle $\gamma \ll 1$ is of special interest. Under this condition, each joint makes a small contribution to the flexibility, making the chain contour very smooth. One can expect that this limiting case approaches the persistent chain model. Indeed, according to Eq. (2.5), the persistent length $\tilde{l} = 2b/\gamma^2 \gg b$ at $\gamma \ll 1$. We suggest that the reader verify that for $\gamma \ll 1$ Eq. (3.10) turns into Eq. (3.6). A hint: for $\gamma \ll 1$, $(\cos \gamma)^N = \exp(N \ln \cos \gamma) \cong \exp(-N\gamma^2/2)$.

3.5. The rms end-to-end distance of a long polymer chain with fixed valence angles and independent potentials for internal bond rotation can be calculated using matrix methods.

Consider a more complicated model of a polymer chain. Suppose it consists of stiff segments of length b with a fixed valence angle between them and an internal, non-zero potential $U(\varphi)$ similar to that shown in Figure 1.4b. Assume that the potentials for individual bonds $U(\varphi)$ are independent and that $U(\varphi) = U(-\varphi)$ (Fig. 1.4b). This model provides a further approximation of the properties of real chains with rotational-isomeric flexibility (*see* subsection 2.1). Let us calculate $\langle R^2 \rangle$ for this model.

The given model is rather complicated in terms of geometry. Its investigation calls for the introduction of vector and matrix notation.

In this case, the expression (3.9) for $\langle R^2 \rangle$ remains correct. To calculate $\langle \cos \theta_{i,i+k} \rangle$, we shall introduce a local coordinate system for each of the links. The axis x_i will be directed along the i-th link, and the axis y_i will lie in the plane formed by the bonds i and $i-1$ so that the angle between the axes x_{i-1} and y_i is acute. The axis z_i will be directed so as to make the Cartesian coordinate system right-handed. One can easily see that the components of unit vectors $(n_x)_{i+1}$, $(n_y)_{i+1}$, $(n_z)_{i+1}$, lying parallel to the corresponding axes of the $(i+1)$-th coordinate system, are expressed in the i-th coordinate system as

$$(n_x)_{i+1} = \begin{bmatrix} \cos \theta_i \\ \sin \theta_i \cos \varphi_i \\ \sin \theta_i \sin \varphi_i \end{bmatrix};$$

$$(n_y)_{i+1} = \begin{bmatrix} \sin \theta_i \\ -\cos \theta_i \cos \varphi_i \\ -\cos \theta_i \sin \varphi_i \end{bmatrix};$$

$$(n_z)_{i+1} = \begin{bmatrix} 0 \\ \sin \varphi_i \\ -\cos \varphi_i \end{bmatrix}, \qquad (3.13)$$

where θ_i and φ_i are the spheric angles in the i-th coordinate system.

Let v denote the arbitrary vector written in the i-th coordinate system as

$$v = \begin{bmatrix} v_y \\ v_z \\ v_x \end{bmatrix}.$$

Then, by virtue of Eq. (3.13), the transition from the $(i+1)$-th to i-th coordinate system transforms the components of the vector v as

$$v' = \hat{T}_i v,$$

where $\hat{T}_i$ is the orthogonal matrix

$$\hat{T}_i = \begin{bmatrix} \cos \theta_i & \sin \theta_i & 0 \\ \sin \theta_i \cos \varphi_i & -\cos \theta_i \cos \varphi_i & \sin \varphi_i \\ \sin \theta_i \sin \varphi_i & -\cos \theta_i \sin \varphi_i & -\cos \varphi_i \end{bmatrix}$$

and v' is the column vector with components v'_x, v'_y, v'_z.

Now, it should be noted that $\cos \theta_{i,i+k} = ((n_x)_i ; (n_x)_{i+k})$. To calculate this scalar product, we transform the vector $(n_x)_{i+k}$ by successive transitions [from the $(i+k)$-th to the $(i+k-1)$-th coordinate system, then from the $(i+k-1)$-th to the $(i+k-2)$-th system, and so on] to the i-th coordinate system. In this system, the vector $(n_x)_{i+k}$ equals

$$\hat{T}_i \hat{T}_{i+1} ... \hat{T}_{i+k} (n_x)_{i+k},$$

and consequently

$$\cos \theta_{i,i+k} = [100] \hat{T}_i \hat{T}_{i+1} ... \hat{T}_{i+k} \begin{bmatrix} 1 \\ 0 \\ 0 \end{bmatrix} = (\hat{T}_i \hat{T}_{i+1} ... \hat{T}_{i+k})_{11}, \qquad (3.14)$$

where the lower sign 11 denotes the corresponding matrix element and

$$\langle \cos \theta_{i,i+k} \rangle = \langle (\hat{T}_i \hat{T}_{i+1} ... \hat{T}_{i+k})_{11} \rangle$$

$$= \frac{\int (\hat{T}_i \hat{T}_{i+1} ... \hat{T}_{i+k})_{11} \Pi_{j=i+1}^{i+k-1} \exp\left[-\frac{U(\varphi_j)}{T}\right] \sin \theta_j d\theta_j d\varphi_j}{\int \Pi_{j=i+1}^{i+k-1} \exp\left[-\frac{U(\varphi_j)}{T}\right] \sin \theta_j d\theta_j d\varphi_j}$$

$$(3.15)$$

The integrations in Eq. (3.15) separate to yield

$$\langle (\hat{T}_i \hat{T}_{i+1} ... \hat{T}_{i+k}) \rangle = \langle \hat{T} \rangle^{k+1},$$

where

$$\langle \hat{T} \rangle = \frac{\int \hat{T} \exp\left[-\frac{U(\varphi)}{T}\right] \sin\theta \, d\theta \, d\varphi}{\int \exp\left[-\frac{U(\varphi)}{T}\right] \sin\theta \, d\theta \, d\varphi}$$

$$= \begin{bmatrix} \cos \gamma & \sin \gamma & 0 \\ \sin \gamma \langle \cos \varphi \rangle & -\cos \gamma \langle \cos \varphi \rangle & 0 \\ 0 & 0 & -\langle \cos \varphi \rangle \end{bmatrix};$$

$$\langle \cos \varphi \rangle = \frac{\int \cos \varphi \exp\left[-\frac{U(\varphi)}{T}\right] d\varphi}{\int \exp\left[-\frac{U(\varphi)}{T}\right] d\varphi}.$$

While averaging the matrix $\hat{T}$, we took into account that $\langle \sin \varphi \rangle = 0$ for the symmetric potential shown in Figure 1.5b, when $U(\varphi) = U(-\varphi)$.

Thus, the relation (3.9) can be rewritten as

$$\langle R^2 \rangle = Nb^2 + 2b^2 \left(\sum_{i=1}^{N} \sum_{k=1}^{N-i+1} \langle \hat{T} \rangle^k \right)_{11}$$

$$= Nb^2 + 2b^2 \left(\sum_{k=1}^{N} (N-k+1) \langle \hat{T} \rangle^k \right)_{11}$$

$$= Nb^2 ((\hat{E} + \langle \hat{T} \rangle)(\hat{E} - \langle \hat{T} \rangle)^{-1} - (2\langle \hat{T} \rangle / N)(\hat{E} - \langle \hat{T} \rangle^N)(\hat{E} - \langle \hat{T} \rangle)^{-2})_{11}$$

$$\cong Nb^2 ((\hat{E} + \langle \hat{T} \rangle)(\hat{E} - \langle \hat{T} \rangle)^{-1})_{11} \qquad (3.16)$$

where $\hat{E}$ is the unit matrix ($\hat{E}v = v$). The third equality is obtained considering that the geometric progression of matrices obeys the same formulas as that of scalar quantities. In fact, all of the elements of the matrix $\langle \hat{T} \rangle$ are less than unity; therefore all elements of $\langle \hat{T} \rangle^N$ are much less than unity. This is because $N \gg 1$. The quotient $\hat{X}$ of the matrix $\hat{E} + \langle \hat{T} \rangle$ and matrix $\hat{E} - \langle \hat{T} \rangle$ is found from the equation $\hat{E} + \langle \hat{T} \rangle = \hat{X}(\hat{E} - \langle \hat{T} \rangle)$ and leads to a system of two linear equations for the elements X_{11} and X_{12}. Its solution is

$$X_{11} = \frac{1 + \langle \cos \varphi \rangle}{1 - \langle \cos \varphi \rangle} \frac{1 + \cos \gamma}{1 - \cos \gamma}.$$

Thus, we finally obtain

$$\langle R^2 \rangle = Nb^2 \frac{1 + \langle \cos \varphi \rangle}{1 - \langle \cos \varphi \rangle} \frac{1 + \cos \gamma}{1 - \cos \gamma}. \tag{3.17}$$

As expected, Eq. (3.17) predicts an infinite growth of $\langle R^2 \rangle$ as $\varphi \to 0$ and $\cos \varphi \to 1$; in the latter case, the chain conformation produces a planar *trans*-zigzag (*see* subsection 2.1). It is interesting that according to Eq. (3.17), $\langle R^2 \rangle / N \to 0$ at $\cos \varphi \to -1$. In this case, it can easily be shown that the size of the chain conformation remains finite even at $N \to \infty$.

3.6. *The size of more complicated chains with rotational-isomeric flexibility is calculated by approximation of discrete rotational-isomeric states.*

The assumption made in the previous subsection that internal rotation potentials for different bonds are independent is rarely satisfied in practice. The interaction of neighboring chain links, or in other words, the interdependence of the potentials of internal rotation around the neighboring bonds, requires the application of methods of the statistical physics of cooperative one-dimensional systems. In Sec. 6, we present these methods in a form that is convenient for further applications; here, we should note that their application to systems with a continuous set of states (defined by continuous variation of the internal rotation angle) comes up against some technical problems. To overcome these, M. V. Volkenstein, T. M. Birstein, and O. B. Ptitsin, as well as P. Flory, developed in the 1950s the formalism of the so-called rotational-isomeric approximation (see Refs. 2–4). The crux of the theory is based on the fact (already mentioned in subsection 2.1) that links reside most of the time in conformations characterized by the minima of the internal rotation potential $U(\varphi)$. Consequently, the links possess a practically discrete set of rotational isomeric states. In more detail, one can examine the rotational-isomeric approximation together with appropriate mathematic methods and results of the studies of specific polymers in Refs. 2–4.

4. GAUSSIAN DISTRIBUTION FOR AN IDEAL POLYMER CHAIN AND THE STANDARD MODEL OF A MACROMOLECULE

4.1. *The statistical distribution of the end-to-end vector of an ideal polymer chain is Gaussian.*

In addition to the mean square of the end-to-end vector $\langle R^2 \rangle$ of the coiled chain characterizing the mean size of the coil, there is a more specific quantity describing the coil, $P_N(R)$. This is the probability distribution function that the end-to-end vector of the chain consisting of N links equals R.

For the freely jointed chain, the vector R equals the sum of N independent, randomly oriented contributions u_i. According to the central limit theorem of probability theory, such a quantity has (at $N \gg 1$) the Gaussian distribution (*see* subsection 4.3):

$$P_N(R) = (2\pi N l^2/3)^{-3/2} \exp[-3R^2/(2Nl^2)]. \qquad (4.1)$$

The factor $(2\pi N l^2/3)^{-3/2}$ is found from the normalization condition $\int P_N(R) d^3R = 1$. The Gaussian function (4.1) decays at the distance of order $R \sim N^{1/2} l$, which agrees with Eq. (1.3) defining the size of the freely jointed chain. Evidently, an accurate calculation of the mean square using the general formula $\langle R^2 \rangle = \int R^2 P_N(R) d^3R$ would yield precisely the result (1.3).

Other ideal chain models with different flexibility mechanisms and without free joints are more complicated, because their consecutive elementary segments are not oriented independently. However, the orientational correlations diminish with distance very rapidly, in fact, exponentially, as Eq. (2.2) predicts. One can expect, and this can be proved, that the central limit theorem also is valid for the exponential decay of correlations. Then, treating any ideal polymer as an effective freely jointed chain of Kuhn segments, we can obtain the correct result for the statistical distribution of the end-to-end vector

$$P_N(R) = (2\pi \langle R^2 \rangle/3)^{-3/2} \exp[-3R^2/(2\langle R^2 \rangle)], \qquad (4.2)$$

where $\langle R^2 \rangle$ is given by Eq. (3.3).

Note that the components of the vector R also obey the Gaussian distribution. The Gaussian distribution is probably the most significant and distinctive property of the ideal polymer coil. In this regard, the coil itself and even the ideal chain are called Gaussian.

4.2. *Owing to the statistical independence of conformations of the different sections of an ideal polymer chain, this chain can be mathematically described as a Markov chain.*

We have seen that not only the mean square $\langle R^2 \rangle$ of the Gaussian coil but also the probability distribution $P_N(R)$ (4.2) depend on only one quantity specified by the chemical structure of a specific polymer: the Kuhn segment length l. We now pass to a more detailed method of describing chain conformations and discuss the statistical distribution of all possible spatial forms (conformations) of the polymer chain.

Choose some N points of the chain (Fig. 1.6a) and trace the chain conformation by registering positions of the chosen points via their radius vectors x_t, where $t = 1, 2, \ldots, N$ are the point numbers. By increasing (or decreasing) the number N, we increase (or decrease) the accuracy of the conformation description. We shall deal with the quantity $P(x_1, x_2, \ldots, x_N) = P_N\{x_t\}$, the probability density of the set of chosen point coordinates or of the polygonal line modeling

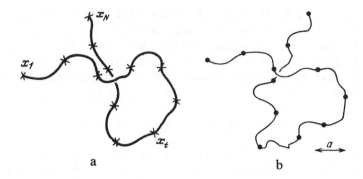

FIGURE 1.6. (a), Polymer chain conformation specified by radius vectors of selected points x_t. (b), Standard bead-on-a-filament model of a polymer chain.

the polymer chain with the chosen accuracy (Fig. 1.6a).

Suppose that initially we have a portion of the chain comprising $t-1$ sections (points $x_1, x_2, \ldots, x_t$). Then we add one more section with the point x_{t+1} at the end of the main chain. When the points x are spaced widely on the ideal chain, the statistical distribution of the coordinates of the point x_{t+1} certainly depends on x_t but not on the coordinates of the foregoing points $x_{t-1}, x_{t-2}, \ldots, x_1$. More precisely, denoting the conditional probability of finding the end of the additional section at the point x_{t+1}, provided its beginning (i.e., the end of the foregoing section) is fixed at the point x_t by $g(x_t, x_{t+1})$, we can say that the value of $g(x_t, x_{t+1})$ is independent of the conformations of the foregoing sections of the chain $x_{t-1}, x_{t-2}, \ldots$[a] Being a probabilistic quantity, g satisfies the normalization condition

$$\int g(x', \ x)d^3x = 1. \tag{4.3}$$

Similar conclusions are valid for all points x_t of the chain, from the beginning x_1 to the end x_N. Thus, the probability P takes the form

$$P(x_1, \ x_2, \ \ldots, \ x_N) = g(x_1, \ x_2)g(x_2, \ x_3)\ldots g(x_{N-1}, \ x_N). \tag{4.4}$$

It should be pointed out that such structure of the function $P(x_1, x_2, \ldots, x_N)$ is conditioned by the linear memory of an ideal polymer. There are "linear" interactions of sections along the chain but no volume interactions. This latter fact accounts for different sections of the ideal chain being statistically independent. The relevant probabilities appear in Eq. (4.4) as factors of a simple product.

---

[a]Strictly speaking, the value of g depends not only on the coordinates x_{t+1} and x_t but also on the chain orientations at these points. For the sake of brevity, we omit the designations of the corresponding variables.

Systems with the probability distribution of type (4.4) were investigated by A. A. Markov and are called *Markov chains*. An ideal polymer chain thus belongs to this class.

Next, provided the points $\{x_t\}$ are chosen so rarely that the distance between them exceeds the Kuhn segment length l, the probability for the point x_{t+1} is independent of the chain orientation at the point x_t. That is, the value of g in this case depends only on the distance between the points:

$$g(x', x) = g(|x'-x|), \quad \int g(|y|)d^3y = 1. \tag{4.5}$$

Choosing, for example, the points $\{x_t\}$ at each joint location of a freely jointed chain, we obtain for this model

$$g(x', x) = [1/(4\pi l^2)]\delta(|x-x'|-l). \tag{4.6}$$

The last relation expresses the trivial fact that in the case of a freely jointed chain, the point x_{t+1} can be found with equal probability at any point of the sphere of radius l and center at x_t.

Knowing such a detailed characteristic as the probability distribution $P(x_1, x_2, \ldots, x_N)$ (4.4), one can certainly find any less specific parameter, for example, the distribution of the end-to-end distance $P_N(R)$. In this case, the end-to-end distance $x_N - x_1 = R$ must be fixed and then the probability $P(x_1, x_2, \ldots, x_N)$ integrated over all conformations having the given coordinates of the ends, that is, over the coordinates of all intermediate points $x_2, \ldots, x_{N-1}$:

$$P_N(R) = \int P(x_1, x_2, \ldots, x_{N-1}, x_N = x_1 + R)d^3x_2...d^3x_{N-1}$$

$$= \int \delta\left(\sum_{i=1}^{N-1} y_i - R\right) \prod_{i=1}^{N-1} g(y_i)d^3y_1...d^3y_{N-1}, \tag{4.7}$$

where $y_1 = x_{i+1} - x_i$ and the simplifying assumption (4.5) was used.

From Eq. (4.7) follows the Gaussian distribution (4.1) or (4.2), which was derived by a formal reference to the central limit theorem. This will be proved in the next subsection.

*4.3. *The end-to-end distance of an ideal macromolecule on the scale of a polymer coil obeys the Gaussian distribution to a high degree of accuracy.*

To transform Eq. (4.7), it is convenient to make use of the known integral representation of the delta function

$$\delta(x) = (2\pi)^{-3} \int \exp(-ikx)d^3k;$$

Then we obtain

$$P_N(R) = (2\pi)^{-3} \int \exp(-ikR)\widetilde{P}_N(k)d^3k, \qquad (4.8)$$

$$\widetilde{P}_N(k) = g_k^N, \qquad (4.9)$$

$$g_k = \int g(y)\exp(iky)d^3y. \qquad (4.10)$$

Thus, the Fourier transformation of the unknown function $P_N(R)$ has a very simple form $\widetilde{P}_N(k) = g_k^N$. It is easy to see that the value of g_k is equal to unity at $k=0$ [which follows immediately from the definition (4.10) and the normalization conditions (4.3) and (4.5)] and tends to zero as $|k|$ grows. Therefore, the function g_k^N has the form of a very narrow peak at $N \gg 1$. This makes it possible to write $g_k^N = \exp(N \ln g_k)$ and to replace the exponential index by the first terms of the expansion in a power series of k. If we limit ourselves to the square term, the calculation of the integral (4.8) leads to the Gaussian distribution (4.1). Taking into account the first correction term (proportional to k^4), we can find the correction to the Gaussian distribution

$$P_N(R) = (2\pi Nl^2/3)^{-3/2} \exp[-3R^2/(2Nl^2)] \left[1 - \frac{3}{20N} \left(5 - \frac{10R^2}{Nl^2} + \frac{3R^4}{N^2l^4} \right) + \cdots \right].$$
$$(4.11)$$

The coefficients of the correction (in brackets) are calculated for the case of the freely jointed chain (4.6) $g_k = (\sin kl)/kl$. It is seen that at $R^2 \lesssim Nl^2$, the corrections are of order $1/N$ and can be neglected for $N \gg 1$. If $R^2 \gg Nl^2$, the corrections can become substantial; nevertheless, owing to the exponential decrease, the function $P_N(R)$ itself is so small in this region that the deviations from the Gaussian law are quite inessential for most practical problems (*see*, however, subsection 8.3)

4.4. *The distance between any two not-so-close points of an ideal polymer chain, just as the end-to-end distance, obeys the Gaussian distribution.*

According to the previous subsection, the Gaussian distribution follows from the general formula (4.7) at $N \gg 1$ irrespective of the form of the function $g(y)$ (i.e., of the specific structure of the polymer chain). Having written Eq. (4.8) for the arbitrary g, we obtain

$$P_N(R) = (2\pi Na^2/3)^{-3/2} \exp[-3R^2/(2Na^2)],$$

$$\langle R^2 \rangle = Na^2, \qquad (4.12)$$

where a is defined by the relation

$$a^2 = \int y^2 g(y)d^3y \quad (g_k \cong 1 - (ka)^2/6 \text{ at } k \to 0). \qquad (4.13)$$

Obviously, for the ideal polymer chain, Eq. (4.12) equally holds for any pair of points i and j in the chain with $|i-j|$ substituted for N and $x_i - x_j$ for R. This conclusion can be confirmed by the formulas of subsection 4.3.

Evidently, for lengths of the order of a Kuhn segment or less, the distribution is far from Gaussian and depends on the specific chain structure. As shown, however, the statistical distribution of an ideal chain on large length scales becomes Gaussian because of chain flexibility irrespective of the structure. This makes it possible to develop a unified model suitable for description and investigation of large-scale properties of any ideal macromolecule.

4.5. *A chain of "beads" connected by Gaussian filaments provides a versatile standard model for the description of large-scale properties of macromolecules.*

While discussing large-scale (not local) properties, there is no need to trace all sections of the macromolecule. We can study the conformation of a polymer chain in more general terms by choosing the points $\{x_t\}$, characterizing the spatial arrangement (*see* subsection 4.2, Fig. 1.6a), sufficiently wide on the chain. When the length of the chain section separating the neighboring points t and $t+1$ exceeds that of the Kuhn segment, the correlation between these points is Gaussian

$$g(x, x') = (2\pi a^2/3)^{-3/2} \exp[-3(x-x')^2/2a^2]. \qquad (4.14)$$

as was shown in subsection 4.4.

The chosen points of the chain can be pictured as the beads and the rest of the macromolecule as the filament connecting the beads (Fig. 1.6b). In the real case, the filament is material and the beads imaginary. In the theoretic papers, however, the standard "bead-on-a-filament" model is frequently used, in which (conversely) the beads are identified with the real links (or groups of links) and the filament treated as immaterial. Owing to the Gaussian correlations of neighboring beads (4.14), this model is also called the *standard Gaussian model of the polymer chain*. From Eq. (4.14), it follows that in the standard model, the quantity a signifies the rms distance between the neighboring beads in the chain.

The primary advantage of the standard model is associated with its considerable simplification of the mathematic description. For example, Eq. (4.12) is exact for this model, and this can be proven easily by performing the calculations of subsection 4.3. On the other hand, the large-scale properties of polymer coils, for example, the Gaussian distribution $P_N(R)$, are independent of the fine local structure of the chains. Therefore, the choice of the standard model for the description of such properties does not impose any restrictions on generality.

This last statement signifies that for any polymer chain, one can find a standard chain of beads such that their large-scale characteristics will coincide. Indeed, as is shown in the next section, the properties of the coil as a whole do not depend on all chain parameters but only on L and l, namely, on the product $Ll = \langle R^2 \rangle$. For the standard Gaussian model, $\langle R^2 \rangle = Na^2$. Hence, having divided the polymer chain with parameters L and l into sections by N chosen points, it is sufficient to select the parameter a of the bead model so that

$$Na^2 = Ll, \tag{4.15}$$

and all large-scale properties of the coil of beads will coincide with the corresponding characteristics of the initial coil.

Having taken this into account, we examine the properties of Gaussian coils in the next section, using primarily the standard Gaussian model of the polymer chain. Eq. (4.15) is be used to relate the characteristics of the model with the parameters of real chains.

5. PROPERTIES OF A GAUSSIAN COIL

5.1. *A Gaussian coil is characterized by a single macroscopic spatial scale; the radius of gyration, hydrodynamic radius, end-to-end distance, and so on are all of the same order of magnitude and the relative fluctuations of any of these quantities are of the order of unity.*

From the distribution $P_N(R)$, it is possible to calculate all the moments (i.e., the rms values of any even power of R). In terms of the standard model [see Eq. (4.12)],

$$\langle (R^2)^n \rangle = \int (R^2)^n P_N(R) d^3R = (Na^2)^n [1 \cdot 3 \cdot \ldots \cdot (2n+1)/3^n] \tag{5.1}$$

(the averaged values of odd powers of R are of course equal to zero). Remarkably, any characteristic end-to-end distance $R \sim \langle (R^2)^n \rangle^{1/2n}$, irrespective of the moment number n, proves to be proportional to $aN^{1/2}$ (i.e., to the same power of N). This also can be seen directly from the form of the Gaussian distribution (4.1). It features the quantities a and N only in the combination $aN^{1/2}$, and there is no other length except a. Thus, the coil as a whole has only one characteristic size, $aN^{1/2}$.

This conclusion is important, because different experimental methods are used for the measurement of various coil parameters. For example, in subsection 5.5, we show that the investigation of the elastic scattering of light by a dilute polymer solution allows one to measure the mean square of the radius of gyration of the coil[b]

$$s^2 = \frac{1}{2} N^{-2} \left\langle \sum_{i=1}^{N} \sum_{j=1}^{N} (r_{ij})^2 \right\rangle, \tag{5.2}$$

By the method of inelastic light scattering, one can measure the diffusion coefficient, or in the final analysis, the so-called hydrodynamic radius of the coil (*see* subsections 32.4 and 33.4)

---

[b]Generally, the square of the radius of gyration of a body or a system of points is a ratio of a moment of inertia to a mass. For a system of identical particles, we must apparently take the sum of the squares of the distances from all particles to the center of gravity, then divide it by the total number of particles N. We urge the reader to prove that this quantity can be brought to the form (5.2). (The theorem is from Lagrange) (see Ref. 4, Appendix 1).

$$R_D^{-1} = N^{-2} \left\langle \sum_{i=1}^{N} \sum_{j=1(i \neq j)}^{N} \left| r_{ij} \right|^{-1} \right\rangle. \tag{5.3}$$

where r_{ij} is the vector connecting the links i and j. What are the values of these characteristic parameters?

They can be calculated easily in terms of the "bead" model, because for this model, the Gaussian distribution (4.1) is valid for any chain section i, j (see subsection 4.4). Consequently, it is easy to find that

$$\langle r_{ij}^2 \rangle = |i-j|a^2, \quad \langle |r_{ij}|^{-1} \rangle = (6/\pi)^{1/2}|i-j|^{-1/2}a^{-1}. \tag{5.4}$$

The simplest way to calculate the sums (5.2) and (5.3) is to replace them by integrals, which is valid for $N \gg 1$. For $N \gg 1$, the result is written as

$$s = (1/6)^{1/2} \langle R^2 \rangle^{1/2} \quad ((1/6)^{1/2} \approx 0.41), \tag{5.5}$$

$$R_D = (3\pi/128)^{1/2} \langle R^2 \rangle^{1/2} \quad ((3\pi/128)^{1/2} \approx 0.27). \tag{5.6}$$

Both quantities are indeed of the order of $aN^{1/2}$, as was expected.

The expression (5.1) also makes it possible to determine the relative magnitude of the fluctuations of R^2, that is, the square of the end-to-end distance of the chain:

$$\frac{\langle (R^2 - \langle R^2 \rangle)^2 \rangle}{\langle R^2 \rangle^2} = \frac{\langle R^4 \rangle - \langle R^2 \rangle^2}{\langle R^2 \rangle^2} = \frac{2}{3}. \tag{5.7}$$

It follows that the fluctuations of R^2 are of the order of the mean value of this quantity. The analogous conclusion can be drawn for other even powers of R and also for the quantities s, R_D, and others. This implies that the Gaussian polymer coil is a strongly fluctuating system; additional confirmation of this conclusion is given in subsections 5.2 and 5.3.

It should be noted that all results of this subsection obtained in terms of the standard "bead" model are quite appropriate. This is because we look into the properties of the coil as a whole which are not associated with any small-scale properties.

Formally, this follows from the fact that the replacement of the sums (5.2) and (5.3) by the integrals either does not lead to any singularity in the integrand at $(i-j) \to 0$ [for Eq. (5.2)] or that this singularity is integrable [for Eq. (5.3)]. This indicates that the contribution from small scales to the macroscopic quantities s, R_D, and so on is inessential.

5.2. The Gaussian coil is an extremely "loose" system having low density and intense fluctuations.

A smooth decay of the Gaussian exponential (4.2) or (4.12) leads one to picture the coil as a blurred "cloud" of links that is denser at the center and more loose at the periphery. This representation, however, is wrong. Indeed, let us evaluate the mean concentration of links (beads) in the coil. We know from Eq.

(4.12) that the coil size is of the order of $aN^{1/2}$, and consequently, its volume is approximately $a^3 N^{3/2}$. This volume contains N beads, and their mean concentration is thus equal to

$$n \sim N/a^3 N^{3/2} \sim a^{-3} N^{-1/2}; \qquad (5.8)$$

In other words, the mean contour length of the chain contained in a unit volume of the coil is of order $L/(Ll)^{3/2} \sim l^{-3/2} L^{-1/2}$ (i.e., these quantities tend to zero as the chain length grows). Smallness of n signifies that nearly all of the volume of the coil is empty, or rather free of the chain, being occupied by a solvent.

This situation is clearly illustrated in Figure 1.7, where a typical computer-simulated conformation of the ideal coil is shown. The chain fails to take up the volume of the coil.

How, then, does the smooth distribution (4.2) come about? Obviously, it appears only because of averaging over the immense number of possible conformations of the polymer coil.

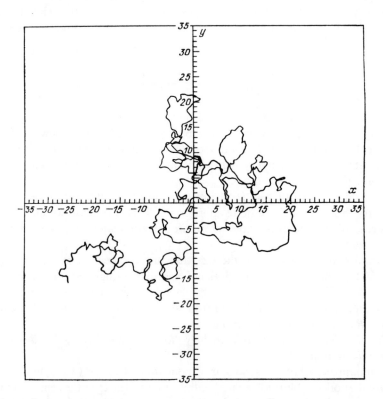

FIGURE 1.7. Conformation of a Gaussian coil for a freely jointed chain of 626 segments, each segment is of unit length. (Courtesy of N. K. Balabayev.)

From this, it follows that the fluctuations of the chain shape in the coil are by no means local in their nature but represent macroscopic pulsations. This fact also is illustrated in the next subsection.

5.3. *The correlation radius of concentration fluctuations of the links in a coil coincides in order of magnitude with the coil size.*

We have seen (*see* subsection 5.1) that the fluctuations of the coil size are of the order of the size itself. Now we describe the fluctuations of the coil structure in more detail. A common approach to the fluctuation analysis involves the study of correlation functions. In terms of the standard "bead" model, the microscopic concentration of the chain links (beads) in the conformation (i.e., a microscopic state) $\Gamma \equiv (x_1, x_2, \ldots, x_N)$ is consistently defined as

$$n_\Gamma(x) = \sum_{i=1}^{N} \delta(x_i - x). \tag{5.9}$$

Now we can readily obtain the average concentration in the vicinity of any spatial point x, that is, the spatial distribution, or the profile, of the average concentration. Assuming for simplicity that the initial link is fixed at the origin ($x_1 = 0$), we get

$$n(x) \equiv \langle n_\Gamma(x) \rangle = \sum_{i=1}^{N} P_i(x).$$

Replacing here the summation by an integration and taking into account that the distribution $P_i(x)$ is Gaussian, we find for the chain consisting of i links:

$$n(x) = (3/2\pi)^{3/2} a^{-3} N^{-1/2} \int_0^1 q^{-3/2} \exp(-3\xi^2/2q) dq, \tag{5.10}$$

where $\xi^2 = x^2/(Na^2)$ and $i/N = q$ is the integration variable. In the range $1/N^{1/2} \ll \xi \ll 1$, the function $n(x)$ diminishes as $1/(a^2|x|)$, while at $\xi \gg 1$, it falls exponentially [as $(N^{1/2}/ax^2)\exp(-3x^2/2Na^2)$]. This ties in with our knowledge of the uniqueness of the characteristic scale $x \sim aN^{1/2}$ (i.e., $\xi \sim 1$) in the coil and with the evaluation (5.8) of the average concentration of links in the coil.

Similarly, one can also find the correlation function

$$\langle n_\Gamma(x_1) n_\Gamma(x_2) \rangle = \left(\frac{3}{2\pi a^2}\right)^3 \frac{1}{N} \int_0^1 dq_1 \int_0^1 dq_2$$
$$\times \frac{\exp[-(3\xi_1^2/2q_1) - (3(\xi_1 - \xi_2)^2/2|q_1 - q_2|)]}{q_1^{3/2}|q_1 - q_2|^{3/2}}.$$

The awkward appearance of this integral should not fluster anyone. What is essential is that the conversion to dimensionless variables ξ and q is possible. This fact is sufficient for one to infer that the correlation radius, at which the correlation function

$$[\langle n_\Gamma(x)n_\Gamma(x')\rangle - \langle n_\Gamma(x)\rangle\langle n_\Gamma(x')\rangle] \Big/ (\langle n_\Gamma(x)\rangle\langle n_\Gamma(x')\rangle),$$

declines, corresponds to the same single scale in the coil, $\xi \sim 1$.

5.4. The Gaussian coil possesses the property of scale invariance.

Uniqueness of the intrinsic macroscopic length scale $\langle R^2\rangle^{1/2} = (Ll)^{1/2} = aN^{1/2}$ lies at the root of scaling invariance of the Gaussian coil, the property whose importance is expounded to a full measure in studies of nonideal polymer systems (*see* Secs. 16, 18, 19, 25, 26). To interpret this property, note that in the transition from a real polymer chain to the standard "bead" model (*see* subsection 4.5), the choice of N dividing points, or the number of beads, is arbitrary. What is important is that the distance a between the neighboring beads must be found from Eq. (4.15) for each choice.

For example, having made a certain choice of N points, we can pass to a choice of half as many points, that is, make the substitution $N \rightarrow N/2$. Then a "bond" in the new bead chain becomes equivalent to a pair of "old" bonds, and the "new" correlation function equals [cf. Eqs. (4.4) and (4.7)]

$$\tilde{g}(x,x') = \int g(x, x'')g(x'', x')d^3x'' = (4\pi a^2/3)^{-3/2} \exp[-3(x-x')^2/4a^2].$$

The correlation function thus remains Gaussian, with a^2 being replaced by $2a^2$, as it should be according to Eq. (4.15). Similarly, it can be checked easily that the addition of new beads between existing ones results in the substitutions $N \rightarrow 2N$, $a^2 \rightarrow a^2/2$, and so on.

Arbitrariness in the choice of N makes sense. It implies that any Gaussian chain section drawn at different scales yields the same pattern (*see* Fig. 1.8) provided the smallest discernible elements are much greater than the persistent length. It is this property that is referred to as *scale invariance*.

5.5. For the Gaussian coil, the statistical structure factor, closely associated with the pair correlation function of concentrations and measured in elastic radiation scattering experiments, is exactly defined by the simple Debye formula.

The correlation parameters that we began considering in subsection 5.3 are studied by widespread research methods associated with elastic (i.e., without frequency change or, in optical terms, Rayleigh) scattering of radiation, light, x-rays, neutrons, and so on. The scattered radiation intensity measured in such experiments is proportional to the quantity

$$G(k) = \frac{1}{N}\left\langle \left|\sum_{n=1}^{N} \exp(ikx_n)\right|^2\right\rangle = \frac{1}{N}\left\langle \sum_{n=1}^{N}\sum_{m=1}^{N} \exp(ik(x_n-x_m))\right\rangle, \tag{5.11}$$

which is called the *static structure factor*, or a *form factor of the system*. Here, k is the scattered radiation vector

$$|k| = (4\pi/\lambda)\sin(\theta/2), \tag{5.12}$$

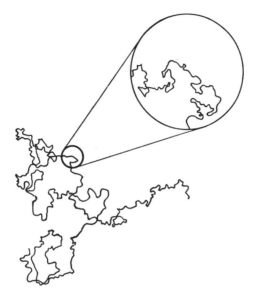

FIGURE 1.8. Explanation of scale invariance of a coil.

θ is the scattering angle, λ is the wavelength, and x_n are the coordinates of scattering centers (*see* Ref. 21 for more detail). Provided we are not interested in small-scale specifics while considering the scattering by a polymer system, the standard Gaussian model of the polymer chain can be applied without restricting generality, with the points $\{x_n\}$ being identified with the beads.

The structure factor is related to the correlation function as

$$G(k) = \frac{1}{N} \int \langle n_\Gamma(x) n_\Gamma(x') \rangle \exp[ik(x-x')] d^3x d^3x', \qquad (5.13)$$

We suggest that the reader prove this relation proceeding from Eq. (5.9) for $n_\Gamma(x)$.

Later, we analyze the structure factors of various polymer systems (*see* subsections 19.6, 23.4, 25.9, 33.3). Now we give one general relation and calculate the structure factor of the Gaussian coil.

A simple general expression for the form factor $G(k)$ of a single polymer chain can be obtained in the limit $|k| \to 0$, corresponding, according to Eq. (5.12), to long-wavelength radiation or scattering through small angles. In this limit, $kx_n \ll 1$, because all of the beads are contained within a restricted volume. Expanding the exponential in Eq. (5.11) in a power series of $|k|$, taking into account that the linear term is zero owing to the symmetry of x_n and x_m and using the definition of the radius of gyration s (5.2), we obtain

$$G(k) \cong N\left(1 - \frac{1}{3}s^2k^2\right) \quad (sk \ll 1). \tag{5.14}$$

It should be noted that this asymptotic expression is applicable both for the ideal Gaussian coil and the real polymer chain with arbitrary volume interactions.

Let us now calculate the structure factor of a Gaussian coil for arbitrary k. For the bead model, the vector $x_i - x_j = r_{ij}$ is distributed according to the Gaussian law for any $i - j$, and we therefore get

$$\langle \exp[ik(x_i - x_j)]\rangle = \int P_{|i-j|}(r)\exp(ikr)d^3r = \exp[-|i-j|(ka)^2/6], \tag{5.15}$$

where the integral is computed by using Eq. (4.12). Substituting this result into the definition of $G(k)$ (5.11) and calculating the sums by replacing them by the integrals, we find that at $N \gg 1$ and $ka \ll 1$

$$G(k) = \frac{12}{(ka)^2}\left\{1 - \frac{6}{N(ka)^2}\left[1 - \exp\left(-\frac{N(ka)^2}{6}\right)\right]\right\}. \tag{5.16}$$

In other words, defining the dimensionless variable

$$x = (ks)^2 = N(ka)^2/6,$$

where s is the rms gyration radius of macromolecule (see Eq. (5.2) or (5.5)), one can rewrite the expression for the scattered intensity per one monomer in the form

$$\frac{G(k)}{N} = g_D(x) = \frac{2}{x^2}[x - 1 + \exp(-x)].$$

It is clearly seen from this expression, that only one macroscopic length scale is involved. The relation (5.16) was obtained by P. Debye in 1944. The function $G(k)$ defined by Eq. (5.16) is called the *Debye scattering factor*.

In the limiting case of short-wavelength radiation and not small angles of scattering, that is, at $N(ka)^2 \gg 1$, the Debye formula yields

$$G(k) \cong 12/(ka)^2 \sim \lambda^2/[a^2 \sin^2(\theta/2)]. \tag{5.17}$$

Because the typical size of polymer coils is approximately 10 to 50 nm, the condition $N(ka)^2 \sim (ka)^2 \gg 1$ can be satisfied only for hard x-rays or neutrons ($\lambda \ll s$) when the asymptotic dependence (5.17) is indeed observed.

As a rule, for the more common case of light scattering, $\lambda \gg s$ $\sim N^{1/2}a \sim (Ll)^{1/2}$. The corresponding asymptotic behavior of the Debye formula coincides, naturally, with the general expression (5.14).

5.6. *The radius of gyration, the contour length, and the Kuhn segment of the chains can be determined from experimental data on elastic scattering by a dilute polymer solution.*

From Eq. (5.14), it follows that the mean square radius of gyration s^2 of the macromolecule can be found by measuring the angular dependence of elastic light scattering by dilute polymer solutions. Note also that according to Eqs. (5.5) and (5.3), $s^2 = Ll/6$ for the ideal macromolecule. The total contour length L of the macromolecule is proportional to its molecular mass M, which can be measured with comparative ease. For example, the value of M can be determined by the elastic light scattering technique mentioned earlier by extrapolating the scattered light intensity to $\theta = 0$ (*see* Ref. 21). Many other methods also are available. The proportionality coefficient M/L, the molecular mass per unit length, can be estimated from the stereochemical data. Thus, the measurements of s^2 and M in dilute solutions of ideal macromolecules (i.e., when volume interactions are absent), allow the Kuhn segment length to be found.

6. AN IDEAL CHAIN AS A RANDOM WALK. MATHEMATICAL ASPECTS

6.1. *A chain in an ideal coil forms a path similar to the random walk of a Brownian particle; the fluctuating orientation of each consecutive link relative to the previous one can be treated as a random walk on a sphere.*

Consider a freely jointed polymer chain whose initial point is located at the origin. Then, according to Eq. (1.3), the mean end-to-end distance of the macromolecule is proportional to $N^{1/2}$. This recalls the following well-known property of Brownian motion: a Brownian particle located at the initial moment of time at the origin would be found after the time interval t shifted from it by a mean distance proportional to $t^{1/2}$. Furthermore, the statistic distribution of spatial positions of the particle is given by the fundamental solution of the diffusion equation, that is, exactly by the Gaussian exponential (4.1). Understandably, this coincidence is not surprising. It testifies that both systems (an ideal polymer chain and a Brownian particle) are described in terms of the same mathematic approach.

Suppose we take successive fixed-length vectors whose orientations are random and statistically independent of one another. The vectors are then summed. In the case of the Brownian particle, these vectors are the displacements occurred during the consecutive time intervals Δt; the sum of $N = t/\Delta t$ vectors yields the total displacement for finite time t. In the case of an ideal freely jointed polymer, the individual vectors denote the segments and their sum the end-to-end vector. Thus, the analogy is perfect.

For more complicated models of a polymer chain (differing from the freely jointed model) one has to employ a more complex random walk of the Brownian particle with the correlation of consecutive displacements taken into account. Nevertheless, the conformations of the ideal polymer chain also are analogous in this case, or rather coincide, with the random walk path of the Brownian particle. The coordinate variation along the chain, or the ordinal number of a monomer link, acts as a time variable.

Together with the ordinary Brownian motion, there is a well-known random variation of orientations that is characteristic of anisotropic microscopic particles. It can be described as a succession of elementary turns in the random direction. If one defines the orientation of microscopic particles by a unit vector fixed at the origin, these turns result in a random walk that the end of this vector performs over the surface of a unit sphere. Using the analogy between the number of a link and the time, it is possible to assert that the orientations of consecutive links in the ideal polymer chain also form a random walk on the unit sphere.

An exact analogy of ideal polymer conformations with random walk trajectories can provide the basis for the mathematical treatment of conformational statistics. We begin with a detailed examination of necessary mathematical tools in the next section on the basis of the standard Gaussian model of a polymer chain located in uniform empty space. This trivial example has already been investigated quite extensively without resorting to any special mathematical apparatus. In the next section, however, we generalize to more complex polymer chains and various external effects. This requires mathematical apparatus that for tutorial purposes are described using the simplest example.

6.2. *A partition function of a polymer chain with fixed ends is called the Green function; it is a fundamental solution of a diffusion-type equation.*

As in any problem of statistical physics, investigation of the conformational statistics of macromolecules begins with the distribution function $\rho(\Gamma)$ and the partition function.

For the simplest standard Gaussian model of a polymer chain, the microscopic state Γ (i.e., the chain conformation) is specified by the set of coordinates of "bead" links: $\Gamma = \{x_0, x_1, \ldots, x_N\}$ (the total number of links in the chain equals $N+1$). The probability of a given microscopic state Γ is given by Eq. (4.4), that is, in this case,

$$\rho(\Gamma) = g(x_0, x_1)g(x_1, x_2)...g(x_{N-1}, x_N), \tag{6.1}$$

where the factors $g(x_i, x_{i+1})$ describe the bonds between neighboring links (i.e., the linear memory); the structure of the formula conforms to the one-dimensional linear connectivity of the ideal polymer chain.

The partition function is obtained by integrating the distribution $\rho(\Gamma)$. It is often convenient to consider a macromolecule whose both terminal links are fixed in space, that is, with the links indexed 0 and N fixed at the given points x_0 and x_N, respectively. The partition function of such a macromolecule depends on x_0 and x_N and is called the *Green function of the polymer chain*, or the *chain propagator*. It is designated as

$$G\left\{\begin{matrix} 0 & N \\ x_0 & x_N \end{matrix}\right\} \equiv \int \rho(\Gamma')\delta(x_0'-x_0)\delta(x_N'-x_N)dx_0'dx_1'...dx_N' \equiv G_N(x_N), \quad (6.2)$$

where the brief notation on the right-hand-side implies that the initial link at the point x_0 is held fixed.

By definition, the Green function (6.2) can be written for the considered simplest case as

$$G_N(x_N) = \int g(x_0, \ x_1)g(x_1, \ x_2)...g(x_{N-1}, \ x_N)dx_1...dx_{N-1}. \quad (6.3)$$

Equation (6.3) is readily interpreted in terms of the analogy with Brownian motion: the Green function

$$G\left\{\begin{matrix} 0 & N \\ x_0 & x_N \end{matrix}\right\}$$

is proportional to the conditional probability that the Brownian particle arrives at the point x_N at the time moment N provided it left the point x_0 at the moment 0. Equation (6.3) expresses this probability in a form that can be called a *path integral*. In this case, the trajectories look like broken lines with vertices $x_1, \ldots, x_{N-1}$.

As a rule, Brownian motion is described mathematically as follows. First, we assume that by the time moment N, a certain statistical distribution $G(x_N)$ of the positions of the Brownian particle x_N has been realized. Then, we find how this distribution varies because of the jumps that the particles perform during the next time interval δN. Proceeding in this fashion, and remembering that the number N links in the chain corresponds to the time variable, we set $\delta N=1$ and write the simple recursive relation for G_{N+1}:

$$G_{N+1}(x_{N+1}) = \int G_N(x_N)g(x_N-x_{N+1})dx_N. \quad (6.4)$$

As the Brownian motion represents a Markovian process and the ideal polymer may also be treated in the same terms, it is natural to identify Eq. (6.4) as a Chapman-Kolmogoroff equation for transitional probabilities. After introducing the "linear memory operator"

$$\hat{g}: \quad \hat{g}\psi(x) = \int g(x, \ x')\psi(x')dx'. \quad (6.5)$$

Eq. (6.4) can be rewritten in the form

$$G_{N+1}(x) = \hat{g}G_N. \quad (6.6)$$

We show later that this equation is of the diffusion type: $G_{N+\delta N} = G_N + \delta N(\partial G_N/\partial N)$ and $g=1+(a^2/6)\Delta$, so that Eq. (6.6) reduces to

$$\frac{\partial G_N}{\partial N} = \frac{a^2}{6} \Delta G_N,$$

where Δ is the conventional Laplace operator. Fixing of the zero-th link at the point x_0 plays the role of an initial condition of the type $G_0(x) = \delta(x - x_0)$ in this equation. For such an initial condition, the solution is

$$G_N(x) = \text{Norm} \cdot \exp[-3x^2/2Na^2]$$

in exact correspondence with Eq. (4.12).

All the results of the previous Sec. 5 can be reformulated in terms of transitional probabilities or diffusion equations; however, this mathematical apparatus is intended for a different purpose. It will help us make some broad generalizations:

1. More complex ideal chains in which the state of a link is specified not only by the position x in space, but also by the orientation u and, possibly, some internal degrees of freedom (e.g., by helicity or adsorption of a small molecule from the solution).

2. Ideal chains subjected to various external influences.

The latter problem (interesting and important by itself) is also necessary for a subsequent investigation of non-ideal polymer systems with volume interactions by the method of the self-consistent field.

6.3. *Many effects experienced by polymer systems can be phenomenologically described in terms of effective external fields.*

Speaking of an effective external field, we do not necessarily mean a real physical (electric, magnetic, and so on) field. We mean only a way to describe a spatial inhomogeneity or anisotropy of the external conditions or to express the dependence of the energy φ of a link on generalized coordinates.

Thus, the compressing external field in which the energy of the link depends on its spatial position $\varphi(x)$ may characterize an attraction of the links to a foreign particle suspended in a solvent, their adsorption on a surface separating solvent phases, the placing of the chain into a microscopic cavity of limited volume, and so on. Figure 1.9 illustrates several typical plots for the spatial profile of the potential $\varphi(x)$, corresponding to typical physical situations.

Subsection 8.2 shows that stretching of the polymer chain by its ends may be regarded as an action on its segments of an external, orienting field in which the energy of the segment depends on its orientation $\varphi(u)$. An analogous (in physical terms) situation is observed in a polymer chain undergoing the helix–coil transition (*see* subsections 40 and 41), where the energy of the link depends on its state (whether the link belongs to helical or coil sections).

Bearing all this in mind, we now develop the required mathematical apparatus.

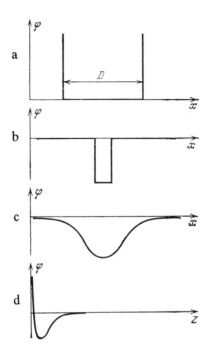

FIGURE 1.9. Typical plots of spatial profiles of the potential $\varphi(x)$, representing various physical conditions at which a polymer chain may exist. (a), Spheric cavity of diameter D (*see* subsections 7.1 and 7.3). (b), Attracting nucleus localized within a small volume (*see* subsection 7.4). (c), Extended potential well (*see* subsection 7.6). (d), Potential well near impenetrable adsorbing wall (*see* subsection 7.7).

6.4. *In the statistics of a polymer chain located in an external field (as in the random walk theory), the Green function satisfies a diffusion-type equation and can be written in the form of a path integral.*

Gradually introducing complications and making generalizations, we now consider the simplest bead model of a macromolecule (with the links specified by the position vectors x_i) placed in the external field $\varphi(x)$. Because the statistical weight of the i-th link at the point x_i equals $\exp[-\varphi(x_i)/T]$, the distribution function $\rho(\Gamma)$ takes the following form different from Eq. (6.1):

$$\rho(\Gamma) = \exp(-\varphi(x_0)/T)\, g(x_0,\ x_1)\exp(-\varphi(x_1)/T)\, g(x_1,\ x_2)$$

$$\times \exp(-\varphi(x_2)/T)\cdots g(x_{N-1},\ x_N)\exp(-\varphi(x_N)/T). \quad (6.1\text{A})$$

As before, the Green function is denoted as in Eq. (6.2), but in its definition via the path integral (6.3), it is convenient to assume that the external field does not act on the initial link, that is, $\varphi(x_0)=0$. Then,

$$G\binom{0\ \big|\ N}{x_0\ \big|\ x_N} = \int \prod_{i=1}^{N} \left[g(x_{i-1},x_i) \exp\left(-\frac{\varphi(x_i)}{T} \right) \right] d^3x_1...d^3x_{N-1}. \quad (6.3A)$$

From Eq. (6.3A) directly follows the next recursive relation, analogous to Eq. (6.4):

$$G\binom{0\ \big|\ N+1}{x_0\ \big|\ x_{N+1}} = \int G\binom{0\ \big|\ N}{x_0\ \big|\ x_N} Q(x_n,x_{N+1}) d^3x_N, \quad (6.4A)$$

where $Q(x',x)$ denotes the Green function for a "polymer" with two links and one bond, namely, a dimer

$$Q(x',x) \equiv G\binom{0\ \big|\ 1}{x'\ \big|\ x} = g(x',x) \exp\left(-\frac{\varphi(x)}{T} \right). \quad (6.5A)$$

By analogy with Eq. (6.6), Eq. (6.4A) may be rewritten in the form

$$G_{N+1}(x_{N+1}) = \hat{Q} G_N(x_N). \quad (6.6A)$$

Here, the action of the transfer operator $\hat{Q}$ is defined by Eq. (6.4).

As in Eq. (6.6), the relation (6.6A) plays the role of a diffusion equation and, in many cases, reduces to a common differential equation. Its formal solution can be written in the form

$$G\binom{0\ \big|\ N}{x_0\ \big|\ x_N} = [\hat{Q}]^N(x_0,x_N) \equiv \int Q(x_0,x_1)Q(x_1,x_2)...Q(x_{N-1},x_N) d^3x_1...d^3x_{N-1}. \quad (6.7)$$

The integral in Eq. (6.7) may be treated as a path integral along trajectories having the form of broken lines connecting the points x_0 and x_N. Thus, in the problems of polymer statistics as in the random walk theory, one can introduce the diffusion equation (6.6) for the Green function, path integral (6.7), and transfer operator $\hat{Q}$.

For an arbitrary model of the polymer chain, the state of a link is specified not only by its coordinate x but also by some other parameters (e.g., the orientation u, the state of helicity, and so on). Let α_i denote the set of variables defining the state of the i-th link. It is easy to see that in the general case, Eqs. (6.1) to (6.6) remain valid provided the substitution $x_i \rightarrow \alpha_i$ is made. In this case, the function $g(\alpha_{i-1},\alpha_i)$ retains the meaning of the conditional probability that the i-th link is in the state α_i provided that $(i-1)$-th link is in the state α_{i-1}; $\varphi(\alpha)$ is the external field acting on the link in the state α (see subsection 6.2).

In solving some problems, we may be interested only in one or a few individual variables of the i-th link from the whole set α_i. Suppose we consider the link orientation u_i, which is essential for describing of chain flexibility. If we determine the functions $g(u_{i-1},u_i)$ and $\varphi(u_i)$ independently of other variables (e.g., of x_i), then Eqs. (6.1) to (6.6) can be written only for the space of link orientations, that is, with the substitution $x_i \rightarrow u_i$ made (cf. the analogy of a change in chain direction with a random walk on the surface of a unit sphere, discussed in subsection 6.1).

6.5. *In terms of the analogy with a random walk, the external field $\varphi(x)$ plays the role of a source (or a sink) of Brownian particles.*

Consider again Eqs. (6.3a) or (6.7). It can easily be seen that the greater contribution to these integrals (i.e., to the Green function) is provided by the trajectories visiting negative potential regions where $\varphi(x) < 0$ (i.e., the potential wells). The potential well augments the value of G, which in terms of a random walk signifies an increased number of Brownian particles in the diffusion cloud. Therefore, in the case of a polymer, the potential well behaves as a source producing diffusing particles.

Conversely, the potential hill, that is, the positive potential region $\varphi(x) > 0$, leads to a decrease in G. In other words, it acts as a sink (absorber, trap) for the Brownian particles.

In particular, while solving the problem of the polymer chain near an impenetrable potential wall using the diffusion equation, the "absorbing" boundary condition must be imposed at the wall edge, that is, $G(x) = 0$ for the points x lying next to the border.

6.6. *The Green function is conveniently written in the form of a bilinear series.*

Employing the transfer operator $\hat{Q}$, it is convenient to use its eigenfunctions $\psi_m(x)$ [in the general case, $\psi_m(\alpha)$] defined by the equation

$$\hat{Q}\psi_m \equiv \int \hat{Q}(x', x)\psi_m(x')d^3x' = \Lambda_m\psi_m(x), \tag{6.8}$$

where Λ_m are the corresponding eigenvalues. The operator $\hat{Q}^+$ conjugate to $\hat{Q}$ has the eigenvalues $\psi_m^+(x)$ found from the equation

$$\hat{Q}^+\psi_m^+ \equiv \int Q(x, x')\psi_m^+(x')d^3x' = \Lambda_m\psi_m^+(x). \tag{6.9}$$

From the definition (6.5) of the operator $\hat{Q}$ and the symmetry of the operator $\hat{g}$ for the standard model [$g(x,x') = g(x',x)$ according to Eq. (4.14)], one can easily show that

$$\psi_m^+(x) = \psi_m(x)\exp(\varphi(x)/T). \tag{6.10}$$

Thus, the eigenvalues Λ_m in Eqs. (6.8) and (6.9) are identical.

Further, it follows from the symmetry of the operator $\hat{g}$ that the eigenfunctions $\psi_m(x)$ form an orthogonal basis, that is, for the appropriate normalization

$$\int \psi_m(x)\psi_{m'}^+(x)d^3x = \delta_{mm'}, \quad \sum_m \psi_m(x)\psi_m^+(x') = \delta(x-x'), \tag{6.11}$$

where δ is the Dirac delta function or Kronecker delta symbol. The summation with respect to m may also imply an integration in the case of a continuous spectrum. In the proper basis, the operator $\hat{Q}$ is diagonal, that is,

$$Q(x',x) = \sum_m \Lambda_m \psi_m^+ (x') \psi_m(x).$$ (6.12)

Substituting the bilinear series (6.12) into the expression (6.7) for the Green function and allowing for the orthogonality conditions (6.11), we obtain

$$G\begin{pmatrix} 0 & N \\ x_0 & x_N \end{pmatrix} = \sum_m \Lambda_m^N \psi_m^+ (x_0) \psi_m(x_N).$$ (6.13)

The presentation of the Green function in the form of the bilinear series (6.13) as well as Eqs. (6.8) to (6.12) are valid in the general case when the state of the link is characterized by the set of variables α if the transition probability g is symmetric: $g(\alpha,\alpha') = g(\alpha',\alpha)$. Then, in Eqs. (6.8) to (6.13), all of the variables x are simply replaced by α. When $g(\alpha,\alpha') = g(\alpha',\alpha)$, it is also possible that Eqs. (6.11) to (6.13) remain valid provided the eigenvalues Λ_m in Eqs. (6.8) and (6.9) coincide. In this book, we only consider models in which the indicated condition is met, and the relations (6.11) to (6.13) therefore are assumed to be valid in the general case as well.

Let us now clarify the physical sense of the conjugate operator $\hat{Q}^+$. The Green function G_{N+1} can be obtained from G_N by adding a link to either the end of an N-link chain or its beginning. The former approach is described by the "diffusion" equation (6.6) with the operator $\hat{Q}$ and the latter by the analogous equation containing the conjugate operator $\hat{Q}^+$.

As for the spectrum of the operators $\hat{Q}$ and $\hat{Q}^+$, the following statement can be made in the general case: because $Q(\alpha',\alpha) > 0$, the spectrum of Λ_m is restricted to a finite interval and the eigenfunctions of the largest eigenvalue are positive: $\psi_0(\alpha) > 0$, $\psi_0^+ (\alpha) > 0$. However, Λ_0 can belong both to the discrete and the continuous spectrum, with the properties of the system proving to be substantially different in these cases. We consider both cases in turn.

6.7. *If a set of possible states for each link is limited, the transfer operator has a discrete spectrum and the correlations diminish exponentially along the chain; this proves in particular the law of exponential decay of orientational correlations along the chain.*

Consider the case when the transfer operator has a discrete spectrum. From the random walk theory, it is known that this situation occurs when the random walk is confined to a limited region and cannot escape to infinity. For example, in describing flexibility, u is the unit vector of direction or the point on a sphere; the random walk on the sphere is always confined (by the surface of the sphere) and characterized by a discrete spectrum. The same is true for a chain "walking randomly" between the helic and the coil states (*see* Chapter 7), a chain compressed within a cavity (*see* Sec. 7), and some other cases.

If the spectrum is discrete, then the largest eigenvalue Λ_0 is separated from the neighboring one by a gap, and as $t \to \infty$, the corresponding term of the series

(6.13) becomes much larger than all the others:

$$G\left(\begin{array}{c|c}1 & t \\ \alpha_1 & \alpha_t\end{array}\right) \cong \Lambda_0^t \psi_0^+(\alpha_1) \psi_0(\alpha_t). \qquad (6.14)$$

In Eq. (6.14), we passed from the standard model to the general case when the state of the link is characterized by the set of variables α. Equation (6.14) corresponds to the so-called ground-state dominance approximation. The expression (6.14) indicates first that no correlation exists between the chain ends. The eigenfunctions $\psi_0^+(\alpha)$ and $\psi_0(\alpha)$ therefore can be interpreted as probability densities for the initial and final links, respectively, which are multiplied in Eq. (6.14) just as are probabilities of independent events.

To examine the correlations along the chain, one more term should be kept in the bilinear series (6.13):

$$G\left(\begin{array}{c|c}1 & t \\ \alpha_1 & \alpha_t\end{array}\right) \cong \Lambda_0^t \psi_0^+(\alpha_1) \psi_0(\alpha_t) + \Lambda_1^t \psi_1^+(\alpha_1) \psi_1(\alpha_t).$$

In this approximation, the correlations are not decoupled. The ratio of the correction term to the main one, however, decays exponentially with the growth of t:

$$(\Lambda_1/\Lambda_0)^t = \exp[-t \ln(\Lambda_0/\Lambda_1)]. \qquad (6.15)$$

In other words, a finite correlation radius exists along the chain

$$t_c = 1/\ln(\Lambda_0/\Lambda_1). \qquad (6.16)$$

Exponential decay of correlations along the chain [see Eq. (6.15)], occurring when the transfer operator has a discrete spectrum, proves the statement in subsection 2.2: orientational correlations decay exponentially for sufficiently large distances along the chain, that is, Eq. (2.2) is valid. Indeed, as already pointed out while describing flexibility, α is the unit vector of chain direction $\mathbf{u}$ at a given point. Because of the finiteness of the set of orientations, the corresponding transfer operator $\hat{g}$ [with the nucleus $g(u',u)$] has *a fortiori* the discrete spectrum.

Note that a discrete spectrum is typical for ordinary one-dimensional models of statistical physics (e.g., the Ising model). Because of this, the method of transfer operators (also called the *Kramers-Vannier method*) is often mentioned in publications on polymer physics with reference to the Ising model analogy.[2-4]

Example. To illustrate the general method, we shall use it to consider the model of a chain consisting of segments of length b with free internal rotation and a fixed valence angle. In this case, $\alpha \equiv u$ is the unit vector of the segment direction and

$$g(\alpha',\alpha) \equiv g(u',u) = g(\cos\theta) = \delta(\cos\theta - \cos\gamma)/2\pi,$$

where γ is the valence angle between neighboring segments (*see* subsection 2.4). The equation for the eigenfunction (6.8) takes the form

$$\int g(\cos\theta)\psi_l(u')d\Omega_{u'} = \Lambda_l\psi_l(u).$$

The largest eigenvalue Λ_0 equals 1, and corresponding eigenfunction is constant [generally, it is always true for $\varphi=0$, which follows from the stochastic nature of the operator $\hat{g}$ (4.3)]. Using the theorem on the summation of Legendre polynomials, it can easily be demonstrated that $\psi_l = P_l(\cos\vartheta)$ and

$$\Lambda_l = 2\pi\int g(\cos\theta)\ P_l(\cos\theta)\ d\cos\theta = P_l(\cos\gamma).$$

According to Eq. (6.16), the persistent length of the considered model is equal to $b/|\ln\cos\gamma|$, which coincides identically with the expression obtained using a simpler technique.

Thus, in studies of the flexibility of macromolecules, the transfer operator method allows the persistent lengths of model chains to be calculated. In reality, however, these calculations are rather complicated, and a simpler calculation of the statistical segment (*see* Sec. 3) is often considered to be sufficient.

6.8. *In the case of a discrete spectrum, the free energy and distribution of links over the states are determined by the largest eigenvalue and the corresponding eigenfunction of the transfer operator.*

The partition function of the chain is derived from the Green function by summing over the states of the ends, that is in the ground-state dominance approximation (6.14) for an N-link chain

$$Z_N = \text{const} \cdot \Lambda^N,$$

and the corresponding free energy

$$\mathscr{F}_N = -T \ln Z_N = -TN \ln \Lambda. \tag{6.17}$$

For convenience, the largest eigenvalue is denoted by Λ instead of Λ_0; the corresponding eigenfunctions is denoted by ψ and ψ^+ to replace ψ_0 and ψ_0^+. In Eq. (6.17), we keep only the leading (thermodynamically additive) term proportional to N.

Recall that the distribution functions for the initial and final links in the ground-state dominance approximation are just equal to $\psi^+(\alpha)$ and $\psi(\alpha)$. Let us now derive the distribution function for a "typical" link removed from both ends of the chain. Consider a "three-point" Green function obtained when in the integral of type (6.7), the state of the intermediate link t is fixed. From Eq. (6.7), it is immediately seen that

$$G\begin{pmatrix} 0 & t & N \\ \alpha_0 & \alpha & \alpha_N \end{pmatrix} = G\begin{pmatrix} 0 & t \\ \alpha_0 & \alpha \end{pmatrix} G\begin{pmatrix} t & N \\ \alpha & \alpha_N \end{pmatrix}. \tag{6.18}$$

If the two parts of the chain, separated by the selected link, are long, then Eq. (6.14) holds for both. Therefore,

$$G\left(\begin{array}{c|c|c} 0 & t & N \\ \alpha_0 & \alpha & \alpha_N \end{array}\right) \cong [\Lambda^t \psi^+(\alpha_0)\psi(\alpha)][\Lambda^{N-t}\psi^+(\alpha)\psi(\alpha_N)]$$

$$= \Lambda^N \psi^+(\alpha_0)\psi(\alpha)\psi^+(\alpha)\psi(\alpha_N). \tag{6.19}$$

As expected, this result is independent of t (i.e., the distributions are identical for all "internal" chain links). Because most links are "internal" in a sufficiently long chain (i.e., the correlations with the end links are decoupled for them), the distribution $n(\alpha)$ of the number of links residing in one or another state is proportional to $\psi^+(\alpha)\psi(\alpha)$ by virtue of Eq. (6.19). By choosing a proper normalization of the function ψ, which can easily be changed without changing the basic formulas and qualitative conclusions of this subsection, one can ensure the equality of these values:

$$n(\alpha) = \psi^+(\alpha)\psi(\alpha). \tag{6.20}$$

Equations (6.17) and (6.20) result from the ground-state dominance approximation. This approximation can be applied when the chain length N is much longer than the correlation radius t_c so that the correlations between the ends are decoupled. According to Eq. (6.16), this happens if the inequality

$$N \ln(\Lambda_0/\Lambda_1) \gg 1, \tag{6.21}$$

holds (i.e., the gap in the spectrum is sufficiently wide).

6.9. *The spatial form of a chain in a Gaussian coil is described by the transfer operator with continuous spectrum; for a continuous spectrum, correlations along the chain extend over its entire length.*

In the two previous subsections, we considered the case of the discrete spectrum of the transfer operator $\hat{Q}$. The results obtained will be used repeatedly below. An example of the opposite situation of a continuous spectrum is considered now. Let us return to the problem of the spatial form of the free ideal coil for the standard Gaussian model of a polymer chain. In this problem, $\varphi = 0$, $\alpha = x$; the operator $\hat{Q}$ is reduced to the operator $\hat{g}$, defined by the equation

$$\hat{g}\psi(x) = \int g(x,x')\psi(x')d^3x', \tag{6.22}$$

where the function $g(x,x') = g(|x-x'|)$ is specified by the relation (4.14). It is easy to check that this operator only has a continuous spectrum and that its eigenfunctions are plane waves $\exp(-ikx)$. Indeed, let us write the equality

$$\int g(|x'-x|)\exp(-ikx')d^3x' = g_k \exp(-ikx), \tag{6.23}$$

which in fact defines the Fourier transform of the function $g(y)$ $(y \equiv x - x')$. It is sufficient to multiply both sides of Eq. (6.23) by $\exp(ikx)$ to see this quite clearly [cf. Eq. (4.10)]. On the other hand, Eq. (6.23) can be treated as an equation for the eigenvalues of type (6.8), with the quantity g_k playing the role of an eigenvalue and the spectrum of these eigenvalues being continuous.

Thus, to use the general relation (6.13), the following redesignations should be made:

$$m \to k, \quad \sum_m \ldots \to (2\pi)^{-3} \int d^3k \ldots, \quad \Lambda_m \to g_k,$$

$$\psi_m \to \exp(-ikx), \quad \psi_m^+ \to \exp(ikx).$$

Therefore, the Green function (6.13) equals

$$G\left(\begin{array}{c|c} 1 & N \\ x_1 & x_N \end{array}\right) = (2\pi)^{-3} \int g_k^N \exp[ik(x_N - x_1)] d^3k. \tag{6.24}$$

Because the nucleus of the operator $\hat{g}$ can be regarded as a probability, the Green function (6.24) is in fact also the probability distribution for the vector $R \equiv x_N - x_0$ connecting the chain ends. It therefore is natural that the result (6.24) coincides with Eqs. (4.8) and (4.9), obtained earlier by a different approach.

After calculating the integral in Eq. (6.24), we obtain the distribution (4.1) [or (4.2)]. Because it cannot be represented as a product of factors that depend individually on x_1 and x_N, one can conclude that in this case, the chain ends are not independent, even in a long chain (i.e., the correlations extend over the whole length of the chain). On the basis of what was stated earlier, it is easy to see that this is the general conclusion for the case of a continuous spectrum for the transfer operator.

*6.10. *The representation of a chain via the standard Gaussian model is equivalent to the representation of the conformational partition function by means of the path integral using the Wiener measure.*

In subsection 4.5, we introduced a standard "bead" model of a polymer chain and noted its validity for any chain, provided that we are considering length scales substantially exceeding the persistent length. We now try to show that the universal nature of the standard Gaussian model has a profound meaning. This model fully displays the analogy between a polymer conformation and a trajectory in conventional diffusive Brownian motion. To make this clear, let us define a chain conformation by radius vectors $x(t)$ as in Sec. 4. If the ordinal numbers t of selected points of the chain (i.e., the coordinates along the chain), are treated as a time variable, then the vector function $x(t)$ can be considered as a trajectory equation. Now return to Eq. (6.7) for the Green function in the form of a path integral; for the standard Gaussian model of a polymer chain, Eq. (6.7) can be rewritten, taking account of the definition (6.3), as

$$G\left(\begin{matrix}0 & t \\ x & x(t)\end{matrix}\right)$$

$$=\mathcal{N}\int \exp\left[-\sum_{\tau=1}^{t}\frac{\varphi(x(\tau))}{T}\right]\exp\left[-\frac{3}{2a^2}\sum_{\tau=1}^{t}(x(\tau)-x(\tau-1))^2\right]\prod_{\tau=1}^{t-1}d^3x(\tau).$$

$$(6.25)$$

where $\mathcal{N}$ is normalizing factor.

If the external field varies only slightly over the length of one link (ensuring the validity of the standard model), one can pass in Eq. (6.25) to the continuous limit. To do this, we write $x(\tau+1)-x(\tau)\approx\dot{x}(\tau)$ (the dot denotes differentiation with respect to the "time" variable t) and obtain

$$G\left(\begin{matrix}0 & t \\ x & x(t)\end{matrix}\right)=\mathcal{N}\int \exp\left\{-\int_0^t d\tau\left[\frac{\varphi(x(\tau))}{T}+\frac{3\dot{x}^2(\tau)}{2a^2}\right]\right\}Dx(\tau), \quad (6.26)$$

where the integral is treated as continual, that is, as an integral over all continuous paths $x(\tau)$. The factor

$$\exp\left[-\int \text{const}\cdot\dot{x}^2(\tau)d\tau\right]$$

resulting from Πg_j defines the statistical weight of an individual path and yields what is known in mathematics as the *Wiener measure*. The expression (6.26) is also well known from the theory of diffusion, describing the probability of finding a diffusing particle at the point $x(t)$ at the time moment t provided that it was located at the initial moment 0 at the point x. It should be recalled that Eq. (6.26) does not describe the random walk of one Brownian particle (or a fixed number of particles) but rather the diffusion of a cloud of such particles randomly appearing or being absorbed in proportion to $\varphi(x)$. The same equation, with an additional factor i in the exponent, defines the Green function (or the transition amplitude) of a quantum particle and underlies the Feynman formulation of quantum mechanics.[22]

Thus, the mathematic meaning of the standard model consists in its partition function having the form of the standard continual integral (6.26). The corresponding diffusion and quantum analogies, as well as the expression (6.26) itself, enhance our physical intuition. In practice, however, the most efficient methods are those associated with the application of the recursive "diffusion" equation (6.6). It is necessary to understand what type of equation the continual version of the standard model [i.e., Eq. (6.26)] corresponds to.

*6.11. *The continuous standard Gaussian model is described by the ordinary diffusion equation (or Schrödinger equation).*

Two preliminary remarks are necessary to derive the sought equation (i.e., the so-called Katz-Feynman formula). First, the operator $\hat{g}$ (6.22) with the Gaussian kernel (4.14) is

$$\hat{g} = \exp[(a^2/6)\Delta], \tag{6.27}$$

where Δ is the conventional Laplace operator. Equation (6.27) is easy to prove by Fourier transformation. To do this, take the arbitrary function $\psi(x)$ and write

$$(\hat{g}\psi)_k = \int \exp(ikx)g(x-x')\psi(x')d^3xd^3x'$$

$$= g_k\psi_k$$

$$= \exp(-k^2a^2/6)\psi_k$$

$$= [\exp(a^2/6)\Delta]_k\psi_k.$$

This result confirms the validity of Eq. (6.27).

Second, the continuous version of the standard model formally corresponds to frequently placed beads (i.e., the operator $\hat{Q}$ must be close to unity). This fact follows from Eq. (6.6), because G_{t+1} is almost equal to G_t. Thus, according to Eq. (6.27),

$$\hat{Q} = \exp(-\varphi/T)\hat{g} = \exp[-\varphi/T + (a^2/6)\Delta] \cong 1 - \varphi/T + (a^2/6)\Delta.$$

Inserting the relation $G_{t+1} = \cong G_t + \partial G_t/\partial t$ into Eq. (6.6), we finally obtain

$$\frac{\partial G_t}{\partial t} = \frac{a^2}{6}\Delta G_t - \frac{\varphi(x)}{T}G_t. \tag{6.28}$$

This equation coincides with the Schrödinger equation in imaginary time for a particle in the external field $\varphi(x)$. This must be expected, because the Feynman amplitude (6.26) is known to satisfy the Schrödinger equation.

6.12. The method of generating functions is convenient for solving many problems.

Still another general method briefly discussed in this subsection is equivalent to the transfer operator method. In some cases, it is more convenient technically and is frequently cited in the literature on statistical physics of macromolecules. A primary concept of this method is a generating function

$$\mathscr{L}(p) = \sum_{N=1}^{\infty} Z_N p^N, \tag{6.29}$$

where Z_N is the partition function of the chain of N links [obtained from the Green function (6.2) by summation over the states of the terminal links]. The generating function may be treated as a large partition function; in this case, $p \equiv \exp(\mu/T)$, where μ is the chemical potential of a link and p is the activity of a link. It is convenient, however, to treat p as a complex variable, then the Z_N values turn out to be coefficients in a power series expansion of the analytic function $\mathscr{L}(p)$. Once the function $\mathscr{L}(p)$ is found, Z_N can be determined using the Cauchy formula

$$Z_N = \frac{1}{2\pi i} \oint \mathscr{L}(p) p^{-N-1} dp, \tag{6.30}$$

where the integration contour should enclose the origin but not any singularities of the function $\mathscr{L}(p)$. From the theory of the functions of a complex variable, the integral (6.30) is known to be determined by the singularities of $\mathscr{L}(p)$. At $N \gg 1$, because of a decrease of the factor p^{-N-1}, only the singularity closest to 0 is substantial. Specifically, if this closest singularity p_0 is a first-order pole, then

$$Z_N \cong p_0^{-N-1} B \text{ Res } \mathscr{L}(p_0),$$

which yields for the free energy

$$\mathscr{F}_N = -T \ln Z_N = TN \ln p_0. \tag{6.31}$$

Thus, all one needs to know about the generating function in the simplest case is the position of the pole p_0 closest to 0.

Using the bilinear expansion of the Green function (6.13), one can formally write the generating function as

$$\mathscr{L}(p) = \sum_m C_m \Lambda_m p / (1 - \Lambda_m p), \tag{6.32}$$

where the coefficients C_m depend on the normalization of ψ functions. Discrete eigenvalues provide the generating function with first-order poles. In particular, the closest pole corresponds to the largest eigenvalue, thus making Eqs. (6.17) and (6.31) identical. Within regions having a continuous spectrum, the poles $\mathscr{L}(p)$ merge to generate more complex singularities: branches, cuts, and so on.

Examples of applications of the method of generating functions are given in subsection 40.9.

7. A POLYMER CHAIN IN AN EXTERNAL COMPRESSING FIELD. ADSORPTION OF AN IDEAL CHAIN; THE SIMPLEST COIL–GLOBULE PHASE TRANSITION

7.1. *An ideal macromolecule, compressed within a limited volume, consists of effectively non-correlated sections or subchains and exerts a pressure on the walls as an ideal gas composed of subchains.*

Examine an ideal macromolecule located in a spheric cavity of diameter D with impenetrable walls (Figs. 1.9a and 1.10). Assuming that the cavity size is much greater than the persistence length, we can use the standard Gaussian model of a macromolecule.

We reason in terms of a random walk. The walk starts from a certain point within the cavity, and it proceeds quite freely until the first collision with the

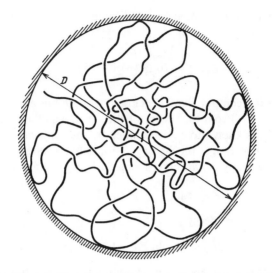

FIGURE 1.10. A polymer chain in a spherical cavity of diameter D.

wall. After reflection, another independent random walk takes place until the next collision, and so on. The length K of the section between two collisions with the wall is easily evaluated:

$$aK^{1/2} \sim D, \quad K \sim (D/a)^2.$$

Thus, the chain consists of $(N/K) \sim N(a/D)^2$ such sections or subchains.

The picture presented here differs drastically from that of a free Gaussian coil. For example, the ends of the macromolecule are separated by the distance of order D and independent of N, because they belong to different subchains for $N \gg K$. The radius of correlation along the chain is of the order K and also independent of N, and so on. Here is the so-called globular state of the polymer chain. The exact definition of this term will be given later.

One can easily evaluate the free energy of the chain in the cavity, in other words, the entropy loss from chain confinement within the cavity. We do this using the simplest reasoning. The free coil is characterized by a single macroscopic scale to be denoted here by $R_{id} \sim aN^{1/2}$ and the cavity by the size D. There is only one quantity in this problem, namely, temperature, having the dimensionality of energy, so that the free energy can only be written in the form

$$\mathscr{F} = T f(aN^{1/2}/D), \tag{7.1}$$

where the function f is not known so far. Because the free energy must be thermodynamically additive that is, proportional to N, $f(x) \sim x^2$, and

$$\mathscr{F} \sim TN(a/D)^2. \tag{7.2}$$

In other words, the entropy loss ($S = -F/T$, as there is no "energy" in this problem $E=0$), is of order unity for each group of K links, that is, for each subchain (each "reflection" from the wall takes away about one degree of freedom from the chain).

The pressure exerted on the walls by the chain is equal to

$$p = -\frac{\partial \mathcal{F}}{\partial V} \sim -\frac{\partial \mathcal{F}}{\partial D^3} \sim TNa^2 D^{-5} \sim \frac{T(N/K)}{D^3}. \tag{7.3}$$

Because D^3 is the volume, Eq. (7.3) obviously expresses the pressure of an ideal gas consisting of N/K particles, or subchains.

7.2. A coil and a globule are two macroscopic phase states of a polymer chain with different fluctuation regimes.

From the example just considered, it is clear that compression transforms a macromolecule into the globular state, which differs drastically from a free Gaussian coil. Condensed globular states of macromolecules are widespread in both animate and inanimate nature, and we discuss them in more detail later (*see* Secs. 21 to 23). These states are often stabilized by volume interactions, that is, by forces attracting links to one another. Now we examine the simplest case of globularization of an ideal chain by an external compressing field. Such a problem acquires immediate physical meaning, for example, in the case of macromolecular adsorption. Besides, the interlink attraction effect can usually be described as resulting from an effective, self-consistent compressing field.

Before proceeding to the theory of a globule, however, we must define in more precise terms the concepts of a globule and a coil. Until now, we have used the term *polymer coil* without its accurate explication. We only implied that a sufficiently long chain assumes not a rectilinear but an extremely entangled spatial conformation; in practice, a long chain in the globular state also possesses this property. As I. M. Lifshitz showed in 1968[30], the difference between the coil and the globule is determined by the character of fluctuations or the fluctuation regime. In particular, the globular state of the polymer chain is a weakly fluctuating state in which the correlation radius of link concentration fluctuations is much smaller than the size of the macromolecule. Conversely, a strongly fluctuating state in which the correlation radius is of the order of the size of the macromolecule itself is referred to as the coil state.

It is evident that a globular state is realized when the compression of the macromolecule is sufficiently high (i.e., if the well is sufficiently deep). On the other hand, as explained earlier, the limited correlations are mathematically expressed via the discrete spectrum of the transfer operator. These two viewpoints turn out to be identical.

Indeed, in the case of the spatial structure of a fluctuating macromolecule, the generalized variable α in the equations of Sec. 6 can be represented by the spatial coordinates of the links, $\alpha \to x$. In this case, Eq. (6.8) for eigenvalues of the transfer operator takes the form

$$\exp[-\varphi(x)/T]\hat{g}\psi - \Lambda\psi = 0. \tag{7.4}$$

It is immediately seen that a decrease in temperature is equivalent, as it should be, to a deepening of the well. Then, one can see that a gap between the highest and the next-to-highest eigenvalues of this equation appears as soon as the temperature drops below a certain critical value (i.e., when the well becomes effectively deep enough). Later, we show this for the limiting cases of an extended and smooth potential, a localized potential, and finally for the general case. The appearance of a discrete spectrum in Eq. (7.4) implies a splitting of correlations, or according to the definition, the formation of a globular state.

It should be noted that when the highest eigenvalue belongs to a discrete spectrum (i.e., the macromolecule is in a globular state), the free energy is described by Eq. (6.17) while the mean link concentration is distributed in space according to Eq. (6.20):

$$n(x) = \psi^2(x)\exp(\varphi(x)/T), \tag{7.5}$$

$$N = \int n(x)d^3x. \tag{7.6}$$

Thus, to explore the behavior of an ideal macromolecule in an external compressing field, one must solve Eq. (7.4) for the maximum eigenvalue Λ. In the presence of a discrete spectrum (globular state), the local link concentration in the globule is distributed according to Eq. (7.5), Eq. (7.6) normalizes the functions $n(x)$ and $\psi(x)$, and the free energy of the globule is given by Eq. (6.17).

Because in this section we are interested primarily in the properties of a polymer chain in an external compressing field only on scales greatly exceeding the persistence length, we shall use further, without any restriction in generality, the standard Gaussian model of the polymer chain.

7.3. *The state of a chain in a smoothly varying external field is described by a Schrödinger-type equation.*

Consider the most significant practical case, when the characteristic scale over which the external field $\varphi(x)$ varies greatly exceeds the distance between neighboring monomer links. In this case, the integral equation (7.4) can be simplified greatly by reducing it to a second-order differential equation. Indeed, taking into account the spheric symmetry of the function $g(y)$ $(y = x - x')$, we obtain

$$\hat{g}\psi = \int g(x-x')\psi(x')d^3x'$$

$$= \int g(y)\psi(x+y)d^3y$$

$$= \int g(y)\{\psi(x) + (y\nabla)\psi + (1/2)(y\nabla)^2\psi + ...\}d^3y$$

$$\cong \psi(x) + (a^2/6)\Delta\psi \tag{7.7}$$

[cf. Eq. (6.27)]. Here, we expand the function $\psi(x+y)$ in powers of y at the point x, keeping only the first three terms of the series. This is justified, because the smooth variation of the external field forces the function ψ to vary smoothly. Consequently, to find how ψ varies over scales of the order a, one can make use of the Taylor expansion. (Note that because of the diminishing of the kernel g with an increase in y, only these variations are significant for the calculation of the integral $\hat{g}\psi$).

From Eq. (7.7), it is seen that when the characteristic scale of variation of the field $\varphi(x)$ much exceeds a, one can make in Eq. (7.4) the substitution

$$\hat{g} \to 1 + (a^2/6)\Delta. \tag{7.8}$$

Moreover, $(a^2/6)\Delta\psi \ll \psi$, because $\psi(x)$ varies over the same scale as $\varphi(x)$. Consequently, rewriting Eq. (7.4) as $\varphi/T + \ln\Lambda = \ln(\hat{g}\psi/\psi)$, we can simplify it still further:

$$\ln(\hat{g}\psi/\psi) \cong \ln[1 + (a^2/6)\Delta\psi/\psi] \cong (a^2/6)\Delta\psi/\psi,$$

and eventually obtain

$$(a^2/6)\Delta\psi = [(\varphi-\lambda)/T]\psi, \tag{7.9}$$

where the designation $\lambda = -T\ln\Lambda$ is used. According to Eq. (6.17), λ is the free energy of a single particle.

Equation (7.9) resembles the steady-state Schrödinger equation for a quantum particle moving in the external potential field $\varphi(x)$. This analogy goes very far. For example, the free energy of a globule is equal to the energy of the discrete ground state (the lowest λ corresponds to the maximum Λ). In the approximation (7.8), the globular density equals

$$n(x) = \text{const} \cdot \psi^2(x), \tag{7.10}$$

that is, is analogous to a quantum-mechanical probability density. The analogy between the theory of polymer globules and conventional quantum mechanics is useful for solving specific problems and refining our intuitive assumptions about the physics of the globular state.

As an example, we apply Eq. (7.9) to the problem of a polymer chain confined in a spheric cavity with impenetrable walls. It is instructive to trace how the qualitative reasoning of subsection 7.1 relates to formal theory. The cavity can be described as a potential well with infinitely high walls: $\varphi(x) = 0$ at $|x| < D/2$, and $\varphi(x) = +\infty$ at $|x| > D/2$ (Fig. 1.9a). The ground-state wave function of such a well is easy to find:

$$\psi \sim \begin{cases} \dfrac{\sin(2\pi x/D)}{2\pi x/D}, & x = |x| \leqslant D/2, \\ 0, & x \geqslant D/2, \end{cases} \tag{7.11}$$

while the corresponding "energy level" is

$$\lambda = 2T\pi a^2/3D^2. \tag{7.12}$$

This last result agrees well with the evaluation (7.2).

Taking into account Eq. (7.10), one can infer from Eq. (7.11) that the local link concentration is distributed inside the cavity as

$$n(x) = (N/\pi Dx^2)\sin^2(2\pi x/D), \tag{7.13}$$

where the proportionality constant was determined from the normalization condition (7.6). Figure 1.11 shows the profile of the local link concentration along the cavity radius. In the case of the ideal chain, the link concentration shows a sharply non-uniform distribution: density in the cavity center is drastically higher than at its periphery. This is certainly a direct consequence of the absence of excluded volume in the ideal macromolecule. By preference, the links congregate at the cavity center, because their entropy rises in this process. Freedom of link motion is not restricted either by the presence of impenetrable walls or other links, because the links do not interact.

Let us also examine the conditions under which Eqs. (7.11) to (7.13) are valid. We have already mentioned that the analogy with the Schrödinger equation is valid only in the case of a smoothly varying ψ function. According to Eq. (7.11), $a^2\Delta\psi/\psi \sim (a/D)^2$ within the well, that is, the condition of smoothness is satisfied for $D \gg a$ (a large well). A more delicate problem pertains to the near-wall region. The function (7.11) shows an inflection at $|x| = D/2$. Analysis indicates that Eqs. (7.11) and (7.13) actually need corrections within a layer of thickness a near the wall. Outside that region, Eqs. (7.11) to (7.13) remain valid.

Next, we should verify whether the expression for globules can be used in this case, that is, whether the ground state of the discrete spectrum of Eq. (7.9) predominates in the partition function. The inequality (6.21) can be rewritten as

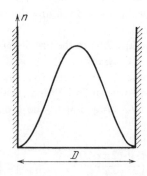

FIGURE 1.11. Link concentration in an ideal polymer chain confined in a cavity of diameter D.

$$N\delta\lambda/T \gg 1. \tag{7.14}$$

It is known that in the case of the Schrödinger equation (7.9) for a particle in an infinitely deep spheric well, $\delta\lambda \sim Ta^2/D^2$. From this, we obtain the following condition for the validity of the relations written above: $Na^2/D^2 \gg 1$, or $R_{id} \gg D$. Thus, the macromolecule forms a globular state only within a cavity whose size is much less than that of the free coil R_{id}. This condition is quite clear. If $D \gtrsim aN^{1/2}$, then the ideal coil essentially moves freely inside the cavity, and its conformation obviously maintains its coil form.

In polymer physics, one frequently comes across the situation when polymer chains are found not in a cavity confined on all sides but rather within a plane slit of width D or within a cylindric tube (pore) of diameter D. Conformational parameters of such chains must be computed when considering the transport of macromolecules through membranes, porous bodies, in chromatography theory, and so on. If the polymer chain can be assumed ideal (e.g., in the case of θ-solvent), then the conformation of the macromolecule in the plane slit or cylindric pore can be described quite similarly to the way it was done earlier (in these cases, the resulting Schrödinger equation will be one- or two-dimensional with cylindric symmetry). In particular, the relation [cf. Eqs. (7.2) and (7.12)]

$$\mathscr{F} \sim TNa^2/D^2, \tag{7.15}$$

is valid for the difference in the free energy between the chain in the slit (or the pore) and in the continuum. This relation can easily be derived by a "scaling" approach similar to that presented in the derivation of Eq. (7.2).

*7.4. *An exact solution of the problem of the interaction of an ideal chain with a point potential well allows a structure analysis of the loops surrounding the globular nucleus.*

Let us direct our attention to the opposite limiting case of a potential well that is concentrated within a small volume; outside that volume, $\varphi = 0$ or $\exp(-\varphi/T) = 1$. If we are not interested in the particulars inside the well region, then we can simply assume the well to be a point, that is, to write

$$\exp(-\varphi(x)/T) = 1 + \beta\delta(x), \tag{7.16}$$

$$\beta = \int [\exp(-\varphi(x)/T) - 1]d^3x. \tag{7.17}$$

Certainly, the value of φ inside the well should be quite large in magnitude for the well to affect noticeably the chain conformation.

At first sight, the point well model seems to have no physical meaning, as the real links possess excluded volume and therefore will not be accommodated in the point well. In the real case of weak absorption of the polymer chain on a small extraneous particle, however, only a few rare links actually stick to this particle, with the free chain sections between them forming long loops. If the size of the particle together with the adsorption layer is much less than the loop size, then the particle can be modeled by a point well. Consequently, the point well model is general enough to analyze the transition between the globular and coil states.

Thus, let us consider Eq. (7.4) for the point field (7.16):

$$(1+\beta\delta(x))\hat{g}\psi=\Lambda\psi(x). \tag{7.18}$$

After Fourier transformation,

$$\psi_k=\int \psi(x)\exp(ikx)d^3x,$$

we obtain an algebraic equation that is easy to solve:

$$\psi_k=\beta\cdot\text{const}/(\Lambda-g_k); \quad \text{const}=(2\pi)^{-3}\int g_{k'}\psi_{k'}d^3k', \tag{7.19}$$

where the values of g_k are defined by Eq. (4.10). If one multiplies Eq. (7.19) by g_k and integrates over k, the integral obtained on the left-hand side will be identical to the one in the numerator on the right-hand side. Consequently, we obtain the closed equation for Λ:

$$1/\beta=(2\pi)^{-3}\int [g_k/(\Lambda-g_k)]d^3k. \tag{7.20}$$

Knowing the temperature, we can find the value of β from Eq. (7.17) and then calculate Λ, the free energy of the globule (6.17), from Eq. (7.20). In this way, we obtain the solution of the problem in closed form. Let us examine it.

Recall that the value of $\psi(x)$ is proportional to the probability density of finding the chain end at point x. Let us normalize it so that it equals that probability (as mentioned before, the normalization condition for the function ψ can be changed without changing the basic formulas of Sec. 6); hence,

$$1=\int \psi(x)d^3x=\psi_{k=0},$$

and therefore,

$$\psi(x)=(2\pi)^{-3}\int [\exp(-ikx)\cdot(\Lambda-1)/(\Lambda-g_k)]d^3k. \tag{7.21}$$

The integrand can be interpreted as a sum of a geometric series in powers of g_k/Λ. This yields the following simple result:

$$\psi(x)=\sum_{m=0}^{\infty} [(\Lambda-1)/\Lambda^{m+1}]P_m(x). \tag{7.22}$$

Here, $P_m(x)$ is the distribution function for the end of the m-link Gaussian coil (4.1). Equation (7.22) has a very plain physical meaning. Again, we reason in terms of a random walk. After leaving the point O, the trajectory walks randomly in free space as a Gaussian coil until an accidental arrival back at the point O. Such a section is, in a certain case, analogous to the subchain introduced in subsection 7.1 for the ideal macromolecule in the cavity. Thus, the globular nucleus located at the point O has a fringe consisting of Gaussian loops and a Gaussian "tail" near the chain end. If in the given conformation the "tail" comprises m links, then the distribution function of its end is $P_m(x)$. Consequently, the expression in the square brackets of Eq. (7.22) is the probability w_m that the "tail" has m links:

$$w_m = (\Lambda - 1)\Lambda^{-m-1}.$$

The average length of the "tail" equals

$$\langle m \rangle = \sum_{m=0}^{\infty} m w_m = (\Lambda - 1)^{-1}. \tag{7.23}$$

It can be shown that the mean length of loops is the same.

Note that the loops forming the fringe around the small (point) nucleus of the globule (Fig. 1.12) are quite similar to the subchains discussed in subsection 7.1. From this analogy, we can infer, and even verify by calculation, that the average length $\langle m \rangle$ of the loop is the correlation radius along the chain.

Generalizing, it can be said that a typical picture of the globular state includes sections (loops, subchains, and so on) of finite (independent of N at $N \to \infty$) length, with the chain remaining Gaussian within the limits of each section and different sections independently located in the same volume. We have already mentioned that the point well model attracts special interest when the loops and the tail are long. From Eq. (7.23), it is seen that this is realized when the value of Λ is close to unity. The free energy of the globule $\mathscr{F} = -NT \ln \Lambda$ is in this case small, while the correlation length along the chain is large.

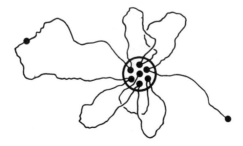

FIGURE 1.12. Conformation of a small globule: a chain globulized by attraction of links to a potential well of small size (surrounded by a *heavy solid line*).

*7.5. *The globule–coil transition preceded by a gradual increase in the thickness of the globular fringe is a second-order phase transition.*

One can expect to observe a specific behavior of the globule considered in the previous subsection in the temperature region where $(\Lambda-1)\ll1$. This is a pretransitional region for the globule–coil transformation. Indeed, the free energy (6.17) is by definition the difference between the energies of the globule and a free ideal coil; therefore, the value $\Lambda=1$ corresponds to the transition. At $(\Lambda-1)\ll1$, the quantity $\psi_k\sim(\Lambda-g_k)^{-1}$ [see Eq. (7.21)] has a narrow peak near $k=0$, because $g_{k=0}=1$ [see Eq. (4.10)]. This means that the function $\psi(x)$ "spreads"; at $|x|\gg a$, it equals

$$\psi(x)\big|_{|x|\gg a}\cong\text{const}\cdot\int\exp(ikx)\cdot\left(\Lambda-1+\frac{k^2a^2}{6}\right)^{-1}d^3k$$

$$=\frac{\text{const}}{x}\exp\left\{-\frac{x}{a}[6(\Lambda-1)]^{1/2}\right\}. \tag{7.24}$$

As we see, the intrinsic size of the fringe is proportional to $\sim a(\Lambda-1)^{-1/2}$. The same result would be obtained from the density calculation, because $n(x)\sim\psi^2(x)$ outside the well. This result agrees with Eq. (7.23): because the loops and the tail of the fringe are Gaussian, the fringe thickness is of order $a\langle m\rangle^{1/2}$.

How does one determine the globule–coil transition temperature, and how does the free energy of the globule behave near the transition? Consider Eq. (7.20). The value of $\beta=\beta_{\text{tr}}$ corresponding to the transition temperature can be found directly from Eq. (7.20) by taking $\Lambda=1$:

$$1/\beta_{\text{tr}}=(2\pi)^{-3}\int[g_k/(1-g_k)]d^3k. \tag{7.25}$$

For example, for the standard Gaussian bead model, $g_k=\exp(-k^2a^2/6)$ [see Eq. (4.14)]. Therefore, $\beta_{\text{tr}}\approx a^3\cdot1.16$. The actual temperature at which β becomes equal to β_{tr} fully depends on the form of the potential $\varphi(x)$. Nevertheless, we can attach more generality to our reasoning by using the quantity β. The temperature dependence of β is smooth. Therefore, $(T-T_{\text{tr}})\sim(\beta-\beta_{\text{tr}})$ near the critical temperature, and for the sake of simplicity, we deal with the deviation of β, but not temperature, from the critical value.

Comparing Eqs. (7.20) and (7.25), we obtain

$$\frac{1}{\beta_{\text{tr}}}-\frac{1}{\beta}=(2\pi)^{-3}\int\frac{(\Lambda-1)g_k}{(\Lambda-g_k)(1-g_k)}d^3k.$$

The main contribution to this integral is provided by the region of small values of $|k|$, because at $\Lambda=1$, the integral diverges at $k=0$. Consequently,

$$\frac{a^3}{\beta_{tr}} - \frac{a^3}{\beta} \cong (2\pi)^{-3} \int_0^\infty \frac{a^3(\Lambda-1) \cdot 4\pi k^2 dk}{[\Lambda-1+(ka)^2/6](ka)^2/6} = (3\sqrt{3}/\sqrt{2}\pi)(\Lambda-1)^{1/2}.$$

Hence, the free energy of the globule (6.17) near the point of its transition to the coil equals

$$\mathcal{F}/(NT) \cong -(\Lambda-1) \cong (2\pi^2 a^6/27)\beta_{tr}^{-4}(\beta-\beta_{tr})^2 = \text{const} \cdot \tau^2, \quad (7.26)$$

$$\tau = (T-T_{tr})/T_{tr}. \quad (7.27)$$

From this relation, one can draw the following conclusion. The capture of a long polymer chain by a potential well is a second-order phase transition. Near the transition, the globule gradually swells, and according to Eq. (7.24), the fringe thickness

$$D \sim a(\Lambda-1)^{-1/2} \sim a/\tau. \quad (7.28)$$

The correlation length along the chain grows as $1/\tau^2$ [see Eq. (6.16)], and the fraction of particles in the globular nucleus diminishes as τ^2.

Although these results were obtained for a point well, they in fact reveal a more general behavior. This is discussed in the next subsection.

*7.6. *A potential well of arbitrary, finite size acts on a polymer chain near the capture temperature (i.e., the coil–globule transition temperature) as a point well.*

This subsection is a formal proof that in a potential field of arbitrary form, satisfying only the condition of rapid decrease (or tending to zero) of the potential at infinity, a globule transforms into a coil via a second-order phase transition, just as in the point well. A central issue of the proof resides in the statement that the wave functions spread near the transition point [see Eq. (7.24)] in a similar manner irrespective of the actual shape of the well. Actually, the function $\psi(x)$ varies slowly far from the potential well, where $\varphi(x) \approx 0$, and the integral operator $\hat{g}$ in Eq. (7.4) can be replaced by the differential operator (7.8) so that Eq. (7.4) in this region is reduced to

$$(a^2/6)\Delta\psi = (\Lambda-1)\psi(x). \quad (7.29)$$

The solution of this equation takes the form

$$\psi(x) = \frac{\text{const}}{x} \exp\left[-\frac{x}{a}(6(\Lambda-1))^{1/2}\right], \quad (7.30)$$

in exact correspondence with Eq. (7.24). Thus, in the general case, the globular structure near the transition point ($\Lambda \to 1$) also resembles Figure 1.12: long Gaussian loops fluctuate around a localization region of the field $\varphi(x)$.

According to Eq. (7.30), the characteristic loop size is approximately $a(\Lambda-1)^{-1/2}$, that is, because the chains are Gaussian, their characteristic length is $m \sim (\Lambda-1)^{-1}$, and the number of loops in the chain is approximately $N/m \sim N(\Lambda-1)$. The free energy of the globule (6.17) $\mathcal{F} \cong -NT(\Lambda-1)$, that is, equals, as always (cf. subsection 7.1), approximately T per each loop.

It now remains to be proven that the value of $(\Lambda-1)$ varies with temperature according to Eq. (7.26) as $(\Lambda-1)\sim\tau^2$ as the globule approaches the transition point.

Consider first the change in Λ resulting from a small temperature change δT in the globular region $(\Lambda>1)$. This change can be found by the following perturbation technique: if the temperature varies from T to $T+\delta T$, then the following substitutions must be made in Eq. (7.4):

$$\exp(-\varphi(x)/T)\to\exp[-\varphi(x)/(T+\delta T)]\equiv\exp(-\varphi(x)/T)+\rho(x),$$

$$\rho(x)\cong\exp(-\varphi(x)/T)\varphi(x)\delta T/T^2,$$

$$\psi(x)\to\psi(x)+\delta\psi(x),\quad \Lambda\to\Lambda+\delta\Lambda.$$

Substituting these relations into Eq. (7.4), and keeping the linear terms in all perturbations, we obtain the non-uniform equation for $\delta\psi(x)$

$$\exp(-\varphi(x)/T)\,\hat{g}\delta\psi-\Lambda\delta\psi=-\rho(x)\hat{g}\psi+\delta\Lambda\psi(x). \tag{7.31}$$

On the left-hand side of Eq. (7.31), there appears the same operator acting on $\delta\psi$ as in Eq. (7.4). Consequently, according to the Fredholm theorem, the non-uniform equation (7.31) has the solution only for such value of $\delta\Lambda$ that makes the right-hand side of Eq. (7.31), weighted by $\exp(\varphi(x)/T)$, orthogonal to the eigenfunction of the uniform equation (7.4), that is, $\psi(x)$[c]

$$\int \exp(\varphi(x)/T)\psi(x)[-\rho(x)\hat{g}\psi+\delta\Lambda\psi(x)]d^3x=0.$$

From the definition of $\rho(x)$, we obtain

$$\frac{1}{\Lambda}\frac{d\Lambda}{dT}=\frac{\int \exp(\varphi(x)/T)\psi^2(x)\varphi(x)d^3x\,T^{-2}}{\int \exp(\varphi(x)/T)\psi^2(x)d^3x}$$

Near the transition point with $\Lambda\to1$, the integral in the denominator grows infinitely, because the ψ function (7.30) spreads and $\exp(\varphi/T)\cong1$ as $|x|\to\infty$.

---

[c]This theorem can be proved simply: according to Eq. (7.4), the left-hand side of the Eq. (7.31) is orthogonal to ψ at any $\delta\psi$, that is,

$$\int \exp(\varphi/T)\psi[\exp(-\varphi/T)\hat{g}\delta\psi-\Lambda\delta\psi]d^3x=\int [\hat{g}\psi-\Lambda\psi\exp(\varphi/T)]\delta\psi d^3x=0.$$

As to the weight factor $\exp(\varphi/T)$, it appears because of the non-Hermitian character of transfer operator in our notations (6.5). To transform it into a clearly symmetric Hermitian form, the potential energy φ of a link should be divided into two equal parts for previous and next bonds of the chain, or in other words, we must write

$$\tilde{Q}=\exp(-\varphi/2T)\hat{g}\exp(-\varphi/2T),$$

This procedure can be reduced simply to redefinition $\psi\to\tilde{\psi}\exp(-\varphi/2T)$, which means exactly the appearance of the weight factor in the scalar products.

At the same time, the integral in the numerator remains finite, because $\varphi(x) \cong 0$ at large $|x|$. Using Eq. (7.30), one can easily evaluate the denominator and obtain

$$d\Lambda/dT = \text{const} \cdot (\Lambda - 1)^{1/2}.$$

The solution of this differential equation is $\Lambda - 1 \sim (T - T_{tr})^2$. This concludes the proof.

Now let us discuss the validity of the results obtained. We assumed that the ground-state predominates; thus, the condition (6.21) must be checked. There is only one discrete eigenvalue near the transition point; therefore, Λ_1 in Eq. (6.21) corresponds to the edge of the continuous spectrum, that is, $\Lambda_1 = 1$. Then, Eq. (6.21) takes the form

$$N \ln \Lambda \gg 1. \tag{7.32}$$

This can be rewritten as $|F| \gg T$. The physical meaning of such a condition is clear. As soon as the free energy difference between the coil and globular states of a macromolecule of N links becomes equal to the temperature, the fluctuation probability of the emergence of a thermodynamically unfavorable state is of order unity, so there is no sense at all in regarding the coil and the globule as different phase states (*see* subsection 21.3). Note also that in accordance with Eq. (7.23), the condition (7.32) can be rewritten as $\langle m \rangle \ll N$ or $D \ll aN^{1/2}$. The physical meaning of these inequalities is also clear. The globule is realized when the chain has many loops (or subchains) and is more compact than the ideal coil.

In the narrow temperature interval $\tau \lesssim N^{-1/2}$, however, the discrete level is so close to the edge of the continuous spectrum that it becomes irrelevant to speak about the globule. As N grows, the width of this interval rapidly diminishes, $\Delta T/T \sim N^{-1/2}$. The finiteness of the transition temperature interval is associated primarily with the finiteness of N. As $\Delta T \to 0$ at $N \to \infty$, the capture of the polymer chain by a potential well is in fact a phase transition (*see* subsection 21.3). Taking into account that this transition proceeds smoothly and continuously, and without any metastable states on the opposite side of the transition point, it can with good reason be treated as a second-order phase transition (*see* subsection 21.4).

7.7. The adsorption of an ideal macromolecule at an attractive surface has the character of a phase transition.

As already mentioned, one of the most significant problems associated with macromolecular behavior in the presence of an external attracting potential $\varphi(x)$ consists of the adsorption of macromolecules on an attracting plane surface. Figure 1.9 illustrates the appropriate potential well. The width is usually very small, being equal to approximately the length of one monomer link.

All general conclusions of this subsection can be applied to a potential well of such shape. In particular, in the case of adsorption, there also exists a critical temperature of capture T_{tr} such that for $T < T_{tr}$, a macromolecule is adsorbed by an attractive surface, whereas for $T > T_{tr}$, it remains free.

Two cases can be identified:

1. Strong adsorption [$\tau \sim 1$, *see* Eq. (7.27)] at which the attraction to the surface is so intense that the chain "flattens" over it.

2. Weak adsorption ($\tau \ll 1$).

In the latter case, the local link concentration profile spreads so greatly that only a very small fraction of links is in the immediate vicinity of the surface. The function $n(x)$ outside the well can easily be found in this case. Introduce the coordinate z in the direction perpendicular to the attractive surface, and at $\varphi(x) = 0$, Eq. (7.4) and its solution can be written, taking into account the symmetry of the problem, in the form

$$(a^2/6)d^2\psi/dz^2 = (\Lambda - 1)\psi,$$

$$n(z) \sim \psi^2(z) \sim \exp[-2(z/a)(6(\Lambda - 1))^{1/2}] \tag{7.33}$$

[cf. Eqs. (7.29) and (7.30)]. We obtain an exponential decay of link concentration with distance from the surface. The characteristic length scale of this decay is equal to the thickness of the adsorption layer D given by Eq. (7.28).

As the temperature approaches the capture temperature ($\tau \to 0$), the adsorption layer keeps swelling. As shown in the previous subsection, the capture of an ideal macromolecule by an adsorbing surface is a second-order phase transition.

8. AN IDEAL POLYMER STRETCHED BY AN EXTERNAL FORCE. ITS RESPONSE TO AN EXTERNAL ORIENTING FIELD

8.1. *A weakly stretched ideal chain exhibits linear elasticity of entropic nature; the corresponding elastic modulus is inversely proportional to the chain length.*

Let us return to the Gaussian distribution (4.1) and (4.2). It may well be treated as follows. The partition function corresponding to a given value of the radius vector R is $Z_N(R) = \mathcal{N}P_N(R)$, where $\mathcal{N}$ is the normalizing factor signifying the partition function for the chain with free ends. It is essential that this factor is independent of R. The free energy of the ideal polymer chain with the given value of R therefore is equal to [*see* Eq. (4.1)]

$$\mathcal{F}(R) = -T \ln Z_N(R) = \text{const} + \frac{3TR^2}{2Nl^2} = \text{const} + \frac{3TR^2}{2Ll}, \tag{8.1}$$

where const is the constant independent of R. Thus, we see that the growing of the end-to-end distance of the chain (e.g., during the stretching of the macromolecule) leads to an increase in the free energy. This implies that stretching of the polymer chain induces an elastic force, inhibiting the stretching. This force is of entropic nature. The expression (8.1) is valid for any ideal macromolecule and also in the absence of any energy interactions (e.g., it is valid for a freely jointed chain model). The elastic force appears because the number of possible conformations with the given end-to-end distance R decreases with stretching. (The totally stretched chain, $R = L = Nl$, has a unique straight conformation.)

It is interesting to note the difference between the free energy loss when the end-to-end distance is fixed [Eq. (8.1)] and when the chain is compressed in the cavity [Eq. (7.2)]: the numerator and denominator interchange.

Consider now the following problem. Suppose that an ideal polymer chain experiences an external stretching force f applied to its end links. In equilibrium, the force f is counterbalanced by the elastic force f_{el} of the chain, $f = -f_{el}$. Let the mean end-to-end vector be R as a result of stretching. Then, $f_{el} = -\partial \mathscr{F}(R)/\partial R$, and therefore

$$f = \partial \mathscr{F}(R)/\partial R = (3T/Ll)R. \tag{8.2}$$

This formula defines the force to be applied to an ideal polymer so that the mean end-to-end vector would equal R. This is a vector equality showing that the chain stretches in the direction of the applied force. The force is seen to be proportional to the displacement. Thus, one can say that an ideal polymer chain obeys the Hooke law.

This mode of expression, however, is quite conditional. The mean value of the vector R for the non-deformed chain equals 0, and we consequently cannot introduce any quantity to be used as a relative deformation, as usually appears in the traditional Hooke law.

The elastic modulus of an ideal chain turns out to equal $3T/(Ll)$. First, note that it is proportional to $1/L$ and therefore is small for long chains. This means that polymer chains are very susceptible to external forces. In the final analysis, it is this circumstance that is responsible for the high elasticity of rubber and similar polymer materials (see Sec. 29). Second, the elastic modulus is proportional to the temperature T (at $l =$ const). This is naturally a direct consequence of the entropic nature of elastic forces.

In the derivation of Eq. (8.2), we used Eqs. (4.1) and (4.2) for the probability distribution of the end-to-end vector. In subsection 4.3, we noted that the Gaussian law is valid for $P_N(R)$ for not-too-large values of $|R|$. Consequently, the Hooke law (8.2) should be modified near the limit of total stretching (see subsection 8.3).

8.2. *The stretching of a chain by its ends can be equated to the action on it by an external orienting field.*

Suppose that under the influence of the force f, the radius vector R changes by the value dR. In the process, the external force performs the work

$$\delta A = f dR = \sum_{i=1}^{N} f dr_i, \tag{8.3}$$

where dr_i is the change of the vector $r_i=x_i-x_{i-1}$ in the given process. On the other hand, considering that the force f transfers along the chain and acts on each link, the polymer chain can be pictured as located in a certain effective stretching potential field $\varphi(f)$ tending to orient each link. In this case, one can write $\delta A = \Sigma d\varphi_i$, where $d\varphi_i$ is the potential energy increment of an individual link. Hence, using Eq. (8.3), we can conclude that

$$\varphi_i = fr_i = fl\cos\vartheta_i, \tag{8.4}$$

where ϑ_i is the angle that a given link makes with the direction of the force f.

8.3. The flexibility of a freely jointed chain can be described exactly at any extension.

In this regard, the partition function of a freely jointed polymer chain being stretched by the ends with the force f can be written as

$$Z = \int \exp\left[-\sum_{i=1}^{N} (fl/T)\cos\vartheta_i\right] \prod_{i=1}^{N} \sin\vartheta_i d\vartheta_i d\varphi_i, \tag{8.5}$$

where φ_i is the azimuth angle of the i-th link. The integration variables in Eq. (8.5) are separated so that after calculating the integrals we obtain

$$Z = ((4\pi \sinh\beta)/\beta)^N, \quad \beta \equiv fl/T. \tag{8.6}$$

Hence, we easily find the free energy $\mathscr{F} = -T\ln Z$ and the mean radius vector R corresponding to the given force f: $R = -\partial\mathscr{F}/\partial f$. Finally, we obtain

$$R \equiv |R| = Nl(\coth\beta - 1/\beta). \tag{8.7}$$

Figure 1.13 shows the quantity $R/(Nl)$ as a function of $\beta = fl/T$. The function $(R/Nl)(\beta)$ is called the *Langevin function*. It was first introduced in the theory of paramagnetism to express the dependence of sample magnetization on magnetic field. This analogy is by no means accidental, because in the both cases, we deal with an orienting field of the same dipole symmetry. For small β ($\beta \ll 1$ or $fl \ll T$) $\coth\beta \sim 1/\beta + \beta/3$, and we get back to the relation $f = 3TR/(Nl^2)$, which is valid in the Gaussian region. If, conversely, $\beta \gg 1$ (or $fl \gg T$), then the asymptotic dependence $f(R)$ takes the form $f = (T/l)(1 - R/Nl)^{-1}$, that is, the value of R approaches saturation corresponding to total chain extension.

Thus, Eqs. (8.1) and (8.2) are applicable provided $fl \ll T$. This condition, with regard to Eq. (8.2), is identical to the inequality $R \ll Nl$, signifying that the chain is still far from total extension.

*8.4. The elasticity of more complex models of macromolecules can be analyzed using the method of transfer operators; in the case of slight stretching, the analysis can be made in general form.

Because the stretching of the chain by its ends is equivalent to the action of an orienting external field, the chain elasticity can be analyzed using the general methods of Sec. 6. It is worth tracing the procedure followed.

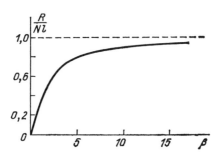

FIGURE 1.13. End-to-end distance of a freely jointed chain as a function of stretching force.

Regarding the definition of the external field (8.4) and the parameter β (8.6), Eq. (6.8) for the eigenfunctions of the transfer operator (6.5a) can be written as

$$\exp(-\beta \cos \vartheta) \int g(u,u')\psi(u')d\Omega_{u'} = \Lambda\psi(u), \qquad (8.8)$$

where ϑ is the angle between u and the external field direction and $g(u,u')$ characterizes the mechanism of chain flexibility.

For example, $g(u,u') = 1/4\pi$ for a freely jointed chain. Integrating Eq. (8.8) over the directions u, we find

$$\Lambda = 4\pi \sinh \beta/\beta.$$

Because the free energy equals $-TN \ln \Lambda$ [Eq. (6.17)], we return to the result (8.6) or (8.7).

More complex models can be analyzed only in terms of a perturbation theory. If $\beta \ll 1$, then Eq. (8.8) can be rewritten as

$$\hat{g}\psi + \hat{h}\psi = \Lambda\psi(u), \qquad (8.9)$$

where $\hat{h} = [\exp(-\beta \cos \vartheta) - 1]\hat{g} = [-\beta \cos \vartheta + (1/2)\beta^2 \cos^2 \vartheta]\hat{g}$ is treated as a perturbation operator. For the unperturbed operator $\hat{g}$, the fundamental eigenvalue equals unity. Linear elasticity corresponds to the second-order perturbation theory. We leave it to the reader to use the perturbation theory formulas taken from any textbook on quantum mechanics to derive the following expression for a chain of rods of length b with fixed valence angle γ (*see* subsection 6.7):

$$\Lambda \cong 1 - (\beta^2/6) \frac{1+\cos \gamma}{1-\cos \gamma};$$

$$\frac{F}{NT} \cong +(1/6)\left(\frac{fb}{T}\right)^2 \frac{1+\cos \gamma}{1-\cos \gamma};$$

$$\langle R \rangle \cong (1/3)f(Nb^2/T) \frac{1+\cos \gamma}{1-\cos \gamma} = (1/3)f(Ll/T). \qquad (8.10)$$

According to Eq. (3.11) it coincides, as expected, with the general result (8.2).

*8.5. *When stretched by the ends, a chain placed in an inhomogeneous medium extends mainly along an optimal trajectory conforming to a potential topography analogous to quasiclassical trajectories of a quantum particle or to the paths of ray optics; such a chain is analogous to the trajectory of a Newtonian, but not a Brownian, particle.*

Assume now that the polymer chain is stretched not in a homogeneous space (considered so far in this section) but rather in an inhomogeneous external field $\varphi(x)$. Such a situation is common in nature; for example, in a star-shaped macromolecule with large number of rays $f \gg 1$, each ray (at least in the central part of the star) being located in a very dense surrounding, stretches out in a great degree. In addition, this situation is of great methodic interest. The globular state of the chain placed in a strong compressing field is analogous to the lowest discrete energy level of a quantum particle. Now, we consider the opposite limiting quasiclassical case.

It is well known that the classical limit is derived in very simple and obvious terms from the Feynman path integral formulation of quantum mechanics.[22] Accordingly, let us turn to the formulation of the path integral in polymer statistics [i.e., to Eq. (6.7), (6.25), or (6.26)]. We analyze the situation when the chain is sufficiently stretched so that a small fraction of the conformations close to one particular trajectory provides a predominant contribution to the partition function or the Green function. What is this particular trajectory, and how is it defined?

From the structure of the path integral, for example, having the form (6.26), it is seen that the largest contribution is provided by the path corresponding to the minimal exponent, that is,

$$\min_{\{x(\tau)\}} \left\{ \int_0^t \left[\frac{3\dot{x}^2(\tau)}{2a^2} + \frac{\varphi(x(\tau))}{T} \right] d\tau \right\}. \qquad (8.11)$$

In this expression, one can easily recognize the least action principle for an ordinary classical particle of "mass" $3/a^2$, velocity $\dot{x}$, and potential energy $U = -\varphi(x)/T$ (the minus sign appears because the action is represented via the Lagrangian function, that is, the difference of kinetic and potential energies; the physical meaning of this sign is clarified later). Varying the action, one can readily derive the equations specifying the selected path in the form of "Newton's second law"

$$(3T/a^2)\ddot{x} = grad\ \varphi \qquad (8.12)$$

or in the equivalent form of the "energy conservation law"

$$(3T/2a^2)\dot{x}^2 - \varphi(x) = E = \text{const.} \qquad (8.13)$$

In this reasoning, however, we neglect for the sake of clearness the external stretching force f (i.e., the orienting field). If this force is taken into account, then the initial or boundary conditions should be defined in the "dynamic" equation (8.12) or the "total" energy in Eq. (8.13). Indeed, if the force f acts on the terminal link t of the chain, then the following equation is to be valid[d]:

$$(3T/a^2)\dot{x}(t) = f. \qquad (8.14)$$

Similarly, for the initial link, $(-3T/a^2)\dot{x}(0) = f$.

For the free energy of the chain, $-T \ln G$ and allowing for the stretching force (8.14), we obtain

$$F = \int_0^t [(3T/2a^2)\dot{x}^2(\tau) - \varphi(x(\tau)) + f\dot{x}(\tau)]d\tau, \qquad (8.15)$$

where $x(\tau)$ is the optimal path defined by Eqs. (8.12) to (8.14). In a certain sense, Eq. (8.15) is analogous to Eq. (6.17) and "Newton's" equations (8.12) to (8.14) to Schrödinger equations (6.6), (6.8), and (6.28).

Let us dwell briefly on a simple physical interpretation of Eq. (8.15). Consider a short section of the chain containing δN monomers, and assume it is stretched so that its end-to-end distance equals δx. According to Eq. (8.1), the free energy of the section equals $3T(\delta x)^2/2\delta Na^2$. Summing the energies of all of the elements δN in the chain, we obtain exactly the first term of Eq. (8.15). Thus, the physical meaning of the minus sign in the definition of the potential energy of an effective Newton's particle $U = -\varphi(x)/T$ becomes clear. The chain is stretched the most (least) in the region of potential wells (hills) of the real field $\varphi(x)$; conversely, the lowest (highest) kinetic energy of the particle corresponds to the potential hills (wells).

In conclusion, we make another observation. The quasiclassical (or ray-optics) approximation is also applicable when the chain is not entirely stretched. The beads on the chain can be chosen with various density (see subsections 4.2 and 5.4). If the chain is slightly stretched, then the approximation of a single optimal path (conformation) is valid only for rare densities of the beads. Later, we see that this corresponds to the condition of extension of the chain of blobs.

9. CONCEPT OF A MACROSCOPIC STATE FOR A POLYMER CHAIN. THE LIFSHITZ FORMULA FOR CONFORMATIONAL ENTROPY

9.1. *A macroscopic state of a polymer chain is specified by the link distribution over the states (coordinates, orientations, and so on), while the properties of the system in a given microscopic state are determined by the linear memory (i.e., by the link connectivity in the chain).*

All ideal chain models admittedly give only a rough description of real properties of macromolecules and polymer systems. In the following chapters, we

---

[d]The end of the chain acts on a confining body with the force $f = -\partial F/\partial x(t) = (T/G)\partial G/\partial x(t)$ $= T\partial(\text{action})/\partial x(t)$; in conformance with mechanics, such a derivative of the action equals the "momentum of a particle" at the moment t, thus yielding Eq. (8.14) (see L. D. Landau and E. M. Lifshitz, Mechanics, Pergamon Press, Oxford, 1988 Sec. 43).

discuss various phenomena caused by volume interactions. The volume interaction effects are more complex and diverse than the ideal chain properties, just as the phenomena in a condensed material are more diverse than in the ideal gas. Nevertheless, models of ideal chains as well as the ideal gas are important, because under some conditions the real systems are closer to ideal systems. In addition, the flexibility causing spatial coiling is no doubt characteristic for any real macromolecule as well as ideal ones. All remaining properties of ideal macromolecules considered previously (in particular, their responses to various external fields) can furnish an answer to the following question: how does the linear memory, (i.e., the very fact of link connectivity in the chain) affect the properties of a statistical system? Or, specifically, in application to an ideal polymer: how do the properties of an ideal gas change on joining its molecules in a chain? The obvious answer is that the change is conspicuous. For example, the density of the ideal gas contained in a cavity is distributed uniformly, whereas the density of the ideal polymer is not (Fig. 1.13). Thus, the ideal polymer does not obey the Pascal law. The susceptibility of the dipoles comprising the ideal gas to an orienting field is constant, whereas in the ideal polymer, it is proportional to the number of particles [see Eq. (8.2)]. The list of examples is easy to continue; however, it is sufficient to draw a principal conclusion that the statistical physics of macromolecules requires a special approach allowing the linear memory or chain structure of molecules (i.e., the "polymer specifics") to be primarily regarded as of paramount importance.

In general, statistical physics reduces the mechanical description of a system in terms of the generalized coordinates of all particles (i.e., in terms of a multidimensional phase space of microscopic variables) to the abbreviated (approximate) description in terms of a relatively small number of specified macroscopic parameters. The price one has to pay for this reduction is appearance of entropy S. The statistical weight $\exp S$ is none other than the number of microscopic states (conformations) possessing the given values of the macroscopic parameters. In the previously mentioned examples, one such macroscopic parameter was the end-to-end vector R. Accordingly, there appeared the quantity $S(R)$, which is the entropy of a macroscopic state with the given value of R.

In the general case, the specific set of macroscopic parameters needed for a meaningful description of a certain physical phenomenon is seldom chosen easily. The choice is not a formal procedure, and it cannot be accomplished by computational means. In most polymer physics phenomena, the density, or more exactly, the distribution of the generalized density of links over their states, serves as a natural macroscopic parameter. It can be, for example, the local concentration of a homogeneous solution, or the spatial distribution of local concentration describing both the fluctuations in the homogeneous solution and interphase boundaries, supra- and intramolecular structures, and so on. It can be the segment orientation distribution function defining the liquid crystal order. More examples can be easily suggested.

Surely, to be a macroscopic parameter, the density distribution $n(\alpha)$ must be smoothed out. For example, the microscopic density (or particle concentration) defined as $n_\Gamma = \Sigma\delta(x-x_i)$ is equivalent to fixing all microscopic variables x_i. Actually, the distribution $n(x)$ should be smoothed out (approximated) over physically infinitesimal volumes, solid angle elements, and so on. If each such element of volume (or solid angle) contains many or at least several particles (links), fixing the smoothed density imposes no restrictions on displacement, rotation, and other local motions of particles within the elementary volumes. The smoothed density therefore yields a truly approximate description of the system and involves fixing a relatively small number of variables. Specifically, attention should be paid so that the distribution $n(x)$ and so on does not exhibit any fast oscillations.

In this manner, we come to the next question, which is one of the basic questions in statistical physics of macromolecules. What is the entropy value of a macromolecule (i.e., of the chain with linear memory) residing in a macroscopic state with a given density distribution? It should be emphasized once again that it is the conformational entropy that determines all of the most specific properties of polymers, both synthetic and biologic.

9.2. *The conformational entropy of a polymer system is independent of the nature of the forces that determine the state of the system; therefore, entropy can be found by considering an ideal chain placed in a fictitious external field that ensures equilibrium for a given macroscopic state.*

Let us return to the abstract designations of Sec. 6, equally suitable for translational and orientational degrees of freedom. Suppose that the link density is distributed as $n(\alpha)$ and the macromolecule resides in a macroscopic state with the given density distribution.

Similar problems in statistical physics are solved by the following technique proposed by M. A. Leontovich in 1938. Let us formally try to find the external field $\varphi(\alpha)$ that ensures the equilibrium distribution $n(\alpha)$ of the ideal chain. (Of course, one should not think how this field might have been realized experimentally). The theory of an ideal macromolecule presented earlier allows the entropy of the chain in the field $\varphi(\alpha)$ to be calculated easily. This solves the problem.

An ideal chain in an external field is described by the set of equations (6.8), (6.10), (6.17), and (6.20). Before, we assumed the field $\varphi(\alpha)$ to be known and looked for $\psi(\alpha)$, $n(\alpha)$, and the free energy $\mathscr{F}$ of the chain; now, the problem is defined differently. The distribution $n(\alpha)$ is known and $\psi(\alpha)$ and $\varphi(\alpha)$ can obviously be found from Eqs. (6.20) and (6.8). To find the entropy, we write the free energy (6.17) in the form

$$\mathscr{F}\{n\} = -T \ln \Lambda = E\{n\} - TS\{n\}.$$

Here, $E\{n\}$ is the energy that the links possess in the field $\varphi(\alpha)$. It is obvious that

$$E = \sum_{\alpha} \varphi(\alpha) n(\alpha),$$

where $\sum_{\alpha}$ should be replaced by the integral, provided the variable α takes on a continual set of values. Consequently, $S\{n\} = \sum_{\alpha} n(\alpha)[\ln \Lambda + \varphi(\alpha)/T]$. One can conveniently eliminate from this relation the fictitious quantity $\varphi(\alpha)$ by using Eqs. (6.5) and (6.8) so that the entropy S is expressed directly as a functional of smoothed density:

$$S\{n\} = \sum_{\alpha} n(\alpha) \ln(\hat{g}\psi/\psi). \qquad (9.1)$$

This result was derived by I. M. Lifshitz in 1968[30].

The function $\psi(\alpha)$ can be expressed directly in terms of $n(\alpha)$ as

$$\Lambda n(\alpha) = \psi^+(\alpha)\hat{g}\psi, \qquad (9.2)$$

which follows from Eqs. (6.20) and (6.8). A specific choice of the value of Λ in Eq. (9.2) is quite unimportant, because it determines only the normalization of ψ function and does not affect the value of $S\{n\}$.

By physical meaning, $\psi^+(\alpha)$ and $\psi(\alpha)$ are the distribution functions of the beginning and the end of the chain. Consequently, the meaning of Eq. (9.2) is very explicit: a chain link is located at the "point" α provided the ends of two sections of the chain approach that point from the two sides. Dealing with a single link, however, we attribute it to (any) one side of the chain, which yields the factor $\psi^+(\alpha)$, while the bond adjoining on the other side provides the factor $\hat{g}\psi$.

It should be noted that in the literature on statistical physics of macromolecules, the distribution function of chain ends $\psi(\alpha)$ is often selected as a macroscopic parameter instead of the density $n(\alpha)$. Then, $n(\alpha)$ and entropy are again determined by Eqs. (9.1) and (9.2).

9.3. *Entropy losses because of the spatial inhomogeneity in concentration of a polymer are determined by the concentration gradient.*

Let us interpret the general results just obtained for such a practically important aspect as the spatial inhomogeneity of a polymer system. We are concerned with local concentration fluctuation as well as equilibrium inhomogeneity, such as a spatial interphase boundary, and so on. In some way, spatial inhomogeneity is unfavorable in terms of entropy, because in the presence of a concentration gradient, the chain bends in certain directions more often than in others. This restricts the conformational set of all bonds (i.e., decreases the entropy).

Thus, suppose there is a spatial distribution of local concentration $n(x)$. Consider only the standard Gaussian bead model. If $n(x)$ varies smoothly, then the substitution $\hat{g} \to 1 + (a^2/6)\Delta$ is valid [see Eq. (7.7)] and $\ln \hat{g}\psi/\psi \cong (a^2/6)\Delta\psi/\psi$. Therefore, $n(x) \cong \text{const} \cdot \psi^2(x)$, or $\psi(x) \sim n^{1/2}(x)$, that is,

$$S\{n\}=(a^2/6)\int n^{1/2}(x)\Delta n^{1/2}(x)d^3x=\text{const}-(a^2/6)\int(\nabla n^{1/2}(x))^2d^3x.$$
$$(9.3)$$

As expected, the local concentration gradient is unfavorable in terms of entropy. This effect is caused by the link connectivity that is formally manifested in $S\{n\}$ being proportional to a^2, the value characterizing chain bonding.

For illustration, it should be noted that Eq. (9.3) allows one to estimate the entropy losses, sustained on placing the ideal chain into a cavity, from another point of view. In this case, $\Delta \sim 1/D^2$, and we immediately arrive at the evaluation (7.2).

Expression (9.3) describes a single chain. When there are many chains (a solution), their conformational entropies add up, that is, $n(x)$ in Eq. (9.3) represents the total local concentration of links of all chains. In addition, one should add in this case the conventional Boltzmann entropy δS of relative translational motion of the chains. Because the link concentration equals $n(x)/N$,

$$\delta S=(1/N)\int n(x)\ln(n(x)/Ne)d^3x. \qquad (9.4)$$

It is essential to recognize that this value tends to zero as N grows. This fact is quite natural, because for a given solution concentration n, the number of independently mixing "particles" (i.e., the chains) diminishes with the polymerization degree. That is why the conformational entropy, which as a rule is insignificant for low-molecular-weight substances compared with the translational entropy, becomes predominant for many properties of polymers and biopolymers.

9.4. Entropy losses because of the orientational ordering of macromolecules can be treated as being caused by "orientational inhomogeneity."

The general formula (9.1) allows one to determine the entropy associated not only with translational but with orientational degrees of freedom as well. We now show specifically how to do this using the example of a persistent model. Let $f(u)$ be the distribution function of chain section orientations; it is normalized so that

$$\int f(u)d\Omega=1. \qquad (9.5)$$

In the isotropic state, $f(u)=1/4\pi$, and any anisotropy is unfavorable in terms of entropy. We discuss the entropy loss in the macroscopic state with the distribution $f(u)$.

Consider the persistent chain using the passage to the continuum limit. Taking a chain consisting of rods of length b with valence angle γ, we let both b and γ tend to 0 so that the Kuhn segment length remains equal to $l=4b/\gamma^2=\text{const}$ (*see* subsection 3.4). In this approach, the anisotropy of the distribution $f(u)$ or $\psi(u)$ clearly should be regarded as smooth, because ψ varies little on turning u

through the angle γ (for $\gamma \to 0$). In complete analogy with the spatial case (7.7), it is easy to show that in this situation

$$\hat{g}\psi \cong \psi + (\pi\gamma^2/2)\Delta_u\psi, \tag{9.6}$$

where Δ_u is the angular part of the Laplace operator, the second term of (9.6) being a small correction. The generalized density of the chain of rods is normalized by the condition $\int n(u)d\Omega = N \equiv L/b$. Therefore, $n(u) = (L/b)f(u)$. Similarly, from Eq. (9.2) we obtain

$$f(u) = \text{const} \cdot \psi^2(u) \quad \text{or} \quad \psi(u) \sim f^{1/2}(u)$$

and for the entropy we obtain

$$S\{f(u)\} = \int \frac{L}{b} f(u) \frac{\pi\gamma^2}{2} \left(\frac{\Delta_u f^{1/2}(u)}{f^{1/2}(u)} \right) d\Omega = \frac{2\pi L}{l} \int f^{1/2}(u)\Delta_u f^{1/2}(u)d\Omega. \tag{9.7}$$

As should be, the auxiliary parameters b and γ are absent in this expression, and only the statistical segment length l appears.

Compare the expression for the orientational entropy of a freely jointed chain. In this model, $\hat{g}\psi = \int \psi(u)d\Omega$ is independent of u, because the foregoing rod has no effect on the orientation of the following rod. Consequently, $n(u) = Nf(u) = \text{const} \cdot \psi(u)$, and

$$S\{f(u)\} = -N \int f(u)\ln[4\pi f(u)]d\Omega. \tag{9.8}$$

As one would expect, this result coincides with the orientational entropy of the ideal gas of rods, because the freely jointed segments are oriented independently.

10. RING AND BRANCHED MACROMOLECULES

10.1. *The gyration radius of an ideal self-crossing (phantom) ring macromolecule is smaller by a factor of $\sqrt{2}$ than that of a linear chain of the same length.*

Along with linear chains, ring macromolecules also exist in nature. The most interesting properties of such macromolecules are determined by the knots that they can form; this is discussed in detail in the next section. Here, we assess the size of an ideal phantom (i.e., freely self-crossing) ring chain.

Let $P_N^{(\text{ring})}(R_{mn})$ denote the probability distribution of the vector connecting the links m and n in an N-link ring macromolecule. Obviously, this value equals the conditional probability $P_N(R_{mn}|R=0)$ so that in the linear N-link chain, the vector between the links m and n equals R_{mn} under the condition that the end-to-end distance equals 0. By definition of conditional probability, we obtain

$$P_N^{(\text{ring})}(R_{mn}) = P_N(R_{mn}|R=0) = \frac{P_N(R_{mn}, R=0)}{P_N(R=0)}, \tag{10.1}$$

where $P_N(R_{mn}| \ R=0)$ is the corresponding joint probability distribution. Because the considered ring chain is assumed to be ideal and phantom, its sections m-n and n-N-0-m are mutually independent. Therefore,

$$P_N(R_{mn},R=0)=P_{n-m}(R_{mn})P_{N-n+m}(R_{mn}). \tag{10.2}$$

Combining Eqs. (10.1) and (10.2) as well as using the Gaussian distribution (4.1), we obtain

$$P_N^{(\text{ring})}(R_{mn})=(3/2\pi\mu l^2)^{3/2}\exp(-3R_{mn}^2/2\mu l^2),$$

$$\mu=|m-n|(N-m+n)/N. \tag{10.3}$$

Thus, the distribution of distances between the links of the phantom ring is Gaussian, but it differs from that of the linear chain (for which there would be $\mu=|m-n|$).

Equation (10.3) allows for an easy calculation of the gyration radius of the phantom ring. As $\langle R_{mn}^2\rangle_{\text{ring}}=\mu l^2$, from the definition of the radius of gyration (5.2) we have

$$\langle s^2\rangle_{\text{ring}}=Nl^2/12=\langle s^2\rangle_{\text{lin}}/2. \tag{10.4}$$

Thus, the ideal phantom ring macromolecule is somewhat more compact (to be exact, by a factor of $\sqrt{2}$) than the corresponding linear chain. The size of such a macromolecule, however, is proportional to $N^{1/2}$, as in the case of a linear macromolecule.

10.2. Ideal branched macromolecules are smaller than linear chains composed of an equal number of links; volume effects are more essential for branched macromolecules.

As mentioned in the Introduction, apart from linear polymers, there exists a large class of branched macromolecules formed by cross-linking linear chain sections or by synthesis in the presence of monomers with three or more free valences. The branched macromolecules are classified into stars, combs, randomly branched, and other types (Fig. 1.14).

Assuming all of the chains composing a branched macromolecule are ideal, we can find the average size of the macromolecule without any major difficulties. Of course, the size of such macromolecules is to be described by the mean square of the radius of gyration $\langle s^2\rangle$, not by the mean square of the end-to-end distance $\langle R^2\rangle$, because the branched macromolecule has many ends. When the branched macromolecule contains N links, one usually writes

$$\langle s^2\rangle=g\langle s^2\rangle_{\text{lin}}=gNa^2/6, \tag{10.5}$$

where $\langle s^2\rangle_{\text{lin}}$ is the mean square radius of gyration of a linear chain composed of the same number of links [see Eq. (5.5)]. The factor g is always less than unity; it characterizes the diminishing size of the average branched macromolecule compared with the linear analogue.

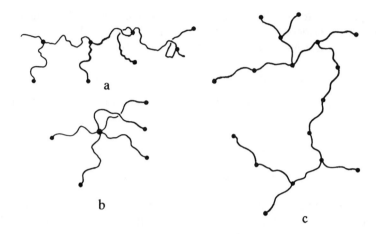

FIGURE 1.14. Branched macromolecules. (a), Comb-like. (b), Star-like. (c), Randomly branched.

For example, for the factor g of a star-like macromolecule with f branches (Fig. 1.14a), each of which contains N/f links, one can readily find:

$$\langle s^2 \rangle = (1/2N^2) \sum_{i,j=0}^{N/f} \sum_{\mu,v=1}^{f} \langle (R_{ij}^{\mu v})^2 \rangle,$$

where $R_{ij}^{\mu v}$ is the distance between the i-th link of the μ-th branch and the j-th link of the v-th branch. The links are enumerated starting from the branching point, and the summation is performed over all link pairs. Obviously,

$$\langle (R_{ij}^{\mu v})^2 \rangle = \begin{cases} (i+j)a^2, & \mu \neq v \\ |i-j|a^2, & \mu = v. \end{cases}$$

Taking into account the last expression, one can easily calculate $\langle s^2 \rangle$:

$$\langle s^2 \rangle = [(3f-2)/6f^2]Na^2, \quad g \equiv 6\langle s^2 \rangle / Na^2 = (3f-2)/f^2. \qquad (10.6)$$

It is seen that for $f \geqslant 3$, $g < 1$, as it should be.

The smaller average size of the ideal branched macromolecule compared with its linear analogue results in corresponding growth in the average link concentration within the macromolecule. It therefore can be expected (as it proves to be in reality) that volume effects, arising primarily from the volume of links, are as a rule stronger for the branched macromolecules than for the linear ones. This is especially important for the randomly branched macromolecules, considered in the next subsection.

10.3. *The size of an ideal randomly branched macromolecule with no cycles varies with the number of links as $N^{1/4}$; the link concentration in such a system grows infinitely with N so that the ideal chain approximation is by no means valid at large N.*

Randomly branched macromolecules are synthesized, for example, from irregular mixture of bivalent and polyvalent monomers, provided an attachment of a polyvalent link to any free valence of the already synthesized part of the macromolecule is equally probable. A wide variety of conditions exists for which the synthesized macromolecules contain practically no cycles. (A *cycle* is a set of macromolecular sections forming a closed contour).

This subsection considers a randomly branched macromolecule without cycles. The size of such a system is obviously of the order of the mean length of a linear chain connecting two randomly chosen ends of a randomly branched macromolecule. If the average number of links between the branching points equals n and the average number of branching points between the two ends v (Fig. 1.15a), we obtain

$$\langle s^2 \rangle \sim vna^2. \tag{10.7}$$

The most difficult task here is how to determine v. To evaluate it, we picture the randomly branched macromolecule as a special network or, strictly speaking, a topologic graph called a *Cayley tree* (or *Boethe cactus*[e]), possessing the same connectivity pattern. Each edge of the graph corresponds to a section between the branching points (Fig. 1.15). The number of edges equals $P \sim N/n$. The unknown quantity v is an average number of graph edges between any two randomly chosen end points; to calculate it, we surround a part of the Cayley tree by a closed contour of length $2P$ (*see* the dotted contour in Fig. 1.15b). Consider this contour as a closed random walk trajectory or a phantom $2P$-link ring macromolecule. With the arbitrary point O chosen as an origin on the Cayley tree, such a macromolecule comprises an equal number of steps in the direction toward the point O and in the opposite direction. This circumstance makes it possible to treat our artificially constructed ring chain as a random walk along a straight line, with steps in the positive and negative directions being consistent with steps toward and away from point O. The size of such a contour conformed to the straight line will equal v. Its value, just as the size of the phantom closed macromolecule in three-dimensional space, is evaluated by means of Eq. (10.4):

$$v \sim (2P)^{1/2}. \tag{10.8}$$

---

[e]The Cayley tree is drawn as follows. Take an arbitrary vertex 0 and draw $z \geqslant 3$ edges (unit sections). By way of illustration in Fig. 1.15b, $z = 3$. From each of z obtained vertices, draw $z - 1$ edges. Then from $z(z - 1)$ new vertices, draw $z - 1$ edges, and so on, *ad infinitum*. All of the vertices of the Cayley tree thus obtained are equal in the sense that z edges emerge from each vertex. The Cayley tree structure is peculiar: at a distance of n steps from any vertex, there are $(z - 1)^n$ other vertices whose number grows exponentially with n. Hence, the Cayley tree cannot be embedded with a finite density in the common three-dimensional (or any finite-dimensional) Euclidean space.

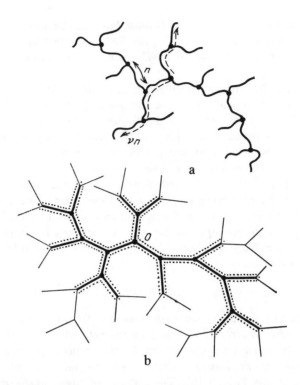

FIGURE 1.15. Randomly branched macromolecule (a) and its layout on a Cayley tree (b). The section between branching points comprises n links and between arbitrary ends vn links. The *dotted line* shows the surrounding contour of length $2P$.

Comparing Eqs. (10.7) and (10.8) and recalling that $P \sim N/n$, we finally obtain

$$\langle s^2 \rangle \sim N^{1/2} n^{1/2} a^2 \qquad (10.9)$$

(B. Zimm, W. Stockmayer, 1949). The size of an ideal randomly branched macromolecule without cycles is seen to grow with N as $N^{1/4}$, far slower than for the linear chain where $s \sim N^{1/2}$. This implies that such a macromolecule for $N \gg n$ must assume a very compact conformation.

According to Eq. (10.9), the volume of the randomly branched macromolecule actually varies as $\langle s^2 \rangle^{3/2} \sim N^{3/4}$ so that the link concentration inside the macromolecule is of order $N^{1/4} n^{-3/4} a^{-3}$ and grows infinitely with N (in contrast to a Gaussian coil, where the concentration tends to zero as N grows). This unnatural result shows that the properties of randomly branched macromolecules with large numbers of links are always affected substantially by volume interactions. Thus, the ideal macromolecule approximation cannot be applied to such macromolecules.

11. TOPOLOGICAL CONSTRAINTS IN POLYMER SYSTEMS

11.1. *Because crossing of a chain section by another is inconceivable, a system of ring polymers "remembers" its topology and cannot alter it.*

A fundamental feature of chain polymer systems is associated with the fact that the macromolecular sections cannot cross one another (Fig. 1.16). Surely, this feature is caused by volume interactions. What is important, however, is that it would remain even if the chains possessed a very small (tending to zero) thickness. This is why self-crossing must forbidden even in the model of an immaterial or ideal polymer.

Obviously, the exclusion of self-crossing by no means manifests itself in equilibrium properties of linear chains. All conformations are feasible, and their realization is only a question of time. In the theory of equilibrium properties, the linear chains can be regarded as phantom, for which any intersections occur without hindrance. (Non-phantom characteristics are essential for the dynamic properties of linear chains discussed in Chapter 6.)

Things are quite different with ring macromolecules. For a non-phantom ring polymer chain or a system of ring chains, the only possible conformations are those that are topologically equivalent to one another, that is, can be obtained only via continuous chain deformation (without break-ups). Thus, exclusion of crossing in the ring chain system results in a topologic restriction on the set of allowable conformations. The topological state of such a system is formed at generation and does not change thereafter. In other words, besides linear memory, ring polymers possess topologic memory as well.

From the standpoint of topology, the ring macromolecule represents a knot (Fig. 1.17). The ring with no knots is termed a *trivial knot.*

Similarly, two or more rings form a link (Fig. 1.18), while a separate ring is characterized by a trivial link. We do not deal here with mathematic problems of the knot and link classification; the interested reader is referred to the monographs[23,24] and review article.[25]

*11.2. *The simplest polymer system with nontrivial topology is an ideal chain with fixed ends lying in the plane with an impenetrable straight line passing across the plane.*

This system is shown schematically in Figure 1.19. A straight line passes through the point O perpendicular to the plane of the drawing. The chain cannot cross the line, and this results in a topologic restriction. The number n of turns

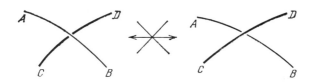

FIGURE 1.16. Mutual crossing of chain sections is impossible.

FIGURE 1.17. The simplest nontrivial knot-trefoil.

of the chain around the point O (in Fig. 1.19, $n=1$) is a topologic invariant that remains constant after the system is prepared (i.e., the chain ends are fixed). The analysis of this schematic model adequately clarifies the essence of the topological problems in polymer physics (S. F. Edwards; G. Prager, H. L. Frisch 1968).

The system geometry is clear from Figure 1.19. The radius vectors R_1 and R_2 of the chain ends are drawn from the point O and make angle θ between them. It is known in advance that a topologic restriction reduces the entropy by decreasing the set of allowable conformations.

Suppose that $W_n(R_1,R_2,\theta,N)$ is the probability that a N-link chain passes from R_1 to R_2, having made n turns around an obstacle. Later, we show that

$$W_n(R_1,R_2,\theta,N)=P_N(R_1-R_2)\varphi_n(\theta,z), \qquad (11.1)$$

where $z\equiv 2R_1R_2/(Na^2)$, $P_N(R_1-R_2)$ is the Gaussian distribution (4.1) for a chain without any restrictions, and

$$\varphi_n(\theta,z)=\exp(-z\cos\theta)\int_{-\infty}^{+\infty}I_{|v|}\exp[i(2\pi n-\theta)v]dv; \qquad (11.2)$$

where I is the modified Bessel function. Now, we demonstrate the derivation of Eqs. (11.1) and (11.2) and analyze them.

For simplicity, we consider the continual standard Gaussian model of a macromolecule (*see* subsections 6.3 and 6.4) by describing its conformation by the vector function $r(s)=\{x(s),y(s)\}$, where s is the length along the chain $[r(0)=R_1$ and $r(s)=R_N]$. First, there arises the purely geometric problem of the determination of the number n for a given conformation. Denote the angle

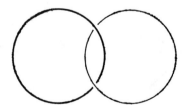

FIGURE 1.18. The simplest topologic entanglement of two ring polymer chains.

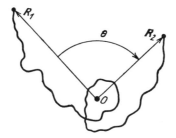

FIGURE 1.19. Entanglement of a polymer chain with an obstacle (point O), chain ends are supposed frozen.

between the current radius vector $r(s)$ and the fixed axis x by $\theta(s)$ $=\arctan[y(s)/x(s)]$. Then, $\dot{\theta}(s)\equiv d\theta(s)/ds$ is the elementary turn of an infinitesimal section ds around the obstacle. Consequently,

$$\int \dot{\theta}(s)ds = \int \frac{x\dot{y}-y\dot{x}}{x^2+y^2}\,ds = 2\pi n + \theta, \qquad (11.3)$$

where a dot over a letter defines differentiation with respect to s. It is useful to rewrite this expression as

$$\int_{r(s)} A(r)dr = 2\pi n + \theta,$$

where the integral is taken over the chain contour in the given conformation $r(s)$ and the vector field $A(r)$ specified by its components

$$A_x = -y/(x^2+y^2), \quad A_y = x/(x^2+y^2).$$

An initial expression for the probability W_n now can be easily obtained. It is sufficient to integrate the ordinary Green's function (6.4) only over the trajectories with the given number n of turns, that is, to insert in the integrand of Eq. (6.4) the appropriate delta-function

$$W_n = \int \delta\left(\int A dr - \theta - 2\pi n\right)\exp\left[-\int \dot{r}^2(s)ds\right]Dr(s).$$

Using for the delta-function the integral relation

$$\delta(\xi) = (2\pi)^{-1}\int \exp(iv\xi)dv;$$

we obtain

$$W_n = (2\pi)^{-1} \int W_v \exp[iv(2\pi n + \theta)] dv,$$

$$W_v = \int \exp\left\{-\int [\dot{r}^2(s) + ivA(r(s))\dot{r}(s)] ds\right\} Dr(s). \qquad (11.4)$$

Recall the analogy between Green's functions of a polymer chain and a quantum particle, discussed repeatedly before. In terms of this analogy, Eq. (11.4) has a very clear meaning: it is the Green's function of a charged particle moving in a magnetic field with vector potential $A(r)$. As a function of the end point coordinates, it satisfies the Schrödinger equation

$$\partial W_v/\partial N - (a^2/6)(\nabla_r - ivA(r))^2 W_v = \delta(N)\delta(r - R_1). \qquad (11.5)$$

Of course, this equation also can be obtained easily from the path integral (11.4) without resorting to the quantum analogy, just as was done with Eqs. (6.26) and (6.28). By making a substitution, one can see that the solution of Eq. (11.5) takes the form (11.1) and (11.2). Let us now examine this solution.

Using the various properties of Bessel functions, it is easy to establish the following:

1. $\sum\limits_{n=-\infty}^{+\infty} \varphi_n(\theta,z) = 1$; if the number of turns[f] around the obstacle is arbitrary, then it is equivalent to the absence of topologic restrictions;

2. $1 > \varphi_n(\theta,z) > 0$; this implies that $W_n < P_N$ (i.e., the onset of topologic restriction reduces the number of possible chain conformations);

3. $\varphi_{n=0}(\theta,z) \to 1$ at $z \gg 1$; this means that a non-entangled ($n=0$) chain becomes free on withdrawal ($z \to \infty$) from the obstacle by the distance exceeding the Gaussian size;

4. $\varphi_{n\neq0}(\theta,z) \to 0$ at $z \gg 1$; an entanglement ($n \neq 0$) with a remote obstacle results in drastic reduction of possible chain conformations;

5. $\varphi_n(\theta,z) \to 0$ at $n \to \pm\infty$; entanglements of high order reduce drastically the conformational set of chains and thus are unfavorable in terms of entropy.

Figure 1.20 shows the approximate dependence of the force that the closed polymer chain ($R_1 = R_2$, $\theta = 0$) exerts on the obstacle, on the distance R between the obstacle, and the given chain link. This force can be calculated by analogy with Eqs. (8.1) and (8.2). Indeed,

$$-T \ln W_n(R_1 = R_2 = R, \theta = 0, N) = \mathscr{F}_n(R^2/Na^2)$$

is the free energy of the given state, with the quantity $\partial \mathscr{F}_n/\partial R$ equal to the sought-after force f_n. As seen from Figure 1.20, at small distances the force is

---

[f]Different signs of n correspond to rounds in opposing (clockwise and counter-clockwise) directions.

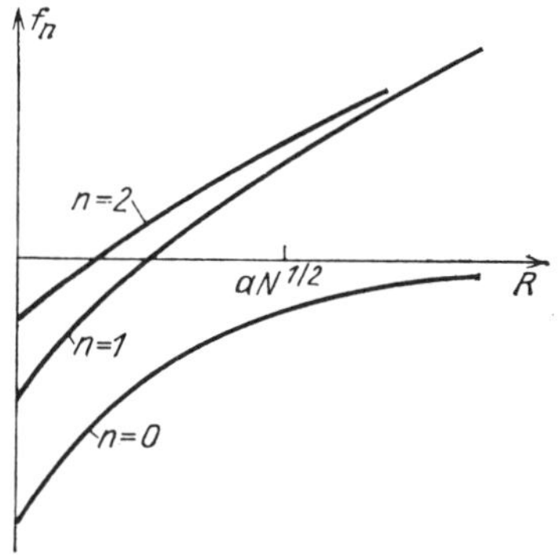

FIGURE 1.20. Force exerted on an obstacle by a polymer ring surrounding it n times as a function of distance between the obstacle and a specified chain link.

negative (i.e., the coil tries to expel the obstacle located in the middle). Moreover, a non-entangled obstacle is expelled ($f_0 < 0$) at any distance. The interaction becomes vanishingly small only for $R \gg aN^{1/2}$. The entangled ring pulls a remote obstacle in ($f_{n \neq 0} > 0$) with a force weakly depending on the order of entanglement.

*11.3. *The development of an analytical theory of polymer knots and entanglements on the basis of known topological invariants meets serious difficulties.*

The previous example makes clear that the investigation of knots and entanglements calls for finding an effective topological invariant, representing a quantity or set of quantities that could easily be determined for any given conformation and would have coinciding values for conformations of one topological type and different values for conformations of different types. For the problem of the entanglement of two closed polymer chains, the so-called Gaussian topological invariant G is widely known: if the two contours C_1 and C_2 are specified by the functions $r_1(s)$ and $r_2(s)$, then

$$G(C_1, C_2) = \frac{1}{4\pi} \oint_{C_1} \oint_{C_2} \frac{[dr_1 \times dr_2] r_{12}}{|r_{12}|^3}, \quad r_{12} \equiv r_1 - r_2. \tag{11.6}$$

Let us prove that G is really a topological invariant. We assume that the contour C_2 is a conductor carrying the current J. Then, according to the Biot-Savart law, the magnetic field induced by the current J equals

$$H(r) = (J/c) \int_{C_2} [(r - r_2) \times dr_2 / |r - r_2|^3].$$

Consequently, the quantity G can be treated as a circulation of the magnetic field along the contour C_1 [because $(a \times b)c = (b \times c)a = (c \times a)b$], and one can use the Stokes formula

$$G = (1/4\pi)(c/J) \oint_{C_1} H(r_1)dr_1 = (c/4\pi J) \int_{C_1} (\nabla \times H(r_1))d\sigma = \int_{C_1} (J/J)d\sigma,$$

where the second integral is a flux of $\nabla \times H$ across the surface spanning the contour C_1. However, $\nabla \times H = (4\pi/c)J$, and the contribution to the flux is provided only by the points at which the stream line (i.e., C_2) intersects with the surface mentioned earlier.

Thus, G is the algebraic (i.e., with the direction taken into account) number of times that a contour crosses the surface stretched over the other contour. This number is clearly a topologic invariant. In fact, it is this invariant that was used in the previous subsection. [As an exercise, the reader may check to see that the Gaussian integral (11.6) yields the expression (11.3) in the specific case when the size of one of the contours tends to infinity]. By analogy with the approach taken in the previous subsection, we could use the Gaussian invariant to analyze the partition function of a chain with a topologic constraint in the form of an arbitrary contour C_2. This would bring us back once again to the Schrödinger equation (11.5), but with a magnetic field of more complex configuration. The real difficulty, however, consists not only (and not so much) in the complicated form of the equation but in the following principal point: there are different entanglements giving equal values for the Gaussian integral; therefore, solution of the equations of type (11.5) corresponds in the general case to a mixture of many topologically different entanglements. The number of turns is a full topologic invariant only for a rectilinear obstacle; therefore, no difficulties arise in only this case. The simplest example of two different (trivial and nontrivial) entanglements with equal (zero) value of the Gaussian integral is shown in Figure 1.21.

It should be noted that no full topologic invariant suitable to form the basis of an analytic theory is known, either for knots or links. In computer-simulation experiments, the most efficient proves to be the application of algebraic (but not integral, as G) invariants, namely, the so-called Alexander polynomials. They coincide only for complex and rarely realized knots.[25] Later, we cite some results of these calculations.

11.4. *A fraction of unknotted chains among ring macromolecules rapidly decreases with an increase in the chain length over several hundreds of segments.*

For a short closed chain, the probability p of a knotted state is negligible. It reaches unity, however, as $N \to \infty$. This is shown in Figure 1.22. It is also evident that the fraction of still more complex knots grows with the chain length. Digressing from the ideal chain model, the probability of knotting decreases dramatically, as $\exp(-23d/l)$, with an increase in the chain thickness d.

11.5. *There is a topological interaction between ring chains; the virial coefficient of the interaction between two non-entangled rings is approximately the cube of the radius of gyration.*

Let us return to Figure 1.20. It shows that the polymer chain that is non-entangled with an obstacle strongly repels it. In the two-dimensional case, the virial

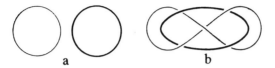

FIGURE 1.21. Two links with identical (0) value of Gauss integral. (a), Trivial. (b), Nontrivial.

coefficient of their interaction is on the order of the square of the Gaussian size Na^2 of the chain.

A quite similar topologic interaction occurs between closed chains in three-dimensional space (M. D. Frank-Kamenetskii *et al.*, 1974). Consider two closed, mutually non-entangled ring macromolecules C_1 and C_2. Provided that the distance between them is large, the chains do not affect one another and each realizes a complete set of allowable conformations. When the chains come closer, some conformations (which are responsible for entanglement of the ring macromolecules) turn out to be forbidden, and the number of feasible conformations is reduced, resulting in a repulsion of entropic nature between the macromolecules. Figure 1.23 shows the computer-simulated repulsion "potential" (i.e., the free energy) of two non-entangled ring macromolecules as a function of distance between their centers of mass. The second virial coefficient of interaction between such macromolecules, which can be determined by osmotic measurements of a dilute solution of non-entangled ring polymers, is of the order of the cube of the coil's radius of gyration. Thus, the situation is qualitatively the same

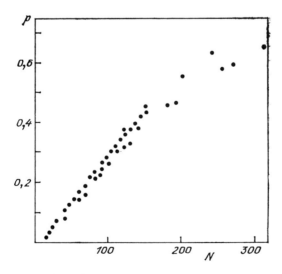

FIGURE 1.22. Probability of formation of a knotted state on random closing of a freely jointed chain of 0 thickness into a ring as a function of the number of segments, N. (Computer simulation data from Ref. 25.)

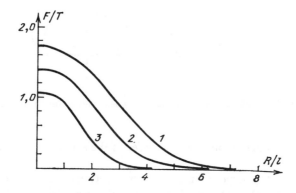

FIGURE 1.23. Free energy of topologic repulsion of two closed, unknotted and non-entangled rings of freely jointed chains of 0 thickness as a function of distance between their centers of mass. Curves 1, 2, 3 correspond to the chains of 60, 40, and 20 segments, respectively. 1—length of a segment. (Computer simulation data from Ref. 25.)

as if the ring chains interacted as impenetrable spheres with radii of the order of the coil size. It should be pointed out that the described interaction is of a topologic nature and continues even for an infinitely thin ideal chain when the proper macromolecular volume equals 0. Similarly, it can be shown that an effective attraction arises between closed macromolecules in the presence of entanglement.

11.6. *To investigate the topological constraints in concentrated polymer systems, the model of "a polymer chain in an array of obstacles" is used; a ring macromolecule, not entangled with the obstacles, takes on a collapsed conformation in the array of obstacles.*

Thus, an analytic approach already proves to be essentially futile for a problem of the topological interaction of two closed macromolecules. The situation grows more complicated in the most interesting cases, when there are many strongly entangled polymer chains in a system, (e.g., in concentrated polymer solutions and melts). Clearly, the topological constraints in such systems can be described effectively only via approximate models.

One such model is "a polymer chain in an array of obstacles." This model assumes that a polymer chain strongly entangled with other chains exists in a certain effective "obstacle lattice," a framework of straight lines forming the edges of a simple cubic lattice (Fig. 1.24). The topologic constraints are emulated by forbidding the chain to intersect obstacles. Figure 1.24b illustrates a two-dimensional version of the model in which the spatial framework of straight lines is replaced by the planar array of points that cannot be crossed by the polymer chain.

Consider, for example, the simple problem of a conformation of the ring macromolecule that is not entangled with obstacles in the obstacle lattice. First, however, return to Figure 1.24 and consider not the chain itself but rather a

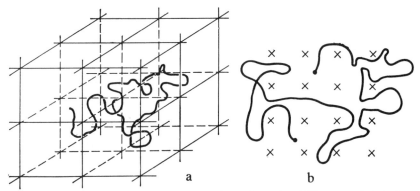

FIGURE 1.24. Polymer chain in three- (a) and two-dimensional (b) arrays of obstacles.

broken line with vertices in the centers of the cells of the obstacle lattice. Clearly, such a broken line can always be chosen (and not in a unique way) so that it is topologically equivalent to the chain (i.e., so that the chain can be transformed into the chosen broken line without crossing the obstacles). The shortest of such lines is called the *primitive path of the given conformation.*

A decisive simplification in the mathematic description of the given model is possible, because the described broken lines can be accommodated on the Cayley tree (cf. subsection 10.3). To each cell of the initial lattice, there corresponds an infinite number of vertices of the Cayley tree. This reflects the fact that each cell can be reached by the infinite number of topologically different ways.

The significance of the introduction of the auxiliary lattice (or tree) becomes clear if one observes that the topologic state of the polymer chain having fixed ends with respect to the array of obstacles is totally specified by the terminal point of the corresponding random walk over the tree. If the terminal points coincide, then the chain trajectories can be transferred continuously from one into another without crossing the obstacle array; in the opposite case, this is impossible. Specifically, the closed (i.e., not entangled with obstacles) trajectory turns into the closed trajectory on the tree. If the closed trajectory is entangled with one or several obstacles, then there is a corresponding non-closed trajectory on the tree.

Understandably, the position of the terminal point of the random walk over the tree plays the role of a topologic invariant in the model of "a polymer chain in an array of obstacles," which can be used to distinguish various topologic states of the chain with fixed ends relative to the array of obstacles. The terminal point of the random walk may be put in correspondence with the shortest path over the tree to the origin of coordinates (i.e., the previously defined primitive path for the given chain conformation). Because the primitive path and its terminal point are in a one-to-one correspondence, one may say that the primitive path for a polymer chain in the obstacle array is also a topologic invariant.

Moving to the problem of the conformation of a closed (not entangled with the obstacles) macromolecule in the array of obstacles, note that such a macro-molecule is depicted by a closed line on the Cayley tree. On the other hand, we know from subsection 10.3 that the conformations of closed random walks over the Cayley tree are equivalent to the conformations of randomly branched macromolecules (Fig. 1.15). In physical terms, the closed chain in the array of obstacles is as if "folded in two," and such a double chain becomes analogous to a randomly branched chain.

Dimensions of the closed chain in the obstacle array therefore can be evalu-ated by Eq. (10.11). In this formula, the number n of links between the branches should be replaced by the number of links in such chain section that can be placed (as a Gaussian coil) in the array of obstacles with spacing c, that is, an $na^2 \sim c$ or $n \sim (c/a)^{1/2}$. Finally, for the radius of gyration of a ring macromole-cule not entangled with the obstacles, we obtain

$$\langle s^2 \rangle = N^{1/2}ca. \tag{11.7}$$

Thus, the size of the closed chain squeezed by the array of obstacles proves to be of order $s \sim N^{1/4}$, that is, much less than $s \sim N^{1/2}$ for a free ideal phantom ring macromolecule (see subsection 10.1).

CHAPTER 2

Polymer Chains with Volume Interactions. Basic Definitions and Methods

In comparison to an ideal macromolecule, the properties of real polymer systems with volume interactions are by far more diverse. These properties are of interest both from theoretical and practical points of view. Volume interactions, however, do not as a rule yield to direct theoretical investigation on the basis of first principles. In this situation (as always in theoretical physics), a choice of successful models of the investigated object and the development of appropriate model concepts (providing in fact a specific language to describe the object) play a decisive role.

12. POLYMER CHAIN MODELS WITH VOLUME INTERACTIONS

12.1. *Many typical volume effects occur on large length scales and therefore are universal (independent of the details of specific chain structure).*

So far, we have used many ideal chain models: freely jointed, fixed valence angle, persistent, and so on. Volume interactions can be introduced into any of these models.

While considering polymer chain models, one always inquires how these models compare with one another and with real chains or how sensitive theoretical results are to the choice of a specific chain model. Clearly, real polymers only partially resemble any schematic model. We return to the answers to these questions many times in our further presentation. Now we note, however, that there is a wide range of conditions under which the behavior of macromolecules is universal, (i.e., independent of their local chemical structure and, consequently, of the choice of model). Thus, in choosing a model, one can proceed from the convenience of the mathematic description. As a rule, the standard Gaussian and lattice models are the best from this viewpoint.

12.2. *Volume interactions in a standard bead model involve strong short-range repulsions and long-range attractions.*

The standard Gaussian bead model shown in Figure 1.6b represents a chain of interacting, spherically symmetric beads strung on an immaterial filament with Gaussian correlations between neighboring link positions [*see* Eq. (4.14)]. In terms of volume interactions, each bead acts as a chain section whose length is of order a. Although the central (i.e., depending exclusively on the distance between bead centers) interaction potential is conditional, it must have the shape shown schematically in Figure 2.1. On small length scales, repulsion is predominant ($\partial u/\partial r < 0$) because of the geometric volume of links, while at longer distances, the beads are attracted by Van der Waals forces ($\partial u/\partial r > 0$).

It should be noted here that in polymer solutions, apart from the interaction between polymer chain links, there is also an interaction between the links and the solvent molecules. In most cases, however, we shall be concerned only with the link interactions, as if ignoring the presence of solvent molecules, while discussing polymer solutions. We shall keep in mind, however, that the link interactions have been effectively renormalized by the presence of a solvent. Consequently, the potential $u(r)$ in Figure 2.1 is a potential of the effective interaction with allowance for the presence of solvent molecules.

There are no doubt some effects that cannot in principle be covered by the standard bead model (e.g., orientational liquid–crystalline ordering of polymer chain segments). Some of these effects are discussed later; as for the rest, the theory of equilibrium volume effects is considered in terms of the standard model as it is adopted in modern polymer physics. For the region of universal behavior of polymer systems, we describe a method of construction (i.e., appraisal of parameters) of the standard Gaussian model, which is equivalent to a given polymer chain model.

12.3. *Volume interactions are macroscopically described by the thermodynamic characteristics of a system of disconnected links.*

It is well known that the thermodynamic quantities of a system of interacting particles as a rule cannot be calculated directly, even in the case of the simplest

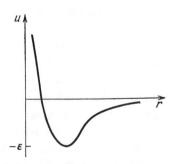

FIGURE 2.1. Typical potential of interaction between links in the standard model.

interaction potential of the type shown in Figure 2.1. Clearly, such a calculation would be futile for polymer systems as well. A way out of this difficulty can be based on the concept of a disconnected link system, a statistical system of particles that are not joined into a chain but act on one another via volume interactions. In the standard Gaussian model, the disconnected link system is merely a system of beads with the interaction potential $u(r)$.

Formally, this can be expressed as follows. The Gibbs distribution for a polymer system is

$$\rho(\Gamma) = \exp(-E(\Gamma)/T) \; \Pi g_j, \qquad (12.1)$$

where g_j are the "bonds" [see Eq. (4.4)] and $E(\Gamma)$ the "interactions," that is, the energy of volume interactions in the microscopic configuration Γ; for example, in the standard Gaussian model, $E(\Gamma) = \Sigma u(x_i - x_j)$. The disconnected link system is a system with the Gibbs distribution

$$\rho'(\Gamma) = \exp(-E(\Gamma)/T). \qquad (12.2)$$

In statistical physics of macromolecules, the volume interactions frequently are described in terms of the thermodynamic parameters of the disconnected link system, which are assumed to be known. To find these parameters is the objective of the statistical physics of ordinary (non-polymer) gases and liquids.

12.4. In polymer systems of moderate concentration, the volume interactions are reduced to relatively rare link collisions and are described by virial coefficients of the disconnected link system.

The density evaluation (5.8) of a Gaussian coil suggests that low density is typical for many polymer systems. As is known from the physics of real gases and dilute solutions (see Ref. 26), thermodynamic functions in low-density systems can be expanded into a power series of the number of particles in a unit volume n (so-called virial expansion). For example, for the free energy F and real-gas pressure p, this expansion takes the form

$$F = F_{id} + F_{int}, \quad F_{int} = NT(nB + n^2C + \dots),$$

$$p = nT(1 + nB + 2n^2C + \dots), \qquad (12.3)$$

where N is the number of particles in the system, F_{id} the part of the free energy corresponding to the ideal gas of N particles, F_{int} the volume interaction contribution to the free energy, and B and C the expansion coefficients (called the *second* and *third virial coefficients*, respectively). These coefficients are defined by the shape of the interaction potential $u(r)$. For example, the following relation is valid for the second virial coefficient B of interaction between point particles:

$$B(T) = (1/2) \int \{1 - \exp[-u(\mathbf{r})/T]\} d^3r. \qquad (12.4)$$

The third and subsequent virial coefficients are given by similar, although awkward, formulas. The terms proportional to the coefficient B in equalities (12.3) are responsible for the contribution of binary collisions (with the participation of two particles) to the thermodynamic parameters of the system, the terms proportional to C for the contribution of triple collisions, and so on.

In accordance with subsection 12.3, polymer systems with relatively small values of n [because of evaluation (5.8), this class comprises a wide range of real systems] can be described in terms of virial coefficients of the disconnected link system. For the examination of concentrated polymer systems, other characteristics of the disconnected link system are also used (*see* Sec. 15).

To proceed further, it is useful to evaluate B from Eq. (12.4) at different temperatures for the interaction potential shown in Figure 2.1. At high temperatures ($T \gg \varepsilon$), the attractive (negative) part of the potential $u(r)$ is negligible, and Eq. (12.4) yields $B \sim v$, where v is the so-called excluded volume of the particle.[a] As the temperature lowers, the second term in the integrand of Eq. (12.4) becomes more important, the value of B diminishes tending to zero at a certain temperature θ linearly with the difference $T - \theta$ and turning negative for $T < \theta$:

$$B \sim \begin{cases} v\tau & \text{for } \tau \lesssim 1, \\ v & \text{for } \tau \gtrsim 1, \end{cases} \qquad \tau \equiv (T - \theta)/\theta. \qquad (12.5)$$

12.5. *A lattice model may be convenient for some analytic calculations as well as computer simulations of polymer systems.*

Lattice models were used extensively in the early development of the volume interaction theory applied to polymer systems, in particular in the work by P. J. Flory. Even now, many theoretic and experimental results are formulated in terms of lattice models of the Flory type. In these models, a polymer chain is represented as a random walk path along the edges of a certain spatial lattice (Fig. 2.2). The volume interactions in the simplest case are determined by the self-avoiding walk condition, that is, by the exclusion of a repeat visit of the chain to the same lattice site (repulsion) and assigning the energy $\varepsilon < 0$ to each pair of off-neighbor (along the chain) links separated by one edge of the lattice (attraction).

13. BASIC DEFINITIONS IN THE THEORY OF VOLUME INTERACTIONS IN POLYMER SYSTEMS

13.1. *The simplest manifestation of volume interactions is the swelling or compression of a polymer coil.*

To become acquainted with the physics of volume effects, it is more convenient to begin with a discussion of the elementary theoretic ideas proposed by P.

---

[a]The value of v coincides with the proper volume of the isotropic beads in the standard Gaussian model. For strongly anisotropic molecules (e.g., for rodlike ones), the values of v differ drastically from the geometric volume (*see* subsection 13.8).

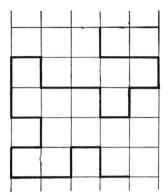

FIGURE 2.2. Lattice model of a polymer chain.

J. Flory as early as 1949 that are related to the basic, simple problem of a single polymer chain. It should be pointed out immediately that the Flory theory is far from rigorous. For many quantities, the predicted values are not accurate; nevertheless, this theory correctly identifies many principal qualitative effects. Being very simple, it allows for many interesting generalizations. Preliminary acquaintance with this theory enables us to present in this chapter the basic notions in a simple and visual way and to characterize the methods of volume interaction theory. In Chapter 3, we re-examine the problem of one chain and set forth an appropriate consistent theory.

The basic characteristic of the volume state of a single polymer chain is its spatial dimensions, for example, the mean square of the radius of gyration s^2. It is convenient to relate this quantity to the analogous size of an ideal Gaussian coil:

$$\alpha^2 = s^2/s_0^2. \tag{13.1}$$

The quantity α directly specifies the role of volume interactions. The coil with $\alpha > 1$ is swollen (extended), and with $\alpha < 1$, it is compressed in comparison with its Gaussian size. Traditionally, the quantity α, irrespective of $\alpha > 1$ or $\alpha < 1$ is referred to as the *swelling parameter of a polymer chain*.

13.2. *According to Flory, the equilibrium swelling parameter of a polymer coil is determined by the balance of the effects of volume interactions and polymer elasticity, which is of entropic nature.*

Following Flory, let us write the free energy $F(\alpha)$ of the polymer coil swollen by a factor of α as a sum of two terms

$$F(\alpha) = F_{el}(\alpha) + F_{int}(\alpha), \tag{13.2}$$

corresponding to entropic elasticity of the chain (F_{el}) and to volume interactions of the links (F_{int}).

The free energy $F_{el}(\alpha)$ of elastic deformation of a coil was calculated in subsection 8.1 for extension and in subsection 7.1 for compression. In accordance with Eqs. (8.1) and (7.2) and omitting the constants and numeric factors, we obtain

$$F_{el}(\alpha)/T \sim \begin{cases} s^2/Na^2 \sim \alpha^2, & \alpha > 1, \\ Na^2/s^2 \sim \alpha^{-2}, & \alpha < 1. \end{cases}$$

It is natural to combine these results by the simple interpolation formula

$$F_{el}(\alpha)/T \sim \alpha^2 + 1/\alpha^2, \tag{13.3}$$

providing the correct asymptotic behavior in both extreme cases: $\alpha \gg 1$ and $\alpha \ll 1$.

As the average link concentration in the coil is low (*see* subsection 5.2), the contribution of volume interactions to the free energy F_{int} of the chain is naturally evaluated by means of the virial expansion into a power series of the concentration (12.3). Assuming the coil to be a dispersed cloud of links distributed within the volume $\sim s^3 \sim a^3 N^{3/2} \alpha^3$ with concentration $\sim N/s^3$, we obtain according to Eq. (12.3):

$$F_{int}\alpha/T \sim s^3 B(N/s^3)^2 + s^3 C(N/s^3)^3 + ... \sim (BN^{1/2}/a^3)\alpha^{-3} + (C/a^6)\alpha^{-6} + ... \tag{13.4}$$

The equilibrium value of the swelling parameter α is determined from the condition of the minimum for the total free energy (13.2) comprising the terms (13.3) and (13.4):

$$\alpha^5 - \alpha = (BN^{1/2}/a^3) + (C/a^6)\alpha^{-3}. \tag{13.5}$$

Equation (13.5) is easy to analyze for all interesting limiting cases.

13.3. *A coil swells if the repulsion between the links dominates, and in this case, the coil size s depends on the chain length N as $s \sim N^{3/5}$; the coil becomes compressed if attraction prevails resulting in $s \sim N^{1/3}$; and near the θ-point, repulsion and attraction counterbalance, the coil becoming Gaussian and $s \sim N^{1/2}$.*

First, consider strong swelling of the coil $\alpha \gg 1$, which is realized for $B > 0$ and $N \gg 1$. In this case, the second terms in both sides of Eq. (13.5) can be neglected, and taking into account the evaluation (12.5), we obtain

$$\alpha \sim (BN^{1/2}/a^3)^{1/5} \sim N^{1/10}\tau^{1/5}(v/a^3)^{1/5}. \tag{13.6}$$

Thus, for the coil size $s \sim \alpha a N^{1/2}$ we obtain

$$s \sim aN^{3/5}\tau^{1/5}(v/a^3)^{1/5}. \tag{13.7}$$

It is essential that for $N \gg 1$, the size of the swollen coil turns out to be much larger than the Gaussian size $s_0 \sim aN^{1/2}$, because $N^{3/5} \gg N^{1/2}$ for $N \gg 1$. Accordingly, if $B > 0$, that is, if there is at least weak repulsion between links (or repulsion slightly prevails over attraction), a sufficiently long chain forms a strongly swollen coil. This is a consequence of coil pliability, with the elasticity modulus of the coil being of the order of $1/N$ (*see* subsection 8.1), namely, being very small. A consistent theory of coil swelling is discussed in Secs. 17–19.

Let us now turn to the opposite limiting case,[b] strongly compressed chain ($\alpha \ll 1$) realized at $B < 0$ and $N \gg 1$. Both terms on the left-hand side of equality (13.5) now can be neglected to obtain the swelling parameter

$$\alpha \sim (-BN^{1/2}a^3/C)^{-1/3} \qquad (13.8)$$

or the chain size

$$s \sim N^{1/3}(-B/C)^{-1/3}. \qquad (13.9)$$

Hence, as soon as $N^{1/3} \ll N^{1/2}$ for $N \gg 1$, one can conclude that a long chain with attractive links is compressed very strongly or, in other words, collapses. Actually, this means (*see* Secs. 20 to 22) that the chain transforms from a coil to a globule. The physical meaning of the result $s \sim N^{1/3}$ [*see* Eq. (13.9)] is very simple. It signifies that the link concentration in the globule $N/s^3 \sim -B/C$ is independent of N (i.e., the globule represents approximately a homogeneous sphere).

Figure 2.3 provides a graphic illustration of computer-simulated conformations of a freely jointed chain of 626 links. The upper left-hand side of the figure (as well as Fig. 1.7) shows an ordinary coil, whereas a typically globular conformation is shown on the lower right-hand side.

Finally, from Eq. (13.5), one can easily find the condition under which the sizes of real and ideal coils coincide. Obviously, $\alpha = 1$ if

$$B \sim -(C/a^3)N^{-1/2}, \qquad (13.10)$$

that is, according to Eq. (12.5), when $|\tau| \sim N^{-1/2} \ll 1$. Recall that τ is the relative deviation from the θ-temperature (i.e., from the temperature at which B turns to 0). Thus, for a long chain ($N \gg 1$), the sizes of real and ideal coils coincide in the immediate vicinity of the θ-temperature (*see* subsection 14.4).

According to the concepts first introduced in the polymer solution theory by P. J. Flory, this last result is valid not only for the overall size but also for other coil characteristics. Close to the θ-temperature, the real polymer chain with interacting links behaves as an ideal Gaussian coil. This conclusion is rather surprising. Indeed, for a common real gas or real solution, a temperature also exists at which the second virial coefficient of molecular interaction vanishes. This temperature is called the *Boyle point*. At the Boyle point, the binary collisions of molecules do not contribute to thermodynamic functions of the gas, for example, pressure (12.3). The gas (or solution) at the Boyle point, however, is

---

[b]In 1949, P. Flory developed the theory of coil swelling given earlier. This theory was later generalized to include compression (O. B. Ptytsin and Yu. E. Eisner, 1965). In the final form, this generalization was obtained by T. M. Birshtein and V. A. Pryamitsin in 1987.

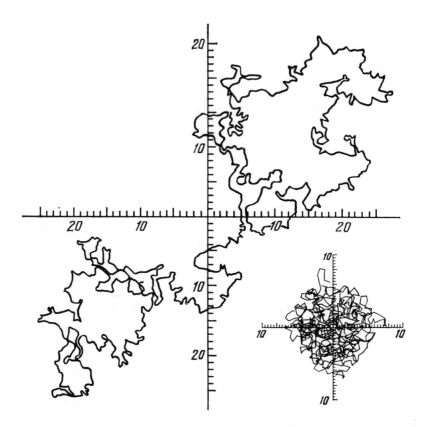

FIGURE 2.3. Typical conformations of a freely jointed chain of 626 links (each link is of unit length) in the coil and globular states obtained by computer simulation. (Courtesy of N. K. Balabayev).

far from ideal, because the triple collisions as well as those of higher multiplicity remain appreciable.

Why, then, does the polymer coil appear ideal under θ-conditions? The possibility of an almost complete neutralization of attraction and repulsion at a certain temperature $T = \theta$ (in the θ-point) is caused by the very fact that the monomer links are connected in a linear chain.

13.4. *Ideal behavior of a real polymer chain is possible at certain θ-conditions, because volume interactions in it are reduced to binary collisions of links.*

To clarify the nature of ideal behavior of a polymer coil in θ-conditions, it is useful to evaluate the number of collisions taking place between the links. We know that the size of a Gaussian coil (*see* subsection 5.1) is of the order of the mean end-to-end distance of the ideal chain, that is, according to Eq. (4.15), $R \sim a N^{1/2}$. The geometric coil volume $V \sim R^3 \sim a^3 N^{3/2}$. However, as mentioned earlier and seen in Figure 2.3, the polymer chain does not fill the coil volume.

With the geometric link volume denoted by v, the volume fraction taken up by links inside the coil is easy to obtain:

$$\Phi \sim Nv/V \sim Nv/(N^{3/2}a^3) \sim N^{-1/2}(v/a^3) \ll 1.$$

Thus, in the long polymer chain $(N \gg 1)$, the volume fraction of polymer in the coil is very small.

To evaluate the number of simultaneous collisions occurring in the volume of a polymer coil, the coil must be regarded as a cloud of free particles or links distributed in volume V. Let us take an instantaneous "spatial photograph" of such a cloud and see how many particles take part at this moment in binary, triple, and other multiple collisions. The number of binary collisions can be estimated as follows. There are N particles near which a partner is located with probability Φ, consequently, the number of binary collisions is of order $N\Phi$. Similarly, the number of triple collisions is of order $N\Phi^2$, and so on. The number of collisions of multiplicity p is

$$\mathscr{N}_p \sim N\Phi^{p-1} \sim N^{(3-p)/2}(v/a^3)^{p-1}. \tag{13.11}$$

The number of simultaneous binary collisions of links in a Gaussian coil is seen to be of order $N^{1/2}$. Although it is small in comparison with N (i.e., each individual link rarely collides), that number is much greater than unity. The number of triple collisions is of the order of unity per a whole coil. As to the collisions of greater multiplicity, their number is very small $(\sim N^{-1/2})$, that is, such collisions are rare and inessential not only for any individual link but for the whole coil.

Now it becomes clear that in the swollen polymer coil (in which the link concentration is lower than in the Gaussian coil), only binary collisions are appreciable. Therefore, the coil swelling parameter (13.6) is determined only by the second virial coefficient of link interaction. It is also clear why the swelling occurs at $B > 0$ (i.e., when repulsion prevails over attraction in binary interactions).

If repulsion and attraction counterbalance each other in binary interactions, that is, $B \approx 0$ [cf. Eq. (13.10)], then an infinite chain turns out to be ideal. This is because a few (of the order of unity) triple collisions cannot substantially affect a very long chain.

Finally, if the coil becomes compressed relative to the Gaussian θ size, the number of triple collisions grows so as to become substantial. Accordingly, properties of a moderately compressed (globular) chain are determined not by one but two virial coefficients of link interaction, the second and the third [cf. Eq. (13.9)].

If the link attraction is great, then a compression of the macromolecule may be so high that the virial expansion (12.3) for F_{int} becomes unwarranted. To describe the globular state in this case, one must know more complex thermodynamic parameters of the disconnected link system besides the virial coefficients (*see* subsection 12.3).

13.5. *The θ-point separates the regimes of good and poor solvents.*

The fundamental notion of the θ-point was clarified earlier for a simple case of single polymer chain. In a system of many chains of not too high concentration (when the volume fraction occupied by links is much less than unity, $\Phi \ll 1$), volume interactions also can be described in terms of corresponding virial coefficients. Such a system is usually realized in a low-molecular-weight solvent. In this case, the character of volume interactions of polymer links (in particular, the values of the virial coefficient) are certainly determined not only by the temperature but also by the composition and the state of the solvent.

If under these conditions $B > 0$ (i.e., link repulsion prevails), individual coils swell. Moreover, different coils tend to keep apart, which promotes mixing of the polymer and solvent. Such a situation is called the *good solvent regime*.

Conversely, if under the conditions $B < 0$ (i.e., link attraction prevails), individual coils are compressed, and different chains tend to stick together, leading to precipitation in a polymer solution of sufficient concentration. This situation is called the *regime of poor solvent or precipitant*.

In simple cases, the good solvent regime is observed at temperatures exceeding θ ($T > \theta$) and that of poor solvent below θ ($T < \theta$). The properties of the solvent itself, however, depend on temperature so that more complex phenomena can appear, leading to an inverse temperature dependence of the solvent quality, several θ-points, etc. The solvent quality can be changed and, in particular, θ-conditions can be reached or passed by varying not only its temperature but also its composition.

To make our reasoning more specific and brief, below we shall treat the θ-point as a θ-temperature and refer to the conditions of good and poor solvent as the cases of high ($T > \theta$) and low ($T < \theta$) temperatures respectively.

13.6. *The properties of a polymer coil in good solvent are universal; with the exception of chain length and stiffness (Kuhn segment), they depend only on one single integral characteristic of volume interactions, the second virial coefficient of link interaction.*

From the earlier considerations, it follows that the conformational characteristics of macromolecules at $T \geqslant \theta$ are independent both of the detailed shape of the effective potential $u(r)$ of volume interactions and the detailed structure of the solvent, and they determined only by one parameter, B. Thus, it can be expected that in good solvent, the volume effects in the coil are universally determined by the value of B independently of the specific nature of the forces responsible for this virial coefficient. On the other hand, according to subsection 4.5, the coil properties that are not connected with volume interactions universally depend on only two quantities, the contour length and effective segment, or in terms of the standard Gaussian model, on the number of links in the chain (N) and the distance between neighboring beads along the chain (a).

Thus, all coil properties are determined by only three parameters: N, a, and B.

Moreover, from dimensionality considerations, it follows that the influence of volume interactions on coil properties can only be expressed by the unique dimensionless combination of parameters B/a^3. Comparing this conclusion with the evaluation (13.11) [see also Eq. (12.5)], it is natural to assume that the coil properties associated with volume interactions must universally depend on the parameter $z \sim \mathcal{N}_2 \sim N^{1/2} B/a^3$, characterizing the number of binary collisions in the coil and ascribing the statistical weight B to each collision. This is exactly what the Flory theory predicts [see Eqs. (13.5) and (13.6)]. The rigorous proof of this assumption is given in Sec. 14.

Any conformational coil characteristic associated with volume effects [e.g., the swelling parameter (13.1)] thus depends only on the combination of parameters z and

$$\alpha^2 = \alpha^2(z), \quad z \equiv 2(3/2\pi)^{3/2} N^{1/2} B/a^3. \tag{13.12}$$

(The numeric factor in this definition of z is chosen for the sake of convenience, as will be clear later.) Similarly, any other analogous characteristic (e.g., $\langle R^p \rangle / \langle R^p \rangle_0$ with $p \neq 2$) is expressed by z.

On the other hand, coil characteristics that are not associated with volume interactions depend on the parameters N and a combined as Na^2 (see subsection 4.5)

By virtue of these factors, the theory of dilute polymer solutions in good solvent is regarded as a two-parameter theory: all macroscopic conformational characteristics are functions of the two parameter combinations Na^2 and $z \sim N^{1/2}B/a^3$. In particular, only these combinations and not the parameters N, a, and B themselves can be experimentally measured (see subsection 13.8).

13.7. The properties of a polymer chain in poor solvent at a moderate distance from the θ-point remain mostly universal: besides a chain length and stiffness (Kuhn segment), they depend on the two integral characteristics, which are the second and third virial coefficients of link interaction.

As shown, the triple collisions become substantial on transition through the θ-point to the poor solvent region, so that the conformational characteristics of the macromolecule are not only specified by the parameters N, a, and B but also by the third virial coefficient C. The polymer properties thus become less universal than in good or θ-solvent. Still, of all volume interaction characteristics [such as the potential $u(r)$, solvent properties, and so on], only the two integral quantities B and C remain substantial, provided the temperature is close to $θ$ (i.e., the link attraction is not too strong).

As the conformational characteristics of the polymer depend on B only via the parameter combination z (13.12), which is proportional to the number of binary collisions and ascribes the statistical weight B to each collision, the third virial coefficient C must appear in the combination C/a^6, which is proportional to the number of triple collisions with weight C. Accordingly, the theory of polymer solutions in poor solvent near the θ-point can be called a three-parameter theory, because all macroscopic conformational characteristics are functions of the three parameter combinations Na^2, $z \sim N^{1/2}B/a^3$ and C/a^6. In particular, only these parameter combinations can be measured experimentally.

13.8. *In the region of universal behavior for each polymer chain, an equivalent standard bead-on-filament model with appropriate parameters can be chosen.*

Universality means that under proper conditions, the behavior of all polymers is identical and, in particular, is the same as for the simplest standard bead model. For example, the size of any polymer chain in good, poor, and θ-solvent is proportional to the molecular mass raised correspondingly to powers 3/5 [Eq. (13.7)], 1/3 (13.9), and 1/2 (4.12).

From a theoretical viewpoint, this signifies that in the universal region, it is sufficient to study only the simplest model of a polymer chain (i.e., the standard bead model) without restricting generality. However, we need to find out how to choose the values of the standard model parameters.

Basically, the answer is clear: the parameters must be such that the values of any observable macroscopic characteristics of the standard bead model coincide with the values of corresponding quantities for the initial real polymer or the initial complex polymer model. It is clear from the previous subsections that all macroscopic conformational characteristics of the bead model are determined by three quantities: Na^2, $N^{1/2}B/a^3$, and C/a^6. Therefore, it is sufficient for these values to be the same, provided of course they can be found for the initial polymer.

In some cases, it may prove to be easier to determine three other macroscopic characteristics for the initial polymer, for example, the size of a single chain in a good, poor, and θ-solvent. The parameters Na^2, $N^{1/2}B/a^3$, and C/a^6 can then be found from the condition of the equality of these macroscopic characteristics for the bead model and the initial polymer.

It should be emphasized that having determined in some way or other the basic values Na^2, $BN^{1/2}/a^3$, and C/a^6, we impose only three conditions on the four parameters N, a, B, and C, that are necessary for a complete specification of the standard bead model. Consequently, one of the parameters can be defined arbitrarily. This reflects the arbitrariness in breaking up the initial chain into elementary links (i.e., in a choice of N). After the definite choice of N, the parameters a, B, and C are found in a unique fashion.

Let us illustrate this by solving the following simple but important problem. Suppose there is a persistent, that is, uniformly elastic (*see* subsection 3.4), model of a macromolecule with the parameters L (total contour length), l (length of effective Kuhn segment), d (thickness or diameter), and τ (relative deviation from θ-point). We must construct the corresponding standard bead model to find its parameters.

We begin by breaking up the chain into links (i.e., by a choice of N). In many applications, it is most convenient to regard a chain section having a length of order d as a link. Then the number of links in the chain L/d and in a unit volume of solution are proportional to the molecular mass and weight polymer concentration (i.e., to easily measured values), respectively, with coefficients defined only by the properties of monomer links but not of the chain. The resulting division of the chain will be referred to as the first division. The second common division, which is convenient in some theoretic calculations, corresponds to the links of length of order l. Thus,

$$N_1 = L/d, \quad N_2 = L/l. \tag{13.13}$$

Because the mean square of the end-to-end distance of the persistent chain equals Ll [see Eq. (3.3)] and of the bead model Na^2 [Eq. (4.12)], the following equalities should be satisfied:

$$Ll = N_1 a_1^2 = N_2 a_2^2,$$

that is,

$$a_1 = (ld)^{1/2}, \quad a_2 = l. \tag{13.14}$$

Let us start with consideration of the first division. As all dimensions of links are of order d, the virial coefficients can obviously be estimated as

$$B_1 \sim d^3 \tau, \quad C_1 \sim d^6. \tag{13.15}$$

Correspondingly,

$$z \sim N_1^{1/2} B_1/a_1^3 \sim (L/d)^{1/2} \tau(d/l)^{3/2}, \quad C_1/a_1^6 \sim (d/l)^3. \tag{13.16}$$

Because in accordance with Eq. (12.5), $B_1 \sim v\tau$, $C_1 \sim v^2$, we obtain

$$v/a^3 \sim p^{-3/2}, \quad p \equiv l/d. \tag{13.17}$$

Hence, it is seen that in the bead model, the parameter v/a^3 specifies the stiffness of the polymer chain: $v/a^3 \sim 1$ (Fig. 2.4a) corresponds to $p \sim 1$ and $l \sim d$ (i.e., to flexible chains), and $v/a^3 \ll 1$ (Fig. 2.4b) corresponds to $p \gg 1$ and $l \gg d$ (i.e., to stiff chains). Of course, reduction of a stiff-chain polymer to a standard

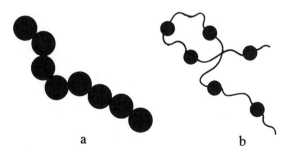

a   b

FIGURE 2.4. Equivalent standard bead model for flexible $v/a^3 \sim 1$ (a) and stiff $v/a^3 \ll 1$ (b) macromolecules.

bead model is possible only in the absence of any effects associated with orientational ordering or liquid-crystalline phases (*see* Sec. 28).

Consider now the second method of chain division. The parameters $N^{1/2}B/a^3$ and C/a^6 must have the same values as in the first division, so one can find B_2 and C_2:

$$B_2 \sim l^2 d\tau, \quad C_2 \sim l^3 d^3. \tag{13.18}$$

Because a persistent chain flexibility manifests itself inappreciably over lengths less than l, one may expect that B_2 and C_2 would coincide in order of magnitude with the corresponding virial coefficients of rod-like macromolecules. This is indeed the case, as we show in Sec. 28.

13.9. *The stiffness parameter of a polymer chain always substantially exceeds unity, or $v/a^3 \ll 1$.*

Let us consider the numeric values of the most significant parameters, p and v/a^3 (13.17), for real polymers. One should realize that these are phenomenological parameters, because, for example, the effective polymer thickness d appearing in the definition $p = l/d$, allows for the role of side groups and solvent in chain flexibility. Nevertheless, it is clear that flexibility cannot manifest itself over a chain section whose length is less than its thickness. Therefore, the inequality $p > 1$ always holds. Actually, the experimental investigations show that even for most flexible polymer chains, $p \approx 3$–4, that is, $v/a^3 \lesssim 0.2$. In stiff chains, the parameter p is much greater. For example, for double helix DNA, $p \approx 50$, that is, $v/a^3 \approx 0.003$. The value of p thus is always appreciably greater than unity, while v/a^3 is appreciably less than unity.

*13.10. *A stricter proof of what was previously stated calls for considering the collisions of chain sections or quasimonomers but not links (monomers).*

Until now, we have not been taking into account the essential circumstance that monomer links are not distributed independently within a coil but are connected in a chain. Consequently, even though the average link concentration in the coil is indeed low at large N, the local concentration of other links around the given link is, as a rule, far from being low and does not diminish with the growth of N. This high local concentration results from links located nearby on the chain. It seems then that it is inadequate to take into account only the second or the second and third virial coefficients of link interaction, so the results presented earlier raise some doubt.

Intuitively, it is clear, however, that the low average link concentration in a coil must lead to a certain universality of coil properties. This actually is the case. If we consider (as we did before) the interaction of chain sections separated from one another in the polymer coil (instead of link interactions as such), the smallness of the average link concentration in the coil signifies that the binary interactions of such chain sections will prevail over their interactions of higher order. Similar to what was shown before, this results in universal behavior that is expressed in terms of a universal dependence not on the second virial coefficient B of link interaction (or B and C) but on the effective renormalized second virial coefficient B^* of chain sections (or B^* and C^*).

We have seen that this universality is connected with the visualization of the polymer coil as a cloud of free, independently colliding particles, analogous to the particles of a rare gas. Therefore, it is natural to assume that in finding the effects of volume interactions on conformational characteristics of the polymer coil, macromolecules can be pictured in the form of a rarefied cloud of N free particles, or "quasimonomers." These quasimonomers differ from monomer links in their interactions, which are characterized by renormalized quantities effectively accounting for the bonding of chain links.

The visualization of a polymer coil as a cloud of quasimonomers is justified, because each chain link is located in a "standard surrounding" of other neighboring links. It is this standard surrounding that renormalizes the properties of each link, so that it appears in the theory of volume interactions in the macromolecule as a quasiparticle whose characteristics do not coincide with those of the initial link.

Thus, all previous conclusions remain valid with B and C being replaced by the renormalized coefficients B^* and C^*. Henceforth, we implicitly assume that this renormalization has already been performed, and we omit the asterisk in the expression for the second and third virial coefficients.

13.11. *Some volume interactions can induce a coil–globule transition, or precipitation, above the θ-temperature.*

Such a transition occurs when the attractive forces, being weaker than the repulsive forces in the case of binary collisions (which corresponds to $B > 0$ or $T > \theta$), prevail in the triple collisions or in those of still higher order. Such a situation is realized, for example, in liquid crystals because of orientational ordering of the segments. In this case, an abrupt condensation at $T > \theta$ may become thermodynamically favorable, because binary collisions are not predominant in the condensed phase, which can be stabilized by the attractive forces via contacts of higher multiplicity. The indicated possibility, however, is rather unusual, so that provided the opposite is not stipulated, we shall be dealing with the common situation when the condensation takes place below the θ point (*see* also Secs. 22 and 44).

14. PERTURBATION THEORY FOR POLYMER COILS WITH VOLUME INTERACTIONS

14.1. *The swelling parameter α of the polymer coil can be expanded in a power series of the parameter $z \sim N^{1/2} B/a^3$, with the term $\sim z^p$ responsible for conformations with p simultaneous binary collisions.*

Embarking on a specific development of the quantitative theory of volume effects, we first try to account for volume interactions as perturbations, treating them as weak interactions.

In subsection 13.1, we defined the square of the swelling parameter as the ratio of the rms radii of gyration of the real and ideal chains [*see* Eq. (13.1)]. An analogous quantity can be considered, which is associated with the ratio of the mean square end-to-end distances $\alpha_R^2 = \langle R^2 \rangle / \langle R^2 \rangle_0$. As $\langle R^2 \rangle \sim s^2$, the behavior of the coefficients α and α_R is qualitatively identical. For simplicity, we calculate the value of α_R by perturbation theory; the result for $\alpha \equiv \alpha_s$ has an analogous structure and is given later.

To calculate the parameter α_R for a coil in good solvent $(T \geqslant \theta)$, we shall consider the Green function of the type (6.2) for the standard bead chain with volume interactions. The value of α_R is expressed via the Green function by the relation

$$\alpha_R^2 = \int R^2 G\binom{0\,|\,N}{0\,|\,R} d^3R \bigg/ \langle R^2 \rangle_0 \int G\binom{0\,|\,N}{0\,|\,R} d^3R. \qquad (14.1)$$

To write the Green function on the basis of Eq. (12.1), it is convenient to use the so-called Mayer function

$$f(x_i - x_j) \equiv \exp[-u(x_i - x_j)/T] - 1. \qquad (14.2)$$

instead of the potential $u(r)$. Then, the expression $\exp[-\varepsilon(\Gamma)/T] \equiv \Pi_{ij}(1 + f_{ij})$ can be expanded in a power series of f to obtain

$$G\binom{0\,|\,N}{0\,|\,R} = \int \delta(x_0)\delta(x_N - R) \prod_{j=1}^{N} g(x_j - x_{j-1}) \bigg\{ 1 + \sum_{1 \leqslant i < j \leqslant N} f(x_i - x_j)$$

$$+ \sum_{i,j,k,l} f(x_i - x_j)f(x_k - x_l) + ... \bigg\} d\Gamma, \qquad (14.3)$$

where the g are Gaussian functions (4.14). The p-th order of perturbation theory corresponds to keeping in Eq. (14.3) the terms containing products of p Mayer functions (i.e., to accounting for p simultaneous binary collisions). Recall that the expansion in the Mayer functions is a standard method of deriving the virial expansion (12.3) for ordinary real gases.

Because in the considered region $T \geqslant \theta$ the value of α^2 is independent of the detailed shape of the potential $u(r)$ and depends only on the second virial coefficient B, the shape of the potential can be chosen arbitrarily. Let us choose it so that the Mayer function (14.2) has a delta shape

$$f(x_i - x_j) = -2B\delta(x_i - x_j) \qquad (14.4)$$

[We suggest the reader calculate the second virial coefficient for this potential according to Eq. (12.4) and see that it is equal to B]. Such a choice for this potential does not restrict generality but rather simplifies the calculations. The expansion in a power series of f in Eq. (14.3) in this case turns (as expected) into a series expansion in powers of B. Using Eq. (14.4), one can easily take all integrals in Eq. (14.3) and obtain from first-order perturbation theory

$$\alpha_R^2 \cong \frac{\int d^3 RR^2 \int d\Gamma \delta(x_0)\delta(x_N-R)\Pi_{j=1}^N g_j\{1-2B\Sigma_{1\leqslant i<j\leqslant N}\delta(x_i-x_j)\}}{Na^2\int d^3 R \int d\Gamma \delta(x_0)\delta(x_N-R)\Pi_{j=1}^N g_j\{1-2B\Sigma_{1\leqslant i<j\leqslant N}\delta(x_i-x_j)\}}$$

$$= \frac{Na^2 - 2B\Sigma_{1\leqslant i<j\leqslant N}[2\pi(j-i)a^2/3]^{-3/2}(N-j+i)a^2}{Na^2\{1-2B\Sigma_{1\leqslant i<j\leqslant N}[2\pi(j-i)a^2/3]^{-3/2}\}}$$

$$\cong \left\{ 1 - 2B \sum_{1\leqslant i<j\leqslant N} \left[2\pi(j-i)\frac{a^2}{3} \right]^{-3/2} \left[1 - \frac{(j-i)}{N} \right] \right\}$$

$$\times \left\{ 1 + 2B \sum_{1\leqslant i<j\leqslant N} \left[2\pi(j-i)\frac{a^2}{3} \right]^{-3/2} \right\}.$$

This is the most important moment. It is clearly seen that the terms proportional to $(j-i)^{-3/2}$ cancel out; the only remaining terms are the ones like $(j-i)^{-1/2}$. This is exactly why the remaining sums, being transformed into integrals, become convergent at $(j-i)\to 0$. In turn, this means that the small-scale local details of the chain structure, and in particular the choice of Mayer function in the form (14.4), become irrelevant. Evaluating the integral, one easily obtains

$$\alpha_R^2 \cong 1+4z/3, \qquad (14.5)$$

where the parameter z is defined by Eq. (13.12). Proceeding with calculations in higher orders of perturbation theory, we obtain the series

$$\alpha_R^2 = \alpha_R^2(z) = 1 + (4/3)z + k_2 z^2 + k_3 z^3 + ..., \qquad (14.6a)$$

where k_i are the numeric coefficients, for example, $k_2 = 16/3 - 28\pi/27 \approx 2.07$.

A similar expansion can be obtained from Eq. (4.11) for the parameter α^2 by somewhat more complicated calculations:

$$\alpha^2 = \alpha^2(z) = 1 + k_1' z + k_2' z^2 + ...; \quad k_1' \approx 1.28, \quad k_2' \approx -20.8. \qquad (14.6b)$$

Positive information that can be extracted from Eqs. (14.6a) and (14.6b) indicates that the swelling coefficient α_R (or α) is a function of one real variable z, in accordance with what was stated in Sec. 13. However, the series (14.6) per se are suitable for the calculation of the functions $\alpha_R(z)$ and $\alpha(z)$ only for $|z| \ll 1$. At the same time, the quantity z contains the large factor $N^{1/2}$, therefore, the inequality $|z| \ll 1$ holds only very close to the θ-temperature (i.e., the temperature at which $B=0$). Thus, despite the low average link concentration in the coil, the effects of volume interactions on the coil size cannot be accounted for in the framework of perturbation theory. As expected, binary collisions of links result in a substantial increase in the coil size in the region $T > \theta$, where repulsive forces prevail. We have already described how the values of α^2 and R can be determined in this region by the Flory method (see subsections 13.2 and 13.3); more refined methods are discussed below.

14.2. *The second virial coefficient of coil interaction in a dilute solution is also given by a power series of z.*

The second virial coefficient for polymer coils reflects the interaction intensity in the presence of rare binary collisions of coils in a dilute polymer solution. This quantity is denoted by A_2. It is also called the *osmotic second virial coefficient*, because it specifies the osmotic pressure π of the polymer in a dilute solution:

$$\pi/T = c/N + A_2(c/N)^2 + \ldots \tag{14.7}$$

[cf. Eq. (12.3)]. Here, c is the solution concentration (i.e., the number of links per unit volume); correspondingly, c/N is the number of chains per unit volume. The value of A_2 can be obtained experimentally from measurements of the osmotic pressure or light scattering in a dilute solution.

The quantity A_2 can be expressed via the interaction potentials of individual links using the relation that generalizes Eq. (12.4) to the case when complex particles with internal degrees of freedom interact:

$$A_2 = (1/2V) \int \{1 - \exp[\varepsilon_1(\Gamma_I)/T + \varepsilon_1(\Gamma_{II})/T$$

$$- \varepsilon_2(\Gamma_I, \Gamma_{II})/T]\} \prod_i g_i^{(I)} \prod_j g_j^{(II)} d\Gamma_I d\Gamma_{II}, \tag{14.8}$$

where Γ_I and Γ_{II} are the sets of link coordinates in the first and second chains, respectively, $\varepsilon_1(\Gamma)$ the energy of one macromolecule in the microscopic state Γ, $\varepsilon_2(\Gamma_I, \Gamma_{II})$ the energy of both macromolecules (including the energy of their mutual interaction); and V the volume containing the macromolecules. (The quantity V must cancel out during the integration).

In the absence of volume interactions, $A_2 = 0$. When the volume interactions are weak, it is worth trying to calculate A_2 by means of perturbation theory. This calculation is carried out in complete analogy with that given in the previous subsection for α^2. The result takes the form

$$A_2 = N^2 Bh(z), \quad h(z) = 1 + k_1'' z + k_2'' z^2 + \ldots, \tag{14.9}$$

where k_i'' are the numeric coefficients; in particular, $k_1'' = (32/15)(2^{5/2} - 7) \approx -2.865$. As z is proportional to $N^{1/2}$, one can conclude (as in the previous subsection) that the expansion (14.9) is suitable for the calculation of A_2 only in a small interval near the θ-temperature.

The physical meaning of Eq. (14.9) is quite clear. Suppose that only one collision between the links of different macromolecules occurs at a given moment of time. Then it can be accomplished in N^2 ways (N possible choices of the link in the first macromolecule and an equal number of choices in the second chain). Each collision gives a contribution B in the expression for A_2, thus forming the factor $N^2 B$ on the right-hand side of Eq. (14.9). As one collision is accomplished (i.e., two links of different macromolecules are brought into contact), the assumption of each consecutive collision yields the additional factor z, as in the perturbation theory expression for α^2.

14.3. *Series terms in the perturbation expansion can be depicted graphically in the form of diagrams; the peculiarity of polymer systems consists in their diagrams directly corresponding to chain conformations.*

Let us return to Eq. (14.3). It can be shown graphically as

$$(14.10)$$

These pictures are called Feynman diagrams. They represent a simple method of writing the awkward integrals in Eq. (14.3). The thick line is the total Green function of a chain with volume interactions. Thin lines represent the ideal Gaussian Green function (4.1). The broken lines show collisions, and each such line gives a factor B. The "internal indices" i, j, . . ., are the ordinal numbers of colliding particles (monomeric links) over which a summation is made. By way of illustration, in Eq. (14.10), together with the zero- and first-order terms [cf. Eq. (14.5)], we showed the second-order terms (*see* the third and fourth diagrams on the right-hand side). As the order of perturbation theory grows, the number of terms increases so that they cannot be enumerated correctly without a diagramatic language.

The series (14.9) for A_2 can be pictured similarly. Because the calculation of A_2 involves a two-chain Green function (i.e., the statistical sum of two chains with four fixed ends), the necessary diagrams take the form of four-vertex figures:

We show here terms corresponding to the zero- and first-order in the series (14.9). More detailed information on diagrams and general methods of perturbation theory can be found in Refs. 5 and 27.

14.4. *Owing to triple collisions of links, the θ-point spreads into a narrow θ-region; the width of the θ-region for chains of length N tends to 0 as $N^{-1/2}$ when $N \to \infty$.*

The dominating role that binary collisions between links play in the properties of a Gaussian or swollen coil results from the fact that the number of collisions of higher order is small [*see* Eq. (13.1)]. Near the θ-point, however, where the binary collision contribution to macroscopic coil characteristics is absent, the role of triple collisions may become appreciable. Indeed, the simple diagram

showing one triple collision, contributes to the expression for the swelling parameter α_R^2, an additional term proportional to the third virial coefficient C, (which for reasons of dimensionality must appear in that term in the combination C/a^6) and to the number of triple collisions [which according to Eq. (13.11) is independent of N]. Retaining also the principal term of the series (14.6) or the first diagram in Eq. (14.10), we obtain

$$\alpha_R^2 \cong 1 + \tilde{k}_1(BN^{1/2}/a^3) + \tilde{\tilde{k}}_1(C/a^6), \qquad (14.11)$$

where $\tilde{k}_1$ and $\tilde{\tilde{k}}_1$ are numeric coefficients. As expected, the triple collision contribution proves to be negligible for finite B and long chains ($N \to \infty$), but it becomes substantial near the θ-point ($B \to 0$).

The calculation of the triple collision contribution to the second virial coefficient of the coil interaction yields the same result:

$$A_2 \cong N^2 B + \tilde{\tilde{k}}_1' C N^{3/2}/a^3 \qquad (14.12)$$

[N^2 is the number of link pairs, and the probability that the third partner interferes in a binary collision is proportional to the average link concentration (5.8) in the coil $a^{-3}N^{-1/2}$; $N^2 a^{-3} N^{-1/2} = N^{3/2} a^{-3}$].

All of this is usually essential for the determination of the θ-point of polymer solution in a real experiment or computer simulations. This point can be defined as a point in which a certain macroscopic coil characteristic takes a value unperturbed by volume interactions. For example, the temperature can be found at which the rms end-to-end distance is the same as for the ideal macromolecule (i.e., $\alpha_R = 1$). According to Eq. (14.11), however, this happens not at $B = 0$ but rather at

$$B = B_\theta^{(\alpha)} = -N^{-1/2}(C/a^3)(\tilde{\tilde{k}}_1/\tilde{k}_1) \qquad (14.13)$$

[cf. the Flory theory result (13.10)]. One can also find the temperature at which the second osmotic virial coefficient of coil interaction is 0. According to Eq. (14.12), this takes place when

$$B = B_\theta^{(A)} = -N^{-1/2}(C/a^3)\tilde{\tilde{k}}_1'. \qquad (14.14)$$

If one recalls that the coefficient B near the θ-point is a linear function of temperature (12.5), it becomes clear that the value of the θ-temperature found by different methods must be slightly different, owing to the difference in numeric coefficients in Eqs. (14.13) and (14.14). The difference in apparent θ-points and their deviations from the true θ-point where $B=0$ is of order $\theta/N^{1/2}$ (i.e., disappears as the chains become longer). In this sense, the θ-region is said to have the width $\Delta\theta/\theta \sim N^{-1/2}$ at finite N.

To summarize the contents of this section, we emphasize that volume interactions can be accounted for within the scope of perturbation theory only in a limited range of problems associated with dilute solutions of polymer coils near θ conditions. In most other cases, different approaches must be taken; these approaches are subdivided into the method of the self-consistent field and the fluctuation theory (or scaling methods). In the following sections, we briefly describe the ideas underlying these methods and the areas of their application. Their specific realization in solving some problems of the statistical physics of macromolecules is discussed in the next chapters.

15. METHOD OF THE SELF-CONSISTENT FIELD

15.1. *The self-consistent field approximation is based on the neglect of fluctuations.*

Let us return to the problem of a single chain and its solution by the Flory method. Recall that in the Flory method, the macroscopic state of the chain is specified by the swelling parameter α and the expression for the free energy $F(\alpha)$ of such a state is sought [*see* Eqs. (13.2) to (13.4)]. It seems that the total free energy can now be obtained by summing over all possible values of α:

$$\mathcal{F} = -T \ln Z, \quad Z = \int \exp(-F(\alpha)/T) \; d\alpha. \qquad (15.1)$$

The quantity $F(\alpha)$, however, has a minimum that is narrow, because $F(\alpha)$ is proportional to N. Therefore, instead of summing over α, it is sufficient to take the largest term of the sum [corresponding obviously to the lowest $F(\alpha)$] or, in other words, to calculate the integral (15.1) by the method of steepest descent:

$$\mathcal{F} \simeq \min F(\alpha) = F(\alpha_{eq}), \qquad (15.2)$$

where α_{eq} is the equilibrium value of the macroscopic parameter at which the function $F(\alpha)$ reaches the minimum. This is exactly how the calculations are performed in Flory theory.

The physical sense of Eq. (15.2) is simple: it indicates that a certain macroscopic state is thermodynamically much more favorable than others. In this case, statistical equilibrium signifies that the system remains in the most favorable state and experiences only insignificant fluctuations near it.

The approximation of the type (15.2), assuming that fluctuations are negligible, is called the *self-consistent field approximation*; the meaning of this name is made clear in subsection 15.4. The Flory theory is a typical theory of a self-consistent field.

Thus, in technical terms, the method of the self-consistent field can be reduced to a minimization of the free energy as a function of a macroscopic parameter. The physical essence of the method is based on the neglect of fluctuations near the most favorable macroscopic state.

It goes without saying that the neglect of fluctuations is not always justified so the fluctuation theory (*see* Secs. 16 to 19) must be applied for the description of some polymer systems. However, we show later that for many polymer melts and concentrated solutions (*see* Secs. 24, 26, and 27), liquid crystals (*see* Sec. 28), globules (*see* Secs. 20 to 22), and so on, the method of the self-consistent field is quite applicable. That is why it should be examined in detail.

In a general formulation of the method of the self-consistent field, as in most typical problems of the statistical theory of macromolecules, a natural macroscopic parameter is provided by the smoothed spatial distribution of the local link concentration $n(x)$ or its coordinate and orientation distribution $n(x,u)$ (*see* subsection 9.1). A natural generalization of Eqs. (15.1) and (15.2) employs the quantity $F\{n\}$, which is the free energy of a macroscopic state with fixed distribution $n(x)$ [or $n(x,u)$] with all remaining degrees of freedom having already reached statistical equilibrium, (i.e., the relaxation having been completed). To calculate the equilibrium free energy of the system, one must perform a summation over all distributions n, that is, the functional integration

$$\mathscr{F} = -T \ln Z, \quad Z = \int \exp(-F\{n\}/T) \, Dn. \tag{15.3}$$

In the self-consistent field approximation [cf. Eqs. (15.1) and (15.2)], the integral (15.3) is obtained by the method of steepest descent:

$$\mathscr{F} \cong \min F\{n\} = F\{n_{eq}\}, \tag{15.4}$$

where n_{eq} is the equilibrium distribution corresponding to the minimum of the free energy.

Significantly, when the method of the self-consistent field is applicable, the free energy $F\{n\}$ of the macroscopic state can be found with less effort. This is shown in the next subsection.

15.2. *The free energy of a polymer system comprises the conformational entropy contribution associated with linear memory and volume interaction contribution expressed via parameters of the disconnected link system.*

Recall that in the simplest Flory theory (*see* subsection 13.2), the free energy $F(\alpha)$ of a macroscopic state is subdivided into two contributions, $F_{el}(\alpha)$ and $F_{int}(\alpha)$ [*see* Eq. (13.2)]. In the general case, when a macroscopic state is specified by the distribution $n(x)$ [or $n(x,u)$], a similar division is ordinarily written as

$$F\{n\} = E\{n\} - TS\{n\}. \tag{15.5}$$

Here, $TS\{n\}$ is an analogue of $F_{el}(\alpha)$, that is, $S\{n\}$ is the conformational entropy of the polymer system. The expression for this quantity, as well as for F_{el}, can be borrowed from the theory of ideal polymers, because as we noted in Sec. 9 (subsection 9.2), the conformational entropy depends on the macroscopic state itself (i.e., on the distribution of concentration n) but is independent of the nature of the forces forming that macroscopic state. Thus, $S\{n\}$ is given by the Lifshitz formula (9.1) derived in Sec. 9.

The quantity $E\{n\}$ from Eq. (15.5) represents the volume interaction contribution to the free energy[c] and is in fact an exact analogue of F_{int} in Eq. (13.2). Like F_{int}, $E\{n\}$ is expressed naturally via the characteristics of a disconnected link system. To show this, first consider a system of standard Gaussian bead chains for which $v \ll a^3$ (Fig. 2.4b). [Recall that this condition corresponds to stiff chains (see subsection 13.8)]. Let us subdivide the volume of the system into the auxiliary volumes ω such that $v \ll \omega \ll a^3$ (Fig. 2.5). If, as usually happens, the forces of volume interaction are short-ranged (Fig. 2.1) and have an interaction radius of order $v^{1/3}$, the volume interaction energies of various volumes ω simply add up. On the other hand, because $\omega \ll a^3$, the link connectivity in a chain practically imposes no restriction on bead motion within the volume ω. Therefore, the volume interaction contribution to the free energy of the volume ω is the same as for the corresponding system of disconnected links (see subsection 12.3).

Because the volume ω is sufficiently small, the free energy of this volume for the disconnected link system can be written as $f(n(x), T)\omega$, where $f(n(x), T)$ is the free energy of a unit volume of the disconnected link system depending on the temperature T and link concentration $n(x)$ at the point where the volume ω is located. The unknown contribution of volume interactions to the free energy of the volume ω equals the difference between $f(n, T)\omega$ and the corresponding value of $f_{id}(n, T)\omega$ for non-interacting links. The expression for $f_{id}(n, T)$ is well known:

$$f_{id} = Tn \ln(n/e), \tag{15.6}$$

because it is a characteristic of an ordinary ideal gas. Summing over all volumes

---

[c]At first sight, Eq. (15.5) is an ordinary division of free energy into entropy $T\partial F/\partial T$ and energy $F - T\partial F/\partial T$ parts. Actually, this is not the case, because the conformational entropy does not coincide with true thermodynamic entropy: $S\{n\} \neq -\partial F/\partial T$. The reason is clear: the conformational entropy $S\{n\}$ as a linear memory characteristic includes the part of internal energy, which is connected with interactions between the links located nearby in the chain, elastic deformations of chain backbone, rotational isomers, and so on. From the statistical standpoint, the quantity $\exp(S\{n\})$ is proportional to the number of chain conformations (or trajectories) comprising the given macroscopic state, with each conformation having the weight defined by chain flexibility. It is this last circumstance that makes $S\{n\}$ different from ordinary entropy. At the same time, $S\{n\}$ acts as an entropy in all processes with constant linear memory. Accordingly, $E\{n\}$ acts in the polymer theory as energy, although strictly speaking, it is only the part of internal energy that is associated with volume interactions.

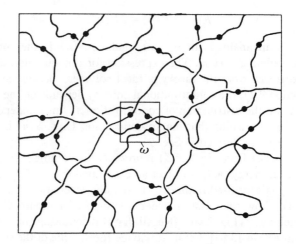

FIGURE 2.5. Explanation of the contribution of volume interactions to the free energy of a polymer system.

ω, we eventually obtain the final expression for $E\{n\}$:

$$E\{n\} = \int [f(n(x), T) - f_{\mathrm{id}}(n(x), T)] d^3x = \int f^*(n(x), T) d^3x,$$

(15.7)

where

$$f^*(n, T) = f(n, T) - f_{\mathrm{id}}(n, T).$$

(15.8)

The physical meaning of Eq. (15.7) is quite clear. The contribution of volume interactions to the free energy of a polymer system is local, and it is determined by the free energy of a unit volume of the corresponding system of disconnected links less the quantity (15.6) responsible for the translational entropy of mutually independent motion of the disconnected links. The latter contribution is excluded (subtracted), because the links connected in the polymer chain cannot move freely.

Equation (15.7) was obtained for the limiting case $v \ll a^3$. From subsection 13.8, it follows that the standard Gaussian chain with $v \ll a^3$ belongs to stiff-chain polymers. For flexible-chain polymers, $v \sim a^3$. The expression (15.7) for $E\{n\}$, however, remains valid in this case as well, provided that $f(n, T)$ denotes the free energy per unit volume of quasimonomers (*see* subsection 13.10) and not of disconnected links. Because we always assume implicitly that the renormalization accounting for the effects of quasimonomers has already been accomplished (*see* subsection 13.10), Eq. (15.7) has to be regarded as valid in the general case.

Thus, the free energy of the polymer system can be represented as a sum of two terms (15.5). The first ($E\{n\}$) is determined only by the properties of the disconnected link system, while the second $(-TS\{n\})$ incorporates all polymer specifics caused by link connectivity.

15.3. *Volume interaction contributions to the free energy of a polymer system can be expressed by the interpolation formulas known in the theory of real gases and solutions.*

To find $E\{n\}$ from Eq. (15.7), one must know the function $f^*(n, T)$ for the disconnected link system. Because this system represents a solution, the function $f^*(n, T)$ cannot be calculated accurately in the whole range of variation of its arguments. On a phenomenologic level, the physical effects, however, can be described by numerous interpolation and semi-qualitative equations of state of real gases and solutions; we give a few examples of this kind. Apart from the expression for $f^*(n, T)$, we also write out the two other quantities to be required later; the contributions of volume interactions to the chemical potential of the link and to the pressure, respectively:

$$\mu^*(n, T) = \partial f^*(n, T)/\partial n, \qquad (15.9)$$

$$p^*(n, T) = n\mu^* - f^*. \qquad (15.10)$$

The well-known Van der Waals equation of state, describing the gas–liquid transition (or separation of a homogeneous solution into dilute and concentrated phases) yields

$$f^* = -nT \ln(1-nv) - n^2 a,$$

$$\mu^* = -T \ln(1-nv) + Tnv/(1-nv) - 2na,$$

$$p^* = Tvn^2[1/(1-nv) - a/Tv], \qquad (15.11)$$

where v and a are the parameters. The analogous Flory–Huggins formulas are more convenient in some respects and are used more frequently:

$$f^* = (T/v)[(1-nv)\ln(1-nv) + nv - \chi n^2 v^2],$$

$$\mu^* = -T[\ln(1-nv) + 2\chi nv],$$

$$p^* = (T/v)[-\ln(1-nv) - nv - \chi n^2 v^2]. \qquad (15.12)$$

χ and v are the parameters. (Eq. (15.12) are derived in the Appendix to Sec. 24.)

In accordance with subsection 12.4, the functions $f^*(n)$, $\mu^*(n)$, and $p^*(n)$ can be expanded in a series of power n:

$$f^*/T = n^2 B + n^3 C + ...,$$

$$\mu^*/T = 2n B + 3n^2 C + ...,$$

$$p^*/T = n^2 B + 2n^3 C + ... \qquad (15.13)$$

[cf. Eq. (12.3)], where B, C, ... are the second, third, and so on virial coefficients.[d] Comparing Eqs. (15.11) and (15.12) with Eq. (15.13), one can obtain the values of the coefficients B and C for the interpolation equations of Van der Waals

$$B=v-a/T, \quad C=v^2/2, \tag{15.14}$$

and Flory-Huggins

$$B=v(1/2-\chi), \quad C=v^2/6. \tag{15.15}$$

A typical dependence of one of the thermodynamic functions (μ^*) on n and T for the equations of state (15.11) and (15.12) is shown in Figure 3.9. Its behavior is governed at low n by the attraction or repulsion prevailing in binary collisions (*see* subsection 12.4), that is, by the second virial coefficient B (15.13). Unlimited growth of μ^* (as well as of f^* and p^*) corresponds to the approximation of dense packing of the particles. For the equations of state (15.11) and (15.12), it happens when $nv \to 1$, so that the particle concentration at dense packing equals $1/v$. Finally, the range of concentrations appearing at lower temperatures, where $\mu^*(n) < 0$, (Fig. 3.9b) corresponds to a domination of attractive forces in particle interactions.

15.4. *In a polymer system the self-consistent field exerted on one monomer link by surrounding links, owing to volume interactions, is defined by the chemical potential of a disconnected link system.*

Everything that was said earlier nevertheless leaves the term *self-consistent field approximation* unclear; we clarify it now. Consider a polymer system with both volume interactions and an external field $\varphi(x)$. For such a system, we can write

$$E\{n\} = \int n(x)\varphi(x)d^3x + \int f^*(n(x), T) d^3x. \tag{15.16}$$

Applying the method of the self-consistent field (15.2) to this system, we must minimize the free energy (15.5), that is, to calculate the derivative $\delta F\{n\}/\delta n$. Thus, we obtain

$$\delta E\{n\}/\delta n = \varphi(x) + \mu^*(n(x)). \tag{15.17}$$

From Eq. (15.17), it is seen directly that the role of volume interactions in this approximation is reduced to the addition to the external field $\varphi(x)$ of a quantity called the *self-consistent field* and that equals the chemical potential of a link:

$$\varphi_{self}(x) = \mu^*(n(x)). \tag{15.18}$$

---

[d]When using the virial expansion (15.13), Eq. (15.8) for $E\{n\}$ is reduced (as it should be) to Eq. (13.4) for F_{int}.

The physical meaning of the expression (15.18) is quite clear. $\varphi_{self}(x)$ denotes the work that has to be accomplished to bring one particle (link) to the point x. During this process, the energy of volume interactions changes exactly by the value $\mu^*[n(x)]$.

15.5. *According to the self-consistent field approximation, the volume interactions in a polymer system proceed as in a dispersed "cloud" of disconnected links.*

In this subsection, we accentuate a simple qualitative consequence of the formulas given previously, because it will be used frequently in later sections. According to Eqs. (15.2), (15.5), and (15.7), the volume interaction contribution to the equilibrium free energy $\mathscr{F}$ in the self-consistent field approximation equals $f^*(n, T)$ per unit volume. This quantity allows a simple virial expansion in series of integral powers of n (15.13). In this expansion, n^2 is the probability of a binary collision in a system of particles with concentration n provided that the particles are free and non-correlated. The value n^3 is the similar probability of a triple collision, and so on. This implies that in the self-consistent field approximation, the volume interactions can be described in terms of non-correlated collisions in a "cloud" of disconnected particles (links) or more precisely, quasimonomers (*see* subsection 13.10).

16. FLUCTUATION THEORY AND SCALING METHOD

16.1. *A low-concentration polymer solution in good solvent is a strongly fluctuating system; there is a qualitative and quantitative analogy between polymers and other systems with developed fluctuations (e.g., magnets near a second-order phase transition point). Strongly fluctuating systems, including polymer ones, are characterized by critical exponents.*

Subsections 5.2 and 5.3 showed by the example of a Gaussian coil that strong fluctuations correlated over long distances are possible in polymer systems. It is also clear that fluctuations can only intensify because of repulsive volume interactions among links. Consequently, the method of the self-consistent field based on the neglect of fluctuations cannot be applied in studies of not-too-concentrated polymer solutions in the region $T > \theta$ when repulsive forces prevail in the volume interactions. This resembles the situation observed in ordinary physical systems, for example, in magnetic systems near a critical point or second-order phase transition where the self-consistent Landau theory does not apply because of fluctuation growth.[28] The indicated analogy becomes more obvious if one compares the universal behavior of polymer coils in good solvent (*see* Sec. 13) to the universal properties of magnets, or other systems, near second-order phase transition points.

In quantitative terms, the universality of second-order phase transitions is expressed by the independence of so-called critical exponents for each specific system from microscopic details of its structure.[28] The critical exponents define the behavior of various physical quantities near a phase transition point. For example, having designated the critical exponent for the fluctuation correlation radius ξ by v, we must write $\xi \sim [(T - T_C)/T_C]^{-v}$, where T_C is the critical temperature. (In the case of a magnet, this temperature is referred to as the *Curie point*).

In a polymer system, the analogue of the variable $(T - T_C)/T_C$ is the parameter $1/N$. Indeed, we know from subsection 5.3 that in an ideal Gaussian coil (and consequently in a real coil in good solvent), the fluctuation correlation radius ξ is of the order of the spatial dimension R of the coil. Hence, it follows that the growth of the chain length N corresponds to the approach to a critical point. Introducing the critical exponent v for a polymer coil, we write

$$\xi \sim R \sim N^v. \tag{16.1}$$

This relation coincides with the result (13.7) of the Flory theory. Thus, according to Flory, $v \approx 3/5$. Later, we will find other examples of critical exponents and discuss how to calculate them (*see* Secs. 17 and 18).

As for magnetism, the critical exponents for polymer systems are universal, that is, independent of the specific chemical structure of macromolecules and defined only by their most general properties (i.e., by the fact of their chain structure, the presence of branches, and so on). What is essential, the critical parameters and fluctuation behavior of polymers as well as other systems, generally depends on the dimensionality d of space. The common cases are three-dimensional bulk systems ($d = 3$) and thin films adsorbed on two-dimensional surface ($d = 2$). The physics of real systems, however, can be substantially elucidated by tackling the problem in a space of arbitrary dimensionality d that can be non-integer (*see* Ref. 29). In this particular case, no geometric sense is imparted to a fractional value of d (*see*, e.g., subsection 18.3). There are such systems, however, (so-called fractals; *see* Ref. 29) for which the fractional dimensionality has a direct geometric and physical meaning.

16.2. The quantitative formulation of the analogy "polymer-magnetic" allows the general results of the fluctuation theory to be directly applied in polymer physics.

Such a quantitative formulation was found by P. G. de Gennes in 1972. He showed that the statistics of a single long polymer chain in good solvent is equivalent to that of a magnet near the second-order phase transition point in the limit when the number n of components of an elementary magnetic moment is formally made to approach 0 ($n \to 0$). In 1975, J. des Cloizeaux showed that a system of many chains, (i.e., a polymer solution) is equivalent to a zero-component magnetic in an external magnetic field. The derivation of the analogy "polymer-magnetic" is presented in Chapter 10 of Ref. 8. These results initiated a rapid development of the statistical physics of polymer coils. By the time they were obtained, the fluctuation theory of second-order phase transitions in magnets had been thoroughly developed, and many results in this area could be transferred to polymer physics by formally assuming $n = 0$ in the appropriate formulas (so-called "$n = 0$ method"). Subsequently, we will not take this purely formal approach, which incidentally has a cardinal drawback: it is applicable only to systems with an equilibrium distribution of chain lengths, which is modified with changing external conditions.

The basic ideas of a consistent fluctuation theory applied to polymer problems will be clarified by means of the so-called decimation method proposed by P. G. de Gennes in 1977. The essence of the method is discussed in Sec. 18.

16.3. *The scaling method often helps in obtaining physical results without resorting to awkward field theory calculations.*

The concept of scaling to be discussed explicitly below can be substantiated on the basis of a consistent fluctuation theory. Application of this concept in practical calculations, however, requires only the acknowledgment of the fact that the radius of correlation is the only macroscopic characteristic length in the fluctuating system. For example, in subsections 5.1 to 5.3, it was shown that all characteristics of microscopic dimensions of a Gaussian coil are of the same order, that is, all of them are really of the order of the radius of correlation. In fact, uniqueness of a characteristic scale is both a basis and primary manifestation of universality in fluctuating systems. By some examples, we later show how one can make significant conclusions from this only assumption using scaling evaluations.

APPENDIX TO SEC. 16. POLYMER-MAGNETIC ANALOGY

Whenever in the literature on polymer physics the polymer-magnetic analogy is mentioned, the word *magnetic* implies, in fact, a simple theoretic model to describe the second-order phase transition, that is, a system with Ginzburg-Landau Hamiltonian:[28,48]

$$H = \int d^d x \left\{ \epsilon \sum_{\alpha=1}^{n} u_\alpha^2(x) + a^2 \sum_{i=1}^{d} \sum_{\alpha=1}^{n} [\partial u_\alpha(x)/\partial x_i]^2 + h u_1(x) \right.$$
$$\left. + B \left[\sum_{\alpha=1}^{n} u_\alpha^2(x) \right]^2 \right\}. \tag{16.2}$$

Here, $u_\alpha(x)$ is the n-component order parameter, d the dimensionality of space, h the external magnetic field, ϵ the deviation of temperature from the Curie point, and a and B the coefficients.

It should be emphasized again that n and d are different quantities. For example, one can easily conceive both a thin layer of magnetic ($d=2$) with an arbitrary orientation of magnetic moments ($n=3$) and a bulk of magnetic ($d=3$) with a plane of easy magnetization ($n=2$).

The quantitative conception of the polymer-magnetic analogy is:

$$\Omega(\mu_m, \mu_p) = [F_{magnet}(\epsilon = \mu_m, \ h = \exp(\mu_p/2)) - F_{magnet}(\epsilon, 0)]\big|_{n=0}. \tag{16.3}$$

Here, $F_{magnet}(\epsilon, h)$ is the free energy of the magnetic with the Hamiltonian (16.2), that is,

$$F_{magnet}(\epsilon, h) = -\ln \int \exp[-H\{u\}] D\{u_\alpha(x)\}. \qquad (16.4)$$

$\Omega(\mu_m, \mu_p)$ is the thermodynamic potential of the polymer system with parameters d, a, B and chemical potentials of links (μ_m) and chains (μ_p); $\partial\Omega/\partial\mu_m$ and $\partial\Omega/\partial\mu_p$ are the numbers of links and chains in the system, respectively, with the ratio of these quantities being equal to the mean length of the chain. [The derivation of Eq. (16.3) can be found in Chap. 10 of Ref. 8.]

The formulation of the polymer-magnetic analogy given here allows diverse generalizations. For example, one can take into account the triple collisions of links by introducing the additional terms $C[\Sigma_{\alpha=1}^{n} u_\alpha^2(x)]^3$ into the field Hamiltonian. One can also tackle polymer solutions by studying a system of two interacting fields, and so on.

A basic disadvantage of the "field" approach to polymer statistics, however, should be mentioned: It allows only an investigation of the systems with the equilibrium (and consequently, variable with a change of conditions) distribution over the chain lengths.

CHAPTER 3

Single Macromolecule with Volume Interactions

Having formulated the fundamental concepts and methods of the statistical physics of polymer chains with volume interactions, we now proceed to a comprehensive analysis of specific systems. This chapter is devoted to the statistical conformational properties of a single (individual) macromolecule. This situation is realized experimentally in dilute polymer solutions in which the individual chains do not overlap.

17. SWELLING OF A POLYMER COIL IN GOOD SOLVENT (EXCLUDED VOLUME PROBLEM)

17.1. *Volume interactions in good solvent reduce to excluded volume effects (i.e., forbidding conformations with self-crossing of the chain).*

Consider a single macromolecule swollen in good solvent far from the θ temperature ($T > \theta$). In this case (as follows from subsection 12.4), the attractive part of the link interaction potential (Fig. 2.1) is negligible and the second virial coefficient $B \sim v$. Thus, each link has the inherent volume v, which is excluded for other links because of short-range repulsive forces. In this case, one can easily see that the spatial shape of a polymer chain becomes analogous to the trajectory of a self-avoiding random walk. Finding the conformation of a single macromolecule swollen in good solvent, or in other words, the problem of the statistical properties of a self-avoiding random walk, is called the *excluded volume problem*.

Subsection 14.1 showed the perturbation theory to be suitable to the analysis of equilibrium swelling of the coil only in a close vicinity of the θ point, because the parameter z (13.12) is of the order of $z \sim BN^{1/2}/a^3 \sim v\tau N^{1/2}/a^3$, that is, proportional to a large value $N^{1/2}$. If the θ point is not close, then $\tau \sim 1$ and $z \gg 1$. The equilibrium swelling of a coil with excluded volume therefore can be treated mathematically as the problem of the asymptotic behavior of the function $\alpha^2(z)$ for $z \gg 1$.

It should be emphasized that the excluded volume problem, as a kind of a touchstone, played an extremely important part in the development of the statistical physics of macromolecules. The Flory theory, the earliest attempt to solve the excluded volume problem, was discussed in Sec. 13 (*see* subsections 13.1 to 13.3). Recall that this theory is based on the self-consistent field method applied in its most simplified form. We know that the coil is a strongly fluctuating system; therefore, the results of the self-consistent field theory cannot be regarded as correct and must be checked. We shall see, however, that some results of this theory are fairly accurate, so it would be useful to begin the description of the excluded volume problem with the self-consistent field method.

17.2. *An equable application of the self-consistent field method to the excluded volume problem confirms the Flory result for the critical exponent of the coil size* $v=3/5$.

The value $v=3/5$ was obtained by S. F. Edwards in 1965 in the following way. A given link is assumed to experience the action of all remaining links through the effective field $\varphi_{\text{self}}(x)$. This quantity is obviously proportional to the link concentration $n(x)$ at the given point. Indeed, according to Eqs. (15.18) and (15.13)

$$\varphi_{\text{self}}(x) = T B n(x).$$

Because this is a repulsive field having the shape of a hill rather than a well, the diffusion equation for the Green function will not have a discrete spectrum in this field, and $\varphi_{\text{self}}(x)$ should be substituted for $\varphi(x)$ in the "non-stationary Schrödinger equation" (6.28). For brevity, we omit the analysis of this nonlinear equation and quote only the final result:

$$R^2 = \int G_N(x) x^2 d^3x \left(\int G_N(x) d^3x \right)^{-1} \sim N^{6/5}.$$

This coincides with the Flory result (13.7), as $R \sim s$ for the coil. Accordingly, the self-consistent field method is usually applied to various generalizations of the excluded volume problem in the simplified form suggested by P. Flory.

17.3. *Although the Flory result for the critical exponent v is fairly exact, the accuracy of the Flory interpolation formula for coil swelling for arbitrary z (i.e., at arbitrary distances from the θ point) is low.*

The critical exponent v defines a dependence of the coil size not only on the chain length $R \sim N^v$ but also on temperature and the stiffness parameter v/a^3. In fact, because $R \sim a N^{1/2} \alpha$ and α depends only on the parameter combination $z \sim N^{1/2} B/a^3$ (13.12), then $\alpha(z) \sim z^{2v-1}$ and for $z \gg 1$

$$R \sim a N^v \tau^{2v-1} (v/a^3)^{2v-1} \tag{17.1}$$

[$B \sim v\tau$; *see* Eq. (12.5)]. All of the results of the Flory theory corresponding to the value $v=3/5$ agree well with the data obtained both experimentally and by computer simulation. This allows one use the Flory theory to derive the dependence $\alpha(z)$ for any z, not only for the asymptotic $z \gg 1$.

Equation (13.5) is convenient for giving a common interpolation throughout the whole region from a strongly swollen coil at $z \gg 1$ to a globule at $z \ll -1$. It cannot, however, claim high accuracy in the intermediate region. In fact, in the expression $F_{el}/T \sim \alpha^2 + \alpha^{-2}$ (13.3), the term α^{-2} (responsible for chain compression) has no physical meaning in the region of extension ($\alpha > 1$) that we are discussing.

The interpolation for $F_{el}(\alpha)$ used by Flory himself is written as

$$F_{el}(\alpha)/T = (3/2)\alpha^2 - 3 \ln \alpha, \qquad (17.2)$$

where the second term can be interpreted as an entropy of placing the chain ends of a Gaussian coil within the volume $\alpha^3 a^3 N^{3/2}$ for a swollen coil instead of $a^3 N^{3/2}$ for a Gaussian coil; $\ln(\alpha^3 a^3 N^{3/2}) - \ln(a^3 N^{3/2}) = 3\ln\alpha$. Minimizing $F(\alpha) = F_{el} + F_{int}$ and taking into account Eq. (17.2) for $F_{el}(\alpha)$ and Eq. (13.4) for $F_{int}(\alpha)$, Flory obtained

$$\alpha^5 - \alpha^3 = \text{const} \cdot z = (134/105)z. \qquad (17.3)$$

The numeric factor 134/105 is found from the additional condition that Eq. (17.3) at $z \ll 1$ should give a result coinciding with the perturbation theory at least in the first order of z.

Despite the fact that Eq. (17.3) at $z \gg 1$ yields the almost correct asymptotic behavior $\alpha \sim z^{1/5}$, a derivation of the formula implies that it also is incorrect. Indeed, the experimental deviations from the dependence defined by Eq. (17.3) turn out to be quite essential for intermediate $z \sim 1$ (with the relative deviations being of the order of unity).

17.4. The Flory method allows a simple generalization for a chain with excluded volume in a space of arbitrary dimensionality.

As noted in subsection 16.1, it is helpful to consider theoretically the fluctuating systems in a space of variable dimensionality d. This often helps to clarify the general situation. In addition, the basic technique of calculating the critical exponents proposed by K. Wilson and M. Fisher in 1971 (i.e., the ε expansion method) leads to a power expansion in $\varepsilon = d_{cr} - d$. (The critical dimensionality in many cases, including the problem of excluded volume, equals $d_{cr} = 4$.) It therefore is useful to obtain from the self-consistent theory the dependence $\nu(d)$ for the purpose of comparison. We obtain this by generalizing the Flory method.

Regarding the elastic free energy $F_{el}(\alpha)$ [see Eq. (13.3)], the only consequence of a transition to a space of arbitrary dimensionality would be replacement of the factor 3 by d in Eq. (17.2). The free energy $F_{int}(\alpha)$ for arbitrary d is written as

$$F_{int}(\alpha) \sim TN^2 B/R^d \sim TN^{2-d/2}B/(a^d \alpha^d) \sim Tz/\alpha^d, \qquad (17.4)$$

where

$$z = (d/2\pi a^2)^{d/2} B N^{(4-d)/2} \qquad (17.5)$$

is an explicit generalization of Eq. (13.3) to a space of arbitrary dimensionality d.

Minimizing the total free energy $F_{el}(\alpha) + F_{int}(\alpha)$ with respect to α, we obtain

$$\alpha^{d+2} - \alpha^d = \text{const} \cdot z. \tag{17.6}$$

Depending on the value of d, two qualitatively different cases are clearly seen.

When $d > 4$, the quantity z contains the large parameter N raised to a negative power so that for $N \gg 1$, z is close to zero. In this case, consequently, Eq. (17.6) yields $\alpha \cong 1$ (i.e., coil swelling because of the excluded volume of its links is absent). In the repulsive link interaction region $T \geqslant \theta$, the coil is always ideal, and its size depends on N in the usual way: $R \sim aN^{1/2}$, that is, $\nu = 1/2$.

When $d < 4$, the parameter z is much greater than unity for large N (i.e., coil swelling is substantial). In this case, $\alpha \sim z^{1/(d+2)}$, asymptotically, that is, with Eq. (17.5) taken into account, $\alpha \sim N^{(4-d)/2(d+2)}$, or $R \sim \alpha R_0$ $\sim N^{(1/2)+(4-d)/2(d+2)} \sim N^{3/(d+2)}$.

Hence, the self-consistent theory of the Flory type in a space of arbitrary dimensionality d yields

$$\nu = 3/(d+2), \quad d \leqslant 4; \quad \nu = 1/2, \quad d \geqslant 4. \tag{17.7}$$

Formula (17.7) shows that in treating polymer coil swelling, the dimensionality $d = 4$ is quite special; this is a critical dimensionality, above which the coil is always ideal. This formula furnishes some answer to the question why the critical exponent is naturally sought in the form of an expansion in a power series of the deviation from the critical dimensionality $\varepsilon = 4 - d$.

*17.5. *The Flory theory predicts a fairly correct value for the critical exponent ν for all integer dimensionalities; regarding other physical quantities, the accuracy of the Flory theory is not so high.*

For $d = 1$, the result $\nu = 1$, which follows from Eq. (17.7), is trivial and accurate: a self-avoiding random walk along a line allows no returns (i.e., $R \sim N$). The essentially important case $d = 2$ may arise, for example, when a macromolecule is strongly adsorbed by an attractive surface. For $d = 2$, we obtain $\nu = 3/4$ from Eq. (17.7), that is, $R \sim N^{3/4}$. This result agrees well with computer-simulation data and is exactly correct according to some current theoretic studies. When $d \geqslant 4$, the Flory result $\nu = 1/2$ is also exact; this is obvious as the perturbation theory parameter (17.5) is small.

The case $d = 3$ is surely the most important one for which the correct value of the exponent ν is unknown. Using the "polymer-magnetic" analogy, one can obtain the best estimate from the second-order phase transition theory. Assuming $n = 0$ in the corresponding general formula,

$$\nu = 1/2 + \varepsilon(n+2)/[4(n+8)] + \varepsilon^2(n+2)(n^2+23n+60)/[8(n+8)^3] + \ldots$$

and $\varepsilon = 4 - d = 1$, we obtain[a]

$$v = 1/2 + \varepsilon/16 + 15\varepsilon^2/512 + ... \approx 0.592. \qquad (17.8)$$

Because $\varepsilon = 1$ is not small, one can of course inquire why the series is interrupted to keep only those terms whose order is not higher than ε^2. Generally, the determination of the accuracy of the ε expansion is quite complicated, and it lies beyond the scope of this book. Nevertheless, according to current considerations, the error of the second ε approximation for the exponent v does not exceed 1%. In addition, the value (17.8) agrees with computer-simulation results to within 1%.

Thus, for a three-dimensional coil, the Flory result $v = 3/5 = 0.6$ is not accurate, but it is very close to the correct value [cf. Eq. (17.8)]. In fact, the accuracy of the estimation $v \approx 0.6$ is sufficient for any real chain. Nevertheless, a consistent calculation of critical exponents is of great interest and is discussed in the next section. There we present some important methods that have many other applications in polymer physics.

To conclude this section, we should emphasize that the success of the Flory theory in calculating the exponent v for a coil with excluded volume does not prove that the Flory theory is correct. On the contrary, as a self-consistent field theory, it is by no means correct, and this fact reveals itself in studies of other physical quantities, some examples of which are discussed later.

18. RENORMALIZATION GROUP AND ε-EXPANSION METHODS APPLIED TO THE EXCLUDED VOLUME PROBLEM

18.1. *The general idea of the renormalization group method consists in a multiple enlargement of elementary scale (i.e., for polymers) in a gradual transition to treating the chains in terms of enlarged blocks of links.*

For a magnet, the transformation of the renormalization group consists in a transition from the initial picture of the magnet as a system of elementary magnetic moments (spins) located at the lattice sites to a new one in which the role of an elementary spin is played by the magnetic moment of a block composed of several neighboring spins of the previous step (L. P. Kadanov, 1968). To characterize the magnet, recursion relations are written at each step. When written in an appropriate form, they yield a stationary point in the iteration process. Universality of critical behavior is expressed mathematically via universal properties of recursion relations near the stationary point. Analysis of the obtained relations allows the critical exponents to be calculated.

In 1977, proceeding from this general idea, P. G. de Gennes proposed a procedure to realize the renormalization group method directly in terms of

---

[a]Note that the Flory formula (17.7), when expanded in a power series of ε, gives $v = 1/2 + \varepsilon/12 + ...$ (i.e., differs from the correct ε expansion already in the first order). Consequently, the Flory formula is inaccurate for an arbitrary d.

polymer physics. This procedure is referred to as a *decimation procedure* (Fig. 3.1). Suppose we have a standard Gaussian polymer chain of N_0 links with distance a_0 between neighboring links and a second virial coefficient B_0 of link interaction. We pass from the initial division of the macromolecular chain into N_0 links to a new one in which the chain consists of $N_1 = N_0/g$ blocks, each composed of g consecutive links. In the new description, these new blocks act as effective links. Similar to subsection 13.8, we can compose from the blocks an equivalent standard Gaussian chain with the parameters $N_1 = N_0/g$, a_1 and B_1, and we can repeat the block enlargement procedure with this equivalent chain many times (Fig. 3.1). Such procedures of scale enlargements are called *renormalization transformations*. In mathematic terms, they form a so-called renormalization group. The physical idea of the renormalization group method consists of as follows. Because the fluctuations are determined by a length scale of the order of the correlation radius, which in turn is of the order of the coil size (i.e., much greater than the link size), the influence of microscopic structure of the initial chain should disappear with the growth of the block of monomer links during the iterations. In terms of the standard model, the microscopic structure is characterized by the only dimensionless parameter, B_0/a_0^d (*see* subsection 13.5). Accordingly, the renormalization group method is based on the assumption that with the growth of a length scale of the block monomer link (i.e., with an increase of number p of iterations), the quantity B_p/a_p^d no longer depends on B_0/a_0^d and tends to a definite universal limit (a stationary point). To prove this assumption to be valid and to analyze the resulting consequences, one must learn

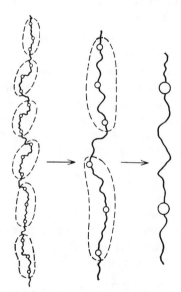

FIGURE 3.1. Explanation of decimation procedure: two renormalizations with $g=3$.

how to find the parameters a_{p+1} and B_{p+1} from a_p and B_p.

*18.2. Recursion relations for parameters of block monomer links are obtained by a perturbation theory.

Suppose (without a proper substantiation so far) that a perturbation theory can be applied to obtain the required recursion relations. It is made clear later why the perturbation theory is valid for this purpose.

We reason in the following way. Consider a chain of g links with parameters a_0, B_0. Let a_1 denote its end-to-end distance in the coil state and B_1 the second virial coefficient of interaction between two such chains. The next step is to take a chain of g^2 initial links (i.e., of g block links of the first "generation"). The chain parameters become a_2, B_2, and so on. After this description, it becomes clear that the quantities a_{p+1}, B_{p+1} are the size and virial coefficient of the coil of g links with the characteristics a_p, B_p. In the absence of volume interactions, they would be expressed by the formulas $a_{p+1}^2 = g a_p^2$, $B_{p+1} = g^2 B_p$, whereas in the real case with volume interactions, one should use the perturbation theory results (14.6) and (14.9) in accordance with the assumption made earlier. These results, however, should be generalized preliminarily in two respects.

First, Eqs. (14.6) and (14.9) are written for the three-dimensional case ($d=3$), and the similar formulas have to be derived for the arbitrary value of d. In this case, the expansion parameter is of the order $(B_p/a_p^d)g^{(4-d)/2}$ [see Eq. (17.5)], and even though the calculations of the coefficients are too bulky to be presented here, they are fairly simple conceptually [see Eq. (14.5), which is easily generalized].

Second, Sec. 14 discussed the perturbation theory for a N-link chain with $N \gg 1$. Here, we are interested in a g-link section in which g is not necessarily large. In first-order perturbation theory, this condition is accounted for by the substitution of $g^{(4-d)/2} - 1$ for $g^{(4-d)/2}$, because the perturbation tends to zero at $g=1$. Eventually, we obtain

$$a_{p+1}^2 \cong g a_p^2 \{1 + k_1(d)\beta_p[g^{(4-d)/2} - 1]\}, \tag{18.1}$$

$$B_{p+1} \cong g^2 B_p \{1 + k_1''(d)\beta_p[g^{(4-d)/2} - 1]\}, \tag{18.2}$$

$$\beta_p \equiv B_p/a_p^d, \tag{18.3}$$

$$k_1(d) = (d/2\pi)^{d/2} \cdot 4/[(4-d)(6-d)],$$

$$k_1''(d) = -(d/2\pi)^{d/2} \cdot 32\{1 - [2^{(8-d)/2} - 2]/(8-d)\}/[(d-2)(4-d)(6-d)]. \tag{18.4}$$

For simplicity, we limit ourselves to first-order perturbation theory. As shown later, this corresponds to the first order of the ε expansion.

Using Eqs. (18.1) and (18.2), one can readily obtain the recursion relation for the dimensionless quantity $\beta_p \equiv B_p/a_p^d$ as well. It is convenient to write it as

$$\beta_{p+1}=\beta_p g^{(4-d)/2}/\{1+\beta_p[dk_1(d)/2-k_1''(d)][g^{(4-d)/2}-1]\}. \quad (18.5)$$

Understandably, β_p tends to β^* (as expected) in the process of multiple repetition of the recursive procedure (i.e., for $p\to\infty$). Taking in Eq. (18.5) $\beta_{p+1}=\beta_p=\beta^*$, we find

$$\beta^*=1/[dk_1(d)/2-k_1''(d)]$$

$$=(2\pi/d)^{d/2}(d-2)(4-d)(6-d)(8-d)[2d(d-2)(8-d)+32(10-d)$$

$$-2^{(18-d)/2}]^{-1}. \quad (18.6)$$

18.3. *The invariant excluded volume corresponding to the stationary point of the renormalization group is small if the dimensionality of space is close to four.*

Moving to the analysis of Eqs. 18.1 to 18.6, we can regard d as an arbitrarily changing parameter, neglecting temporarily its actual geometric sense. From Eq. (18.6), one can immediately see that $\beta^*=0$ for $d=4$ and $\beta^*\ll1$ for $\varepsilon=4-d\ll1$:

$$\beta^*\cong\varepsilon\pi^2/32 \quad\text{for }\varepsilon\ll1. \quad (18.7)$$

At this stage, it is worth returning to the assumption formulated at the beginning of subsection 18.2 and emphasizing that it is the smallness of β^* that allows use of the perturbation theory to write the recursion relations (18.1) and (18.2). In this connection, it should be made clear that the high accuracy of Eq. (18.6) in comparison with Eq. (18.7) is deceptive. The result (18.7) can be refined only by applying second-order perturbation theory to Eqs. (18.1) and (18.2).

18.4. *A diagram of flows is investigated to analyze the renormalization results.*

The recursion relations are convenient to write in the form of differential equations. By assuming the parameter g to be close to unity[b]: $g-1\ll1$.

Let s denote the number of initial monomer links in one block at the p-th renormalization step. The block monomer link of the next $(p+1)$-th step incorporates $s+\Delta s$ initial links. In this notation,

$$g=(s+\Delta s)/s=1+\Delta s/s,$$

and consequently, $g\to1$ corresponds to $\Delta s/s\ll1$. Then,

$$\beta_p\equiv\beta(s), \quad \beta_{p+1}\equiv\beta(s+\Delta s)\cong\beta(s)+\Delta s\frac{d\beta}{ds}.$$

Substituting these results into Eq. (18.5), we reduce it to the form

---

[b]It should be pointed out that the renormalization transformation with non-integer g is logically quite feasible although it cannot be interpreted graphically as in Fig. 3.1. It should be understood in the following way: suppose we have a real polymer chain and choose the standard model for its consideration (see Sec 12). Because of freedom in choosing the number of beads (monomers) in the standard model, we can start with the one of some N_1 and then transform to some $N_2=gN_1$. Even N_1 and N_2 are not necessarily integers, and g is not as well.

$$\frac{2}{\varepsilon} s \frac{d\beta}{ds} = \beta(s) - \frac{1}{\beta*} \beta^2(s).$$

(18.8)

This equation can be solved easily [because it is linear with respect to the variable $1/\beta(s)$], but its properties are illustrated more graphically by the phase portrait in Figure 3.2. When $d\beta/ds > 0$, $\beta(s)$ grows, and when $d\beta/ds < 0$, $\beta(s)$ diminishes with an increase in the length scale s. This is illustrated by the arrows in Figure 3.2. Thus, we see that there is a stable stationary point $\beta = \beta* > 0$ and a non-stationary one $\beta = 0$.

If the solvent is good, $\beta_0 > 0$ from the very beginning, and as the length scale s grows, we arrive at the stationary point $\beta*$, that is, the universal (independent of a specific value of β_0) behavior of the coil with excluded volume. If the solvent corresponds to the θ point (i.e., $\beta_0 = 0$), then $\beta(s) = 0$ for all scales. This stationary point is called Gaussian, because it characterizes the Gaussian regime of the coil. Finally, if the solvent is poor (i.e., $\beta_0 < 0$), $\beta(s) \rightarrow -\infty$ with the growth of the length scale. (The physical meaning of this regime is discussed in Sec. 20.)

This is illustrated by a diagram of flows of the renormalization group transformation (Fig. 3.3), where $\beta(s)$ is shown as a function of the scale s for different β_0 (i.e., for different conditions). Figure 3.3a corresponds to Eq. (18.8), whereas Figure 3.3b describes a more accurate equation that we do not present here for the sake of brevity and that accounts for triple collisions. In terms of physics, it is clear that in this refined diagram, the separatrix, which leads to the Gaussian regime $\beta(s) \rightarrow 0$ as $s \rightarrow \infty$, must begin from a small negative value of β_0, because a small attraction because of binary collisions is needed to compensate a repulsion in the case of triple collisions (*see* subsection 14.4).

*18.5. *The critical exponents are easily found by solving the renormalization group equation; the exponents are universal because of the existence of a stationary point in renormalization transformations.*

Write Eq. (18.1) in the form of a differential equation $(g \rightarrow 1)$:

$$2 \frac{s}{a(s)} \frac{da(s)}{ds} = 1 + \frac{\varepsilon}{2} \beta(s) k_1.$$

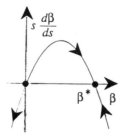

FIGURE 3.2. Phase portrait of the renormalization group equation.

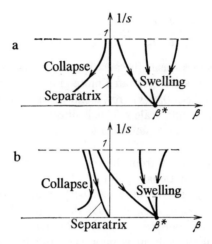

FIGURE 3.3. Diagram of flows of the renormalization group.

Knowing the initial condition $a(s=1)=a_0$, this equation is easy to solve:

$$a(s)=a_0 \exp\left\{\int_1^s [1+(\varepsilon/2)\beta(s')k_1]ds'/2s'\right\}.$$

If $s \gg 1$, then $\beta(s)$ converges to the stationary point β^*, the function $\beta(s')$ can be replaced in the integrand by the constant β^* over the greater part of the integration interval to yield

$$a(s)=a_0 s^{1/2+(\varepsilon/4)\beta^* k_1}. \tag{18.9}$$

Being interested in the properties of the N-link chain as a whole, we should extend the renormalization up to the scale $s=N$ to turn the whole coil into one block monomer link. As a result, we automatically obtain the coil size $R=a(s)\big|_{s=N}$. Consequently, Eq. (18.9) has the expected structure $R \sim N^\nu$ which [after substituting Eqs. (18.4) and (18.7)ᶜ] allows us to find the expression for the critical exponent ν:

$$\nu \cong 1/2+\varepsilon/16.$$

As expected, the first term of ε expansion investigated yields the result coinciding with Eq. (17.8) obtained by the polymer-magnetic analogy.

Analyzing the derivation of Eq. (18.10), one easily notices that the universality of the exponent ν (i.e., its independence of the initial value of β_0) follows from the fact that consecutive renormalizations lead to the initial conditions being gradually "forgotten," that is, $\beta(s)$ tends to the stationary point. In addi-

---

ᶜRecall that the quantity $k_1(d)$ should be substituted not in the accurate form (18.4) but (as all the other quantities) in the form of the first-order expansion in $\varepsilon=4-d$.

tion, if β^* corresponds to the second (Gaussian) stationary point $\beta^*=0$ (Fig. 3.2), then Eq. (18.9) yields the expected result $\nu=1/2$ (Gaussian statistics).

*18.6. *A stationary point of the renormalization group corresponds to scale invariance of coil structure and to a unique characteristic macroscopic length scale.*

Because the function $\beta(s)$ tends to the stationary point β^* with the growth of the length scale s, the coil properties no longer depend on the scale. This is inherent for the scaling invariance. (This concept was interpreted in detail in subsection 5.4 in connection with the standard Gaussian model.)

The scaling invariance persists with the growth of s up to the greatest possible value $s=N$; therefore, $a(s=N)$ proves to be the only macroscopic length characteristic for the coil. Any dimensional characteristic of the chain as a whole should coincide with that length by the order of magnitude (and in particular, by the form of the dependence on N for $N \to \infty$).

The existence of a unique characteristic length scale is to be specially emphasized, because this fact underlies the scaling concept that is extensively used later.

19. PROPERTIES OF POLYMER COILS WITH EXCLUDED VOLUME (THE SCALING CONCEPT AND THE NOTION OF A BLOB)

19.1. *While discussing the properties of nonideal coils, one can consider for simplicity flexible chains in an extremely good solvent.*

In this section, we assume that only repulsive forces act between the links because of their mutual impenetrability. Then according to Eq. (12.5), the second virial coefficient $B \sim v$, where v is the excluded volume of a monomer link. Further, we suppose that the chains are flexible. In this case (as shown in subsection 13.8), we should set $v \sim a^3$ for the equivalent standard chain (Fig. 2.4). The conformational chain characteristics that we study in this section thus will depend on only two parameters: the number N of links in the chain, and the distance a between neighboring links in the chain. The second virial coefficient of link interaction $B \sim v \sim a^3$. For this case, according to Eq. (17.1), the size of the polymer coil $R \sim aN^\nu \approx aN^{3/5}$. The results for stiff chains ($v \ll a^3$) or for temperatures close to the θ-temperature ($B \ll v$) can be generalized in a simple way.

In the region where repulsive forces prevail in volume interactions of link, polymer coils possess the property of universality (*see* subsection 13.6), and application of one or another specific model of a polymer chain to derive macroscopic conformational properties causes no restrictions on generality. In particular, this section uses both the standard Gaussian model of a polymer chain with $B \sim v \sim a^3$ and the lattice model in which a polymer chain is visualized as a self-avoiding walk over a simple cubic lattice with spacing a (Fig. 2.2). If in this walk we neglect attraction between the links, this model will be equivalent to the standard Gaussian macromolecule with $B \sim v \sim a^3$, and its parameters N and a

for the lattice chain will coincide in order of magnitude with those in the equivalent chain.

19.2. The second virial coefficient of interaction of two swollen coils in good solvent is of the order of their volume, just as if they were solid spheres.

This immediately follows from the uniqueness of an intrinsic macroscopic size in the coil and the uniqueness of the scale. Indeed, from dimensionality considerations, we have $A_2 \sim R^3$, where R is of the order of the coil size. Still, this fact should be discussed in more detail.

Suppose that two polymer coils approach one another at some moment of time so close that their overlap volume is of the order of the coil volume $V \sim R^3 \sim a^3 N^{3\nu}$. Then, the link concentration in the overlap volume equals: $n \sim N/V \sim a^{-3} N^{1-3\nu}$. Evaluate the free energy F_{int} of interaction of such coils, and then derive the virial coefficient A_2 from Eq. (14.8). Clearly, $F_{int} \sim Tk$, where k is the number of simultaneous collisions of monomer links of the two coils. (The characteristic energy is of order T, because the interaction of the links is reduced to their mutual impenetrability.) The value of k is determined by a product of the number of links in one of the chains within the overlap volume $nV \sim N$ by the probability collision w for one link (i.e., $k \sim Nw$). Following the Flory theory, we should consider the coil as a cloud of independent particles with density n to obtain $w \sim na^3$, that is $F_{int} \sim TN^{2-3\nu} \approx TN^{1/5} \gg T$ (P. J. Flory, W. Krigbaum, 1950). Actually, when the two links approach one another, the correlative "clouds" of neighboring links also become involved, so the probability of a contact turns out to be somewhat less than na^3. In subsection 19.4 (*see* also subsection 25.5), we show that

$$w \sim (na^3)^{1/(3\nu-1)} [1/(3\nu-1) \approx 1.25].$$

Consequently, the correct estimation of F_{int} is

$$F_{int} \sim TN(na^3)^{1/(3\nu-1)} \sim T.$$

Thus, even though the coils are mutually penetrable as the probability $\exp(-F_{int}/T)$ of their getting into one another is of order unity,[d] the virial coefficient of their interaction is of the order of their volume $A_2 \sim V \sim a^3 N^{3\nu}$ [as the integrand in Eq. (14.8) is of order unity within the volume V].

These views are easily generalized to a space of arbitrary dimensionality d. When $d \geqslant 4$, the estimation yields $F_{int} \ll T$ and Eq. (14.9) is valid, because the perturbation theory parameter [*see* Eq. (17.5)] is small. Conversely, when $d < 4$, $F_{int} \sim T$, so that we eventually obtain

$$A_2 \sim \begin{cases} R^d \sim a^d N^{\nu(d)d}, & d < 4, \\ a^d N^2, & d \geqslant 4. \end{cases} \tag{19.1}$$

---

[d]Note that this probability is much less than unity according to the Flory-Krigbaum evaluation. This leads to the wrong conclusion about mutually impenetrable coils.

19.3. *A model problem on the behavior of a chain with excluded volume in a thin capillary or plane slit provides a chance to see how the scaling concept can be applied in the simplest form and to introduce the notion of a blob.*

Consider a capillary (or a slit) of diameter (or thickness) $D \ll R$, where $R \sim aN^\nu$ is the coil size. On the length scales less than D, a chain located in the capillary (or slit) is insensitive to the imposed constraint, so the size of l-link section of the chain is of the order of al^ν. The maximum length g of the unperturbed section then is defined by the condition

$$ag^\nu \sim D, \quad \text{i.e.} \quad g \sim (D/a)^{1/\nu}. \tag{19.2}$$

The chain section of g links is called a *blob*. As mentioned, the blob size is determined by the condition that the chain remains unperturbed within the blob.

Now we consider a macromolecule as a chain of N/g blobs.[e] According to Eq. (19.1), the excluded volume of a blob is of order D^3. The length of a link in the chain of blobs obviously is of order D as well. Consequently, the macromolecule in the capillary or the slit can be treated as a one- or two-dimensional coil with excluded volume formed by the chain of blobs (Figure 3.4). The size of such a coil is easy to estimate:

$$R_{\text{tube}} \sim \frac{N}{g} D \sim Na \left(\frac{D}{a}\right)^{1-1/\nu} \approx Na \left(\frac{D}{a}\right)^{-2/3}, \tag{19.3}$$

$$R_{\text{sl}} \sim \left(\frac{N}{g}\right)^{\nu_2} D \sim N^{\nu_2} a \left(\frac{D}{a}\right)^{1-\nu_2/\nu} \approx N^{3/4} a \left(\frac{D}{a}\right)^{-1/4}. \tag{19.4}$$

As it should be, size diminishes with growth of D, and when D reaches the size aN^ν of the unperturbed coil, the chain size also turns out to be of the same order of magnitude.

Now let us estimate the free energy ΔF associated with the chain constraint. We do this by the so-called method of scaling estimations. Because T is the only quantity in the problem having the dimension of energy and the coil is characterized by the only length scale aN^ν [so that the quantity D can enter all expressions only in the combination $D/(aN^\nu)$], the free energy ΔF takes the form

$$\Delta F \sim T\varphi(D/aN^\nu), \tag{19.5}$$

where $\varphi(x)$ is the so far unknown function of the dimensionless argument $x = D/(aN^\nu)$. The form of the function $\varphi(x)$ is rather difficult to derive explicitly, but its asymptotic behavior for the strongly compressed chain ($x \ll 1$) can be established easily. Because of the thermodynamic additivity condition, a macromolecule placed in a narrow slit breaks into many independent parts or blobs (Fig. 3.4). Therefore, $\Delta F \sim N$. Hence, $\varphi(x) \sim (x)^{-1/\nu}$, and

---

[e]In the spirit of the renormalization group method (*see* Sec. 18), we bring the renormalization to the length scale $s \sim g$ and regard the blob as a "block" link.

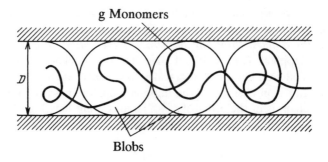

g Monomers

Blobs

FIGURE 3.4. Polymer chain with excluded volume forms a system of blobs in a tube or slit of diameter (width) D.

$$\Delta F \sim NT(D/a)^{-1/v}. \tag{19.6}$$

Comparing this result with Eq. (19.2), we see that the free energy per blob is approximately equal to T. Actually, we see later that this result is of a general character [cf. the estimate (7.2) for the ideal chain in the globular state].

19.4. *A chain with excluded volume placed into a closed cavity forms a globule; its pressure on the walls is higher than that for an analogous globule formed by an ideal chain.*

For the free energy of chain compression in a cavity, Eq. (19.5) is also valid. There is no reason, however, to expect that ΔF will be proportional to N, because the sections of size D are now brought together in space and interact. In the self-consistent field theory for the cloud of links with binary collisions, we obtain $\Delta F \sim TBD^3(N/D^3)^2 \sim N^2$. The correct asymptotic behavior of the function $\varphi(x)$ (19.5) for the strongly compressed chain $x \ll 1$ (and therefore the expression ΔF) is established from the condition that the pressure of the chain on the cavity walls $p = -\partial \Delta F/\partial V \sim -\partial \Delta F/\partial D^3$ must depend only on the link concentration in the cavity $n \sim N/D^3$ but not on N and D independently. This is because the different sections of size D act on one another via the volume interactions defined by the concentration. It is easy to see that the mentioned condition is met, provided the asymptotic behavior of $\varphi(x)$ obeys the power function law. Assuming $\varphi(x) \sim x^m$ and calculating m, it also is easy to find that if $m = 3/(1 - 3v)$, then

$$\Delta F \sim T(aN^v/D)^{3/(3v-1)}, \tag{19.7}$$

$$p \sim Ta^{-3}(a^3N/D^3)^{3v/(3v-1)}, \quad 3v/(3v-1) \approx 9/4. \tag{19.8}$$

The pressure of the real chain on cavity walls is seen to be higher than that of an ideal chain. The ratio of (19.8) to (7.3), $(aN/D)^{1/v(3v-1)}(D/a)^{(2v-1)/v} \gg 1$, provides evidence that the links repel one another. The pressure (19.8), however, is lower than that obtained in the self-consistent theory assuming the independent binary collisions of the links $\sim Ta^{-3}(a^3N/D^3)^2$. (Note that $a^3N/D^3 < 1$, because the volume of links in the cavity is less than the cavity volume.) This is a manifestation of the above mentioned correlation effect (i.e., a decrease in collision probability for any link because of the existence of a cloud of neighboring links). The free energy can be evaluated as $\Delta F \sim TNw$, where w is the collision probability for a given link (*see* subsection 19.2). Comparing this estimate with (19.7), we find $w \sim (na^3)^{1/(3\nu-1)} \ll na^3$ ($n = N/D^3$), which was used in subsection 19.2.

In this subsection, we discuss the compressed state of a macromolecule that is globular in nature. From the local point of view, such a globule (considered on length scales less than D), seems to be composed of many independent chains, and in this sense, it is analogous to the so-called semidilute solution (*see* Secs. 25 and 26). We will show that the correlation radius in such a system is of order $a(na^3)^{-\nu/(3\nu-1)}$, (i.e., much less than D). This proves that the state under consideration is indeed globular.

19.5. *The scaling concept allows one to easily analyze a polymer chain extended by an external force applied to the chain ends; the extension turns to be a nonlinear function of force because of the presence of excluded volume.*

In subsection 8.1, we solved the problem of polymer extension for an ideal chain and showed that the mean end-to-end distance R is connected with the applied force by the linear relation $R = (Na^2/3T)f$ [*see* Eq. (8.2)]. Here, we obtain the analogous dependence (the Hooke law) for a polymer chain with excluded volume.

We know that polymer chain flexibility is caused by a depletion of possible conformations of a macromolecule during its extension, that is, flexibility is a macroscopic property of a polymer coil that is independent of the details of the microscopic structure of the chain. Consequently, the elastic behavior of the coil must depend on the parameters N and a of the polymer chain in the form of the combination $R \sim aN^\nu$ (for an ideal chain, $\nu = 1/2$, and for a chain with excluded volume, $\nu \approx 3/5$), that is, such a behavior must depend on the overall size of the coil but not the microscopic dimensions of individual links. Using other parameters (the force f and temperature T), one can create a unique combination of length dimensionality, (i.e., T/f). Accordingly, the mean end-to-end distance $|\langle R \rangle|$ of an extended chain (this designation was introduced in subsection 8.1 to differentiate this quantity from the coil size R) should be written as

$$|\langle R \rangle| \sim aN^\nu \varphi(x), \quad x \equiv aN^\nu f/T, \tag{19.9}$$

where $\varphi(x)$ is a certain, so far unknown function of dimensionless argument.[f] One can check that Eq. (19.9) is valid for the ideal polymer chain. In this case, $\nu = 1/2$, and according to Eq. (8.2), $|\langle R \rangle| \sim aN^{1/2} \cdot (aN^{1/2}f/T)$, that is,

---

[f]Hereafter, we use the same designations $\varphi(x)$ or $f(x)$ for functions of dimensionless argument, featuring in the method of scaling estimations [cf. Eqs. (19.5) and (19.9)]. It should be remembered that these functions are different in each specific case and are not related.

$\varphi(x) = x$. In the presence of the excluded volume, it is rather difficult to calculate the function $\varphi(x)$ exactly, but if one considers its behavior in the opposite extreme cases of very small and very large values of x, then the comprehensive qualitative representation of the function can be obtained.

First, suppose that $x \ll 1$ (i.e., the stretching force is small). In this case, the coil is only weakly disturbed by the external force, and the "elastic response" $|\langle R \rangle|$ must be linear (i.e., proportional to f). This means that for $x \ll 1$, the function $\varphi(x) \sim x$ or

$$|\langle R \rangle| \sim N^{2\nu} f a^2/T \approx N^{6/5} f a^2/T \text{ when } f \ll T/(aN^{3/5}). \quad (19.10)$$

This result represents the Hooke law for the chain with excluded volume at small extensions. From comparison with Eq. (8.2) obtained for the ideal chain, it is seen that the elasticity modulus in this case turns to be somewhat less (of the order of $1/N^{6/5}$, but not $1/N$ as in the ideal chain). This result is quite natural: the elasticity modulus decreases because of the repulsive forces between the links. The inequality $f \ll T/(aN^{3/5})$, which must be satisfied to provide a linear response, signifies that according to Eq. (19.10), $|\langle R \rangle| \ll aN^{3/5} \sim R$, (i.e., the chain extension is insignificant in comparison with the mean size R of the coil).

Now let us direct our attention to the extreme case of strong stretching $x \gg 1$. In this case, $|\langle R \rangle| \gg R$, and the chain can be pictured as a sequence of blobs (Fig. 3.5). The number of links in a blob g is chosen, as always, such that an external action (in this case, a stretching force) is insignificant inside the blob. In other words, the parameter x of Eq. (19.9) must be of order unity over the length g of the chain within the blob, and therefore, $g \sim (T/fa)^{1/\nu}$. Accordingly, the blob size $D \sim ag^\nu \sim T/f$.

Now consider a chain of blobs. For this chain, the link size is of order D, and the number of links of order N/g. Then, the effective parameter x of Eq. (19.9) substantially exceeds unity, so the blob sequence must be considerably extended in the stretch direction. The dimension of the chain in the longitudinal direction must comprise the dimensions of the individual blobs (i.e., $|\langle R \rangle| \sim ND/g$). Taking into account the estimates $D \sim g^\nu a$ and $D \sim T/f$, we finally obtain

$$|\langle R \rangle| \sim aN(fa/T)^{(1-\nu)/\nu} \text{ when } f \gg T/(aN^\nu). \quad (19.11)$$

The result (19.11) could also be obtained directly, without resorting to the blob picture, from Eq. (19.9) if one assumed that $\varphi(x) \sim x^m$ at $x \gg 1$. The exponent m

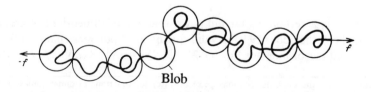

FIGURE 3.5. Polymer chain, stretched by force f by the ends, as a system of blobs.

must be chosen from the obvious physical condition that the value of $|\langle R \rangle|$ is proportional to N at $x \gg 1$.

Note that for $v = 1/2$, we return to the result (8.2) for the ideal chain. For a chain with excluded volume, however, $v \approx 3/5$, and we obtain the nonlinear extension-force relationship $|\langle R \rangle| \sim f^{2/3}$. Thus, we see that the polymer coil with excluded volume behaves under an external stretching force quite differently from the ideal chain.

Clearly, the result (19.11) is valid only when the polymer chain is far from being totally stretched (i.e., $fa \ll T$). Non-universal effects associated with microscopic structure of a specific polymer start showing in the region $fa \sim T$ (cf. subsection 8.3).

Analyzing the last subsections, one notes that they contain the basic assumption that all coil properties are determined by the same size aN^v [see Eqs. (19.5) and (19.9)]. Recall that this scaling assumption, or the scaling invariance hypothesis, is substantiated by the scaling invariance of the coil (see subsection 18.6), and as we saw, this makes it possible to examine the behavior of the polymer coil under various external forces without recourse to bulky calculations. Now we move to a more detailed analysis of properties of the coil itself in the absence of external actions.

19.6. *The behavior of the structure factor over distances shorter than the coil size is defined by the critical exponent v; elastic light and neutron scattering experiments therefore allow the value of v to be measured.*

The notion of the structural factor $G(k)$ [see Eq. (5.11)] was discussed in subsection 5.5 in connection with a Gaussian coil. Recall that the structural factor can be measured directly in experiments on elastic scattering of light, x-rays, or neutrons in dilute polymer solutions. We now show that from such measurements, we also can determine the critical exponent v for coils with excluded volume.

From the definition of the structure factor $G(k)$ (5.11), it follows that in the region of very long wavelengths, where for any links m and n the inequality $|k(x_n - x_m)| \ll 1$ holds, $G(k) = N$. This also confirms the relation (5.14), which is valid in the long wavelength limit irrespective of the presence or absence of the excluded volume of the links.

The structural factor varies with the decrease of the wave vector $|k|$, that is, with the scattering angle [see Eq. (5.12)]. Proceeding from the scaling assumption of uniqueness of the characteristic coil size $R \sim aN^v$, we infer that this variation must be expressed as

$$G(k) = N\varphi(|k|aN^v) \qquad (19.12)$$

[cf. Eqs. (19.5) and (19.9)]. Note that the asymptotic behavior of the function $\varphi(x)$ in the region $x \ll 1$ is known and described by Eq. (5.14); $\varphi \to 1$ for $x \to 0$. The exact expression (5.16), derived in subsection 5.5 for the structure factor of a Gaussian coil (and in this case $v = 1/2$), confirms the scaling formula (19.12).

From the standpoint of the experimental determination of the critical exponent ν for a coil with excluded volume, the most interesting region is that of relatively short wavelengths $|k|aN^\nu \gg 1$ (large-angle scattering). To determine the asymptotic behavior of the function $\varphi(x)$ at $x \gg 1$, it should be noted that in the short wavelength limit, the structural factor must be independent of N, as it follows from the definition (5.11). This is a consequence of the fact that terms with $n \neq m$ in the double sum (5.11) are average values of rapidly oscillating functions, which rapidly diminish as the difference $n - m$ grows (so that the sum $\sum_{m=1}^{N} \langle \exp[ik(x_n - x_m)] \rangle$ remains constant as $N \to \infty$). The fact that $G(k) \sim N^0$ for $|k|aN^\nu \gg 1$ is confirmed by the relation (5.17) derived for the Gaussian coil.

From the condition $G(k) \sim N^0$ at $|k|aN^\nu \gg 1$, it follows that at $x \gg 1$ the function $\varphi(x)$ obeys the power law $\varphi(x) \sim x^{-1/\nu}$. Hence,

$$G(k) \sim (|k|a)^{-1/\nu} \quad \text{at} \quad |k|R \gg 1, \tag{19.13}$$

where R is the size of the coil with excluded volume. Thus, measuring the structural factor in the region of large scattering angles, one can find the critical exponent ν describing the coil with excluded volume. As mentioned, the Flory result $\nu \approx 3/5$ is valid for long polymer chains in good solvent to a high degree of accuracy.

*19.7. *A statistical distribution of the end-to-end distance for a chain with excluded volume differs qualitatively from that of an ideal chain; the behavior of this distribution at long distances is determined by the critical exponent ν and at short distances by the new exponent γ, which describes the end effects and is independent of ν.*

In subsection 4.1, we calculated for the ideal chain the function $P_N(R)$, which is the probability distribution that the end-to-end vector of an N-link chain equals R. This function was shown to have the Gaussian form (4.2) for $N \gg 1$. Let us now calculate the function $P_N(R)$ for the polymer chain with excluded volume.

We know that the size R of the polymer coil with excluded volume determines its unique intrinsic scale. Therefore, the distribution $P_N(R)$ takes the following form in the general case

$$P_N(R) = (1/R^3)\varphi(R/R) \tag{19.14}$$

[cf. the scaling hypotheses (19.5), (19.9), and (19.12)]. The factor R^{-3} appears as a result of the normalization condition $\int P_N(R)d^3R = 1$. In particular, according to Eq. (4.2), the relation (19.14) is valid for the ideal chain, and $\varphi(x) \sim \exp(-3x^2/2)$.

For the polymer chain with excluded volume, a computer simulation of the self-avoiding random walk over spatial lattices yields the function $\varphi(x) \equiv \varphi(|x|)$, as shown in Figure 3.6. (For comparison, this figure also shows by a dashed line the Gaussian function, which is valid for the ideal chain.)

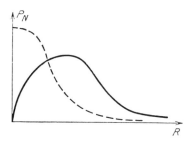

FIGURE 3.6. Statistical distribution of end-to-end distance of a N-link polymer chain with excluded volume. Gaussian distribution for an ideal chain is shown for comparison by the *broken line*.

A consistent analytic calculation of the function $\varphi(x)$ for the chain with excluded volume can be performed only in terms of the fluctuation theory (e.g., using the polymer-magnetic analogy). We shall not do this here; instead, we limit ourselves only to simple evaluations. First, it should be recalled that the function $\varphi(x)$ must be universal and independent of the polymer chain model.

We begin by finding the asymptotic behavior of $\varphi(x)$ at $|x|\gg1$, that is, at long distances $|R|\gg R$. To do this, we use the results of subsection 19.5 pertaining to strong stretching of a chain with excluded volume. The similarity with this problem is as follows: first, $S(R)=\ln P_N(R)$ is the entropy of the chain with the end-to-end distance R; and second, $f=-T\partial S/\partial R$ is the force keeping the chain ends separated by the distance R. Thus, having expressed f from Eq. (19.11) as a function of R and integrated, we can find the entropy S and, consequently, P_N. The result indeed takes the form of Eq. (19.14), and

$$\varphi(x)\sim\exp(-|x|^\delta) \quad \text{when} \quad |x|\gg1, \quad \delta=1/(1-\nu)\approx5/2 \qquad (19.15)$$

(taking into account that $\nu\approx3/5$). Thus, the distribution function $P_N(R)$ diminishes at large values of $|R|$ faster for the chain with excluded volume than for the ideal chain [for which $\delta=2$ according to Eq. (4.2)].

Let us now turn to the opposite limiting case $|x|\ll1$ (i.e., consider the probability of a close approach of the chain ends). Clearly, $P_N(0)=0$, because the chain ends cannot coincide due to excluded volume effects. For small $|R|$, the value of $P_N(R)$ must also be small, because the close approach of the chain ends is hindered by the repulsion of "clouds" of links that are close neighbors to the terminal links of the chain.[g] A similar effect was discussed in subsections 19.2 and 19.4, however, those "clouds" surrounded the internal links of the chain. Obviously, such "clouds" belonging to the terminal monomer links have a different structure, and it turns out that the increase in $P_N(R)$ with growth of $|R|$ cannot be described by means of the critical exponent ν. Therefore, the new

---

[g]This is a purely correlation effect. In the self-consistent field theory, both ends would be independently distributed in the volume of order R^3 and $P_N(|R|\sim a)\sim1/R^3$.

exponent g, independent of ν, must be introduced:

$$\varphi(x) \sim |x|^g \quad \text{when} \quad |x| \ll 1. \tag{19.16}$$

Traditionally, one does not use the exponent g but rather the related exponent $\gamma = \nu g + 1$ (*see* subsection 19.8). Regarding numeric values of g and γ, the fluctuation theory allows one to obtain the ε expansion yielding $\gamma \approx 7/6$ for $d=3$. For $d=2$, we obtain $\gamma \approx 4/3$, and for $d=1$, $\gamma = 1$. Accordingly, $g \approx 5/18 \approx 0.28$.

Using Eqs. (19.14) and (19.16), one easily can estimate the probability of contact for the chain ends, that is, the probability of their approach down to a microscopic distance of order a:

$$P_N(|\mathbf{R}| \sim a) \sim R^{-3} \varphi(a/R) \sim a^{-3} N^{-3\nu - \gamma - 1} \sim N^{-1.97}. \tag{19.17}$$

Thus, this probability essentially diminishes as N^{-2} in distinction to $N^{-3/2}$ for the ideal chain. This result is definitely confirmed by experiment.

*19.8. *The free energy of a swollen coil includes an "end" (associated with chain ends) term, being a logarithmic function of the chain length N and defined by the exponent γ.*

Considering the partition function Z and free energy F of a single polymer coil (both for an ideal coil and with excluded volume), we use the lattice model. Because of the universal behavior of polymer coils, this is by no means a restriction on generality.

For the ideal polymer chain, which can be depicted as a random walk over a spatial lattice with coordination number z, the partition function is obviously $Z_N^{(\text{id})} = z^N$, that is, the free energy $F_{\text{id}} = -NT\ln z$. This result can be rewritten as

$$Z_N^{(\text{id})} = \text{const} \cdot \exp(-\lambda_{\text{id}} N), \tag{19.18}$$

where the quantities const and λ_{id} are independent of N. Obviously, Eq. (19.18) follows directly from the independence of individual parts of an ideal macromolecule. Thus, it is valid for any model of a polymer coil.

Now consider a macromolecule with excluded volume. In terms of a lattice model, its partition function Z_N equals the number of self-avoiding trajectories of length N. To estimate this value, we use our knowledge of the distribution $P_N(\mathbf{R})$. Note that $Z_N P_N(a)$ is the number of such self-avoiding trajectories whose ends are located at neighboring lattice sites. This number equals to a factor of order unity the partition function of a ring chain of $N+1$ links[h]

$$Z_N P_N(a) \sim \widetilde{Z}_{N+1}. \tag{19.19}$$

Regarding the partition function $\widetilde{Z}_{N+1}$, this can be evaluated immediately as

$$\widetilde{Z}_{N+1} \sim (a/R)^3 \exp[-\lambda(N+1)]. \tag{19.20}$$

---

[h]The complications associated with effects of topologic constraint (*see* subsection 11.1) are not taken into account here, so the chains are regarded as phantom even though having excluded volume.

Actually, all links in a ring chain are equivalent, and the main factor therefore must be exponential (corresponding to an additive free energy term linear in respect to N and, of course, with $\lambda \neq \lambda_{id}$). The pre-exponential factor in Eq. (19.20) signifies that the chain ends must approach one another within the volume a^3 instead of the conventional R^3 [cf. Eq. (17.6)] to form a ring. Using Eqs. (19.19), (19.20), and the relation for P_N (19.17), we obtain

$$Z_N \sim N^{\gamma-1} \exp(-\lambda N). \tag{19.21}$$

From the analysis of the estimates made, one can conclude that the power factor $N^{\gamma-1}$ in Eq. (19.21), corresponding to the logarithmic contribution to the free energy, appears because of the end effect, that is, to the fact that the chain ends and their neighboring links are not surrounded with such a thick "cloud" of neighboring links as the internal links are).

Thus, the critical exponent ν describes the typical links, whereas the exponent γ describes the end chain links. All remaining exponents [e.g., g in Eq. (19.16), or δ in Eq. (19.15)] can be expressed via ν and γ.

20. CONDENSED GLOBULAR STATE OF A LONG LINEAR POLYMER CHAIN

20.1. *In a poor solvent, a polymer chain is compressed and evolves to a globular state; globular states are widespread in nature.*

Previous sections considered the volume interactions in good solvent, where repulsive forces act primarily between the links and the macromolecule conforms to a loose fluctuating coil. We now move to a study of the behavior of polymer chains in poor solvent, where attraction plays an essential role in the volume interactions of links.

As noted in subsection 13.3, a polymer coil is compressed in poor solvent; one can say that the polymer chain has "to condense on itself." We now show that such a compression or condensation results in the transformation of the polymer coil into the polymer globule.[i] The coil–globule transition is similar to the phase transition from gas to a condensed state.

An interest in the globules and globule–coil transitions emerged initially in molecular biophysics, when in the 1960s it became clear that the protein

---

[i]That the transition from a good solvent to a poor solvent brings about a cardinal change in the state of a macromolecule can be clarified using the results of investigation of the renormalization group (*see* Sec. 18). We have seen in subsection 18.4 that in the case of poor solvent, when the second virial coefficient of initial monomer interaction is negative ($\beta_0 < 0$), the value of $\beta(s)$, that is, the effective second virial coefficient of interaction of block monomers of length s tends to $-\infty$ as the length scale s grows. In physical terms, this means that sufficiently long blocks of the chain attract one another very strongly and in fact stick together, which implies the transformation of the coil into the globule. In reality, however, the third virial coefficient becomes substantial along with the second as renormalization is performed in the globular state. To investigate the globule, one must introduce a system of two equations of type (18.8), specifying the renormalization of the two indicated quantities. From the formal standpoint, such an approach discloses the analogy between the θ point and the so-called tricritical point.[28] Later, we will study the globular state by simpler methods.

enzymes function in a live cell in the state of dense globules. Specifically, protein denaturation, being an abrupt cooperative transformation occurring at a change of temperature or solvent composition, associated with a conspicuous thermal effect and resulting in a loss of biochemical activity, was at first reasonably interpreted as an unfolding of the protein chain, (i.e., globule–coil transition).

Later, it became clear that one cannot establish a single-valued correspondence between the denaturation of globular protein and the globule–coil transition (*see* Sec. 44). The globular states, however, are very frequent not only in proteins but in other diverse polymer systems, ranging from DNA (*see* subsection 43.1) to macroscopic polymer networks (*see* subsections 29.9 and 30.10).

The contemporary comprehensive appraisal of the role of the globular state in various polymer systems goes back to the pioneer publication by Lifshitz.[30] Following those ideas, we consider in this section the simplest fundamental example of a globule, that is, the condensed state of a long bead chain. (The globule–coil transition for this model is treated in the next section, and many other examples of globules and transitions of the coil–globule type are discussed later.)

20.2. *A large polymer globule consists of a dense homogeneous nucleus and a relatively thin surface layer, a fringe; in equilibrium, the size of the globule settles at such a value that the osmotic pressure of the polymer in the globule nucleus equals zero.*

It is obvious that in a strongly compressed polymer chain, each link accounts for the volume having the order of the proper link volume. According to the estimates made in subsection 13.3, the size of an N-link macromolecule in moderately poor solvent is also proportional to $R \sim N^{1/3}$ [*see* Eq. (13.9)], that is, the volume share of a single link is of order R^3/N and independent of N. It therefore is natural to assume that the equilibrium density of packed links settles throughout all parts of the system independently during compression of the polymer chain (*see* subsection 24.3). This means that the compressed polymer chain represents a globule. Recall that by the definition given in subsection 7.2, the state of a macromolecule in which the correlation radius is much less than the size of the system is called globular; in other words, the fluctuations in a globule (in contrast to a coil) are of a local nature and do not permeate the globule as a whole.

To discuss this in more detail, let us consider a small volume element separated in the globule (Fig. 3.7). In the coil, such an element most probably would contain (according to the estimates of the collision probability made in subsection 13.4) only one chain section, or even no polymer substance at all. Conversely, in the compressed macromolecule, the separated element appears from a local point of view as a system (solution) of many independent chains, representing, in fact, different portions of one macromolecule.[j] The osmotic pressure in such a solution equals $p^*(n)$ [*see* Eq. (15.10)], where n is the link

---

[j]A specific expression for the correlation radius (together with a formal proof of the globular state) is derived in subsection 24.3.

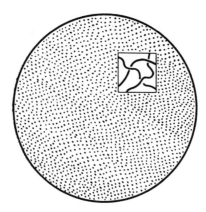

FIGURE 3.7. Elementary volume in a globule appears similar to the region of a solution of independent chains.

concentration in the separated volume element.

It should be reminded that the osmotic pressure of the system of long chains [see Eq. (24.2)] $p^*(n)$ is associated with the pressure of the system of long chains $p(n)$ by the relation

$$p^*(n)=p(n)-nT; \qquad (20.1)$$

Because the polymer links lack the freedom of independent translations, the ideal-gas term nT [cf. Eq. (15.8)] must be subtracted from the polymer pressure.

Further, because the volume elements of the system must be in mechanical equilibrium with one another, the osmotic pressure in the polymer as well as the link concentration must be uniform within the globule: $n(x)=n_0=$ const, at least if the effects of the surface layer of the globule are neglected (see subsection 20.4). Moreover, as the globule coexists with the surrounding solution containing no polymer, the osmotic pressure of the polymer in the globule must equal zero:

$$p^*(n_0,T)=0. \qquad (20.2)$$

This condition specifies the link concentration n_0 in the equilibrium globule. The equilibrium volume of the globule equals $V=N/n_0$, and its size is of order $R\sim V^{1/3}=(N/n_0)^{1/3}\sim N^{1/3}$ [cf. Eq. (13.9)].

The physical meaning of the condition (20.2) is very simple: the link concentration in the globule must settle at such a value that the attractive and repulsive volume interactions counterbalance each other on the corresponding characteristic scale $\sim n_0^{-1/3}$. The globule thus differs from a swollen coil, where the volume repulsion of the links is counterbalanced by the chain bonds.

20.3. *The conformational entropy of a polymer chain in a big globule differs from zero only within the surface layer; the volume approximation in the theory of a globule is based on the neglect of the corresponding surface free energy.*

The discussion presented in the previous subsection leads to the following conclusions:

1. The link concentration distribution $n(x)$ in a large globule has the form shown in Figure 3.8.

2. The fluctuations in a globule are insignificant.

The second conclusion points to the applicability of the self-consistent field theory to the study of the globule (*see* Sec. 15). This allows an easy derivation of the free energy of the globule complying with the equilibrium distribution $n(x)$ of the local link concentration (Fig. 3.8).

We begin with an evaluation of the conformational entropy using the Lifshitz formula (9.3). The local concentration gradient differs from zero only in the surface layer of the globule. Denoting the thickness of this layer by Δ (Fig. 3.8), we can evaluate both its volume as $\sim (N/n_0)^{2/3}\Delta$, where $(N/n_0)^{2/3}$ is the surface area of the globule, and the concentration gradient in the layer as $\sim n_0/\Delta$. Hence,

$$S = \frac{a^2}{6} \int n^{1/2}(x)\Delta n^{1/2}(x)d^3x \sim -a^2\left(\frac{N}{n_0}\right)^{2/3}\Delta\left(\frac{n_0}{\Delta^2}\right) \sim -N^{2/3}a^2 n_0^{1/3}/\Delta.$$

(20.3)

How should one interpret the fact that the entropy of the globule is proportional to $N^{2/3}$ (i.e., to the surface area) while the volume term of the entropy equals zero? We stress that this entropy is much larger than $N^{1/3}$, predicted by the simplified Flory theory (Eq. 13.3). It is to be recalled that the Lifshitz formula (9.1) implies that the entropy is measured relative to the level of the free chain entropy. Thus, the equality to zero of the volume entropy means that the greater fraction of the links in the globule chain, to wit, those links that are located in the homogeneous nucleus of the globule, possess a free (unperturbed) set of permissible conformations. In other words, the chain sections inside the globule nucleus still obey Gaussian statistics. Only in regions where the chain

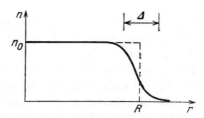

FIGURE 3.8. Distribution of local link concentration in a large polymer globule.

comes to the globule surface is the conformation set of links constrained, because these sections necessarily have the form of loops. In subsection 24.2, we discuss in detail why the statistics of chains or chain segments remain Gaussian when they are homogeneously surrounded by similar chains.

This discussion makes clear that the obvious simple approximation needed to describe the large globule must be associated with the volume contribution to the free energy and the neglect of the surface contribution. In this so-called volume approximation, we totally disregard the conformational entropy; therefore, according to Eq. (15.7),

$$F\{n\}=E\{n\}-TS\{n\}\cong E\{n\}=\int f^*(n(x))d^3x\cong Vf^*(n_0)=Nf^*(n_0)/n_0.$$
$$(20.4)$$

It can easily be shown by direct calculations that the equilibrium link concentration in the globule nucleus, found from the condition $p^*(n_0)=0$ [see Eq. (20.2)], conforms to the minimum of the free energy (20.4):

$$\frac{\partial}{\partial n_0}\left(\frac{f^*(n_0)}{n_0}\right)=\frac{n_0\mu^*(n_0)-f^*(n_0)}{n_0^2}=\frac{p^*(n_0)}{n_0^2}=0. \qquad (20.5)$$

The equilibrium (and also minimum) value of the free energy of the globule is equal in the volume approximation to

$$\mathscr{F}_{\text{vol}}=N\mu^*(n_0). \qquad (20.6)$$

Analyzing the derivation of Eqs. (20.2) and (20.6), one can recognize that in the volume approximation (i.e., for a sufficiently long chain), the equations are valid for any macromolecule and not only for the standard bead model. It is also worth paying attention to the convenient graphic interpretation of these results. To do this, rewrite Eq. (20.2) $p^*(n_0)=0$ in the form

$$0=-p^*(n_0)=f^*(n_0)-n_0\mu^*(n_0)=\int_0^{n_0}[\mu^*(n)-\mu^*(n_0)]dn. \quad (20.7)$$

The geometric meaning of this condition is a coincidence of the shaded areas in Fig. 3.9a. One can see that the equilibrium density n_0 of the globule and the volume free energy $\mu^*(n_0)$ per link are easily found from this illustration.

20.4. To investigate the structure of the surface layer of a large globule and its entropic surface tension, a full system of self-consistent field equations must be formulated.

In accordance with the general principle given in Sec. 15, the equilibrium distribution $n(x)$ of link concentration or the ψ function of the terminal link distribution defined by the relation

$$\Lambda n(x)=\psi(x)\hat{g}\psi \qquad (20.8)$$

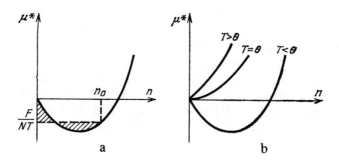

FIGURE 3.9. Interpretation of the conditions defining the density (20.2) and free energy (20.6) of a large globule. (a), Equality of shaded areas. (b), Variation of $\mu^*(n)$ plot with temperature.

[*see* Eq. (9.2)] are determined by minimizing the free energy. According to Eq. (9.1), for the conformation entropy $S\{n\}$ and Eq. (15.7) for $E\{n\}$, free energy can be expressed in the form

$$F\{n\} = \int \{f^*(n(x)) - Tn(x)\ln(\hat{g}\psi/\psi)\}d^3x. \tag{20.9}$$

The minimization must take into account that the number of the links in the chain is constant, that is, the normalization condition

$$\int n(x)d^3x = N, \tag{20.10}$$

should be satisfied. Let us now introduce the indefinite Lagrangian factor λ and write the minimization condition for the free energy (20.9) under the additional condition (20.10) as

$$\delta\left(F\{n\} - \lambda \int n(x)d^3x\right) = 0.$$

By variation,[k] we obtain

$$\mu^*(n(x)) - T \ln(\hat{g}\psi/\psi(x)) - \lambda = 0. \tag{20.11}$$

---

[k]The entropy variation, that is, the variation of the second term in Eq. (20.9), is performed as

$$\delta S = \int \delta n(x)\ln(\hat{g}\psi/\psi)d^3x + \int n(x)\delta \ln(\hat{g}\psi/\psi)d^3x.$$

In the second integral, $n(x)$ is inserted in the form of Eq. (20.8).

$$\wedge \int n(x)\delta\ln\frac{g\psi}{\psi}d^3x = \int \psi\hat{g}\psi\left[\frac{\delta\hat{g}\psi}{\hat{g}\psi} - \frac{\delta\psi}{\psi}\right]d^3x = \int \{\psi\cdot\delta\hat{g}\psi - \hat{g}\psi\delta\psi\}d^3x,$$

This value equals zero, because the operator $\hat{g}$ is, first, linear, that is, $\delta\hat{g}\psi = \hat{g}\delta\psi$, and, second, Hermitian, that is, $\int \psi\hat{g}\delta\psi d^3x = \int \hat{g}\psi \cdot \delta\psi d^3x$. Finally, we obtain

$$\delta S = \int \delta n(x)\ln(\hat{g}\psi/\psi)d^3x.$$

For convenience, we choose the value of Λ featured in Eq. (20.8), which defines only the normalization of the ψ function [note that in Eqs. (20.9) to (20.11), the normalization has no effects whatever] equal to

$$\Lambda = \exp(-\lambda/T), \quad \lambda = -T \ln \Lambda.$$

Then, Eq. (20.11) is rewritten as

$$\exp(-\mu^*[n(x)]/T)\hat{g}\psi = \Lambda\psi(x). \tag{20.12}$$

Together with Eq. (20.8), this equation defines the equilibrium distribution of link concentration in the globule. This is a fundamental equation of the theory of globules.

Comparing Eq. (20.12) with Eq. (7.4), specifying the globular-state structure of the ideal chain in an external field, we see that the chemical potential $\mu^*(n(x))$ of the link acts as a self-consistent field in exact conformity with Eq. (15.18):

$$\mu^*(n(x)) = \varphi_{\text{self}}(x). \tag{20.13}$$

Equations (20.12) and (20.8) form the system to determine the two unknown functions $n(x)$ and $\psi(x)$ [together with Eq. (20.10), they define also the unknown number Λ.] It is convenient to eliminate one of the functions, namely, $n(x)$. This can be done by introducing the parametrically defined function $v(\psi)$ as

$$v \equiv \exp(\mu(n)/2T), \quad v = n/\psi. \tag{20.14}$$

The assignment of the dependence $v(\psi, T)$ is totally equivalent to the assignment of an equation of state for a system of links in any other form, for example, in the form of dependences $\mu^*(n,T)$, $p^*(n,T)$, and so on (see subsection 15.3). Indeed, knowing $v(\psi)$ one can easily find, for example, μ^*:

$$\mu^*(n) = T \ln(v/\psi), \quad n = \psi v,$$

or p^*:

$$p^*(n) = T\psi v(\psi) - 2T \int_0^\psi v(\psi) d\psi. \tag{20.15}$$

If the dependence $v(\psi)$ is assumed to be known, the basic equation (20.12) can be turned into the seemingly simple form

$$\hat{g}\psi = \Lambda v(\psi), \quad N = \int \psi(x)v(\psi(x))d^3x \tag{20.16}$$

because $\mu^* = \mu - T \ln n$.

Before proceeding to study Eq. (20.16), we derive the expression for the equilibrium free energy. In the self-consistent field approximation, the free energy $\mathscr{F}$ equals the minimum value of the functional $F\{n\}$ conforming to the equilibrium distribution $n(x)$, which is the solution of Eq. (20.12) or Eq. (20.16). Substituting the value of $T \ln(\hat{g}\psi/\psi)$ from Eq. (20.12) into Eq. (20.9), we obtain

$$\mathscr{F} = \int \{ f^*(n(x)) - n(x)\mu^*(n(x)) + \lambda n(x) \} d^3x = -TN \ln \Lambda - \int p^*(n(x)) d^3x$$
(20.17)

[cf. the similar formula (6.17) for the ideal chain where $p^*=0$].

Let us now show how the result of the volume approximation (20.6) follows from the general formulas. If the value of ψ is almost constant throughout a large volume whose dimensions are much greater than a, then $\hat{g}\psi \cong \psi$, because the operator $\hat{g}$-averages, roughly speaking, over the sphere with radius a [see Eq. (6.22)]. Therefore, the basic equation (20.12) yields $\mu^*(n_0) = \lambda$, and because $p^*(n_0) = 0$, Eq. (20.17) turns into Eq. (20.6).

*20.5. The entropic surface tension of the globule is determined by a distribution of link concentration within its surface layer.

Because the pressure $p^*(n(x))$ equals zero outside the globule as well as inside its nucleus, that is, the value of $p^*(n(x))$ differs from zero only within the surface layer of the globule, the second (integral) term on the right-hand side of Eq. (20.17) is proportional to the surface area. On the other hand, as $-T \ln \Lambda$ in the first term is as we have shown, equal to $\mu^*(n_0)$, the free energy (20.17) can be rewritten as

$$\mathscr{F} = N\mu^*(n_0, T) + 4\pi R^2 \sigma.$$
(20.18)

Here, the radius R of the globule, its volume V and surface area $4\pi R^2$ are defined by the obvious relations

$$V = N/n_0, \quad (4/3)\pi R^3 = V, \quad 4\pi R^2 = (36\pi)^{1/3}(N/n_0)^{2/3}, \quad (20.19)$$

Taking into account that the problem of the surface of the large globule is actually one-dimensional, the surface tension σ of the globule can be derived from Eq. (20.17):

$$\sigma = - \int_{-\infty}^{+\infty} p^*(n(z)) dz,$$
(20.20)

where z is the coordinate normal to the surface. Also, the integration with respect to z can be extended from $-\infty$ to $+\infty$ because p^* decays exponentially as $z \to \pm\infty$ [see Eq. (20.23)].

Replacing the integral operator $\hat{g}$ in Eq. (20.16) by the differential one according to Eq. (7.7),

$$\hat{g} \cong 1 + (a^2/6)\Delta \cong 1 + (a^2/6) d^2/dz^2,$$

we obtain

$$(a^2/6)d^2\psi/dz^2 = \Lambda v(\psi) - \psi. \tag{20.21}$$

This equation is easily integrated, because it features no explicit argument z, thereby permitting a lowering of the order. The result takes the form

$$z/a = -6^{-1/2} \int_{\psi(0)}^{\psi(z)} \left[2 \int_0^\psi \{\Lambda v(\psi) - \psi\} d\psi \right]^{-1/2} d\psi. \tag{20.22}$$

It can easily be demonstrated that the equality (20.22) actually results in a simple profile of the local link concentration of the type shown in Figure 3.8. In particular, for $z \to \pm \infty$, we obtain the simple exponential asymptotic forms

$$\psi(z)|_{z \to +\infty} \cong \text{const} \cdot \exp(-\vartheta_+ z/a), \quad \vartheta_+ = (\Lambda - 1)^{1/2},$$

$$\psi(z)|_{z \to -\infty} \cong \psi_0 - \text{const} \cdot \exp(\vartheta_- z/a), \quad \vartheta_- = [\Lambda(\partial v/\partial \psi)_{\psi_0} - 1]^{1/2}. \tag{20.23}$$

The first of these asymptotic forms complies with the fact that for great values of $|x|$, the local link concentration is very low, and $\mu^*(n) \approx 0$. Therefore, the indicated asymptotic provides the solution of Eq. (20.21) written in the following simplified form

$$(a^2/6)d^2\psi/dz^2 = (\Lambda - 1)\psi. \tag{20.24}$$

Note also that according to the relation (20.23), the thickness of the surface layer Δ of the globule [see Eq. (3.8)] is of the order

$$\Delta \sim a/(\Lambda - 1)^{1/2}. \tag{20.25}$$

Proceeding from the obtained solution (20.22) of the problem of the surface structure and using Eq. (20.15) for p^*, one can easily calculate the surface tension σ of the globule [see Eq. (20.20)]:

$$\sigma = \frac{aT}{6^{1/2}} \int_0^{\psi_0} \frac{2\int_0^\psi v(\psi)d\psi - v\psi}{[2(\psi_0/v_0)\int_0^\psi v(\psi)d\psi - \psi^2]^{1/2}} d\psi, \tag{20.26}$$

where ψ_0 and v_0 are the values of the corresponding functions within the core of the globule. The result (20.26) explicitly expresses σ via the thermodynamic functions of the system of disconnected links.

20.6. General formulas from the theory of a large globule become simpler in the special case of a low-density globule.

We show later that in the vicinity of the θ point, (i.e., near the globule–coil phase transition), the link concentration in the globule is low. In this case, the virial expansions (15.13) are valid for the functions f^*, μ^*, and p^*. Specifically, the condition (20.2) takes the form

$$p^*(n_0) \cong T Bn_0^2 + 2CTn_0^3 = 0;$$

$$n_0 = -B/2C \quad \text{or} \quad V = -2NC/B \qquad (20.27)$$

(It should be recalled that $B < 0$ in poor solvent). It follows from Eq. (20.27) that the value of n_0 is really small near the θ temperature, so the virial expansion can be used to analyze the globule structure. The result (20.27) also shows that the equilibrium volume of the macromolecule in the globular state near the θ temperature settles at such a value that the attraction of links in binary contacts cancels the repulsion in triple collisions. Next, the volume free energy (20.6) and the number Λ equal

$$\mathscr{F}_{\text{vol}} = -NTB^2/4C, \qquad (20.28)$$

$$\Lambda = \exp(-\mu^*(n_0)/T) \cong 1 + B^2/4C. \qquad (20.29)$$

Finally, the thickness Δ of the surface "fringe" [see Eq. (20.25)], surface tension σ (20.20), and surface free energy $\mathscr{F}_{\text{surf}}$ are

$$\Delta \sim aC^{1/2}/|B|, \qquad (20.30)$$

$$\sigma \sim aB^2T/C^{3/2}, \qquad (20.31)$$

$$\mathscr{F}_{\text{surf}} \sim N^{2/3}Ta(-B)^{4/3}/C^{5/6}. \qquad (20.32)$$

Because the derivation of Eq. (20.26) for the surface tension of the globule involves rather unwieldy calculations, it is worth mentioning that this value can be estimated very easily from the expression for the entropy: as $\mathscr{F}_{\text{surf}} \sim -TS$, then substituting n_0 (20.27) and Δ (20.30) into Eq. (20.3) to obtain Eq. (20.32).

21. GLOBULE-COIL PHASE TRANSITION

21.1. *On approach to the θ point from poor solvent, a globule gradually swells, its size becoming closer to that of a coil, as it should be on approach to the second-order phase transition point.*

Previous sections examined both the gas-like coil and condensed globular states of the macromolecule, and we saw their fundamental qualitative difference. Now we consider how a coil-globule conformational transition between these two states proceeds as the external conditions vary. Subsection 20.1 emphasized the significant role that the coil-globule transitions play; many examples given later further clarify that role.

Consider the problem using the simplest volume approximation for the globule, and take advantage of the graphic interpretation of the results of the volume approximation (20.2) and (20.6) shown by the curve $\mu^*(n)$ in Figure 3.9. Figure 3.9b illustrates the simplest and most typical variation of the dependence $\mu^*(n)$ with temperature. As temperature decreases, the second virial coefficient, defining the slope of the plot $\mu^*(n)$ at $n=0$, becomes negative before all other coefficients. In other words, with the temperature decreasing, attraction begins to prevail over repulsion first in binary collisions. Specifically, such a situation is described by the Van der Waals and Flory-Huggins equations [see Eqs. (15.11) and (15.12)].

From the comparison of Figure 3.9a and b, it can be seen that the volume contribution of free energy of the globule is negative, $\mathscr{F}_{\text{vol}} < 0$, at a temperature below θ. However, we measured the free energy of the globule relative to the energy level of the Gaussian coil; therefore, $\mathscr{F} < 0$ is the condition of the thermodynamic advantage of the globular state. Consequently, we find in terms of the volume approximation that the globular state is at equilibrium for $T < \theta$ while the globule–coil transition temperature $T_{\text{tr}} = \theta$.

On approach to the θ point from poor solvent, the link concentration n_0 in the globule lowers (Fig. 3.9), making the globule volume grow (i.e., the globule swells substantially before turning into a coil). This makes possible use of the virial expansion and as shown in subsection 20.6, gives a universal description of the globule in terms of the second B and third C virial coefficients of link interaction.

Recall that near the θ point

$$B(T) \cong b\tau, \quad \tau = (T - \theta)/\theta,$$

$$C(T) \cong C = \text{const} > 0.$$

Hence, according to Eqs. (20.27) and (20.28), we obtain

$$n_0 \cong |\tau| b/2C \quad \text{or} \quad R \sim N^{1/3} |\tau|^{-1/3} (C/b)^{1/3}, \tag{21.1}$$

$$\mathscr{F}_{\text{vol}} \cong -N\tau^2 \theta b^2/4C \sim -N\tau^2 \theta. \tag{21.2}$$

Were the latter result valid up to the θ-point, it would imply that the globule–coil transition is a second-order phase transition.

In fact, on approach to the θ-point, the thickness of the surface layer of the globule grows [see Eq. (20.30)], and the volume approximation needs corrections in this region.

21.2. A globule–coil transition point lies in the θ-region within an interval of order $N^{-1/2}$ below the θ-point, and it is determined by the balance between entropy gain caused by chain extension and loss in energy of volume link attraction.

The volume approximation result of the globule–coil transformation at the θ-point means that as the conditions transform poor into good solvents, the energy gain because as coil compression disappears at the θ-point. In reality, however, the globule must lose its stability and turn into the coil earlier (i.e., when the energy gain from coil collapse still exists but is insufficient to offset the entropy loss). The entropic (surface) effect is, of course, small (see subsection 20.3), but it becomes essential near the θ-point as the energy contribution is also small.

Indeed, the total free energy of the globule can be written according to Eqs. (20.28) and (20.32) in the form[1]

$$\mathscr{F} = \mathscr{F}_{vol} + \mathscr{F}_{surf} = -N\theta\tau^2(b^2/4C)[1 - |\tau_{tr}/\tau|^{2/3}], \qquad (21.3)$$

where

$$\tau_{tr} \equiv (T_{tr} - \theta)/\theta \cong -2.7a^{3/2}C^{1/4}/(bN^{1/2}). \qquad (21.4)$$

The value of T_{tr} (or τ_{tr}) corresponds to the globule–coil transition point, because for $T < T_{tr}$ (or $|\tau| > |\tau_{tr}|$) $\mathscr{F} < 0$, that is, the globule is stable, whereas at $T = T_{tr}$ (or $\tau = \tau_{tr}$), $\mathscr{F} = 0$.

Thus, the deviation of the globule–coil transition point from the authentic θ point, proportional to $N^{-1/2}$, is of the same order of magnitude as the difference of the apparent θ points (*see* subsection 14.4), that is, the transition point lies in the θ region. $T_{tr} \rightarrow \theta$, as expected, as $N \rightarrow \infty$. In practice, a typical deviation of the temperature T_{tr} from θ is of order 10 K.

21.3. The width of the globule–coil transition in an N-link chain is proportional to $N^{-1/2}$ and tends to zero as $N \rightarrow \infty$; therefore, this transition can be treated as a phase transition.

A key role in the coil-globule transition is played by the fact that the number of links N in a real macromolecule, even though large, is by no means infinite, and it is even very small when compared with the number of particles in typical thermodynamic systems. Consequently, the system may pass into a thermodynamically unfavorable state near the transition point by way of thermal activation. The probability of this process differs noticeably from zero, provided that the difference of the free energies of the two states per a whole macromolecule of N links is of the order of the temperature. Clearly, if this is the case, then it is meaningless to talk about the system being in one definite state. The regions of stable existence of definite states are thus separated by the finite interval ΔT, whose width is determined by the conditions

$$\mathscr{F}(T_{tr}) = 0, \quad |\mathscr{F}(T_{tr} - \Delta T)| \sim T_{tr}. \qquad (21.5)$$

From the expression for the free energy (21.3), it is easy to derive

$$\Delta T \sim (\theta - T_{tr})C^{1/2}/a^3 \sim \theta C^{3/4}/(a^{3/2}vN^{1/2}). \qquad (21.6)$$

In conventional macroscopic systems, the width of the phase transition interval is so negligible that the real picture corresponds to the limit $N \rightarrow \infty$, that is, the non-analytic behavior of the thermodynamic potential is observed, such as discontinuities of its derivatives, for example, entropy, heat capacity, and so on. The globule–coil transition in the macromolecule for $N \rightarrow \infty$ also shows an

---

[1]Here, the numeric coefficients cannot be found from our discourse, because the problem of the surface structure stops being one-dimensional as the globule spreads. Also, Eqs. (20.20) to (20.26) become correct only by the order of magnitude. The coefficients in Eqs. (21.3) and (21.4) are found from a numeric solution of the three-dimensional equation (20.12).

abrupt behavior, because as $N \to \infty$, $\Delta T \to 0$. Therefore, it is natural to consider this transition as a phase transition. For real numbers N however, the finite width of the transition (usually amounting to approximately 1 K) is very significant. Recall that the analogous situation occurs in the ideal chain placed in an external field and undergoing the coil-globule transition [see Eq. (7.32)].

It should be pointed out that there are conformational transitions of a different kind that are not phase transitions. These occur within a narrow interval, but its width is independent of N even as $N \to \infty$. An example is a helix–coil transition (see Sec. 40).

21.4. The character of a globule–coil transition essentially depends on the chain stiffness: for stiff chains, the transition is very sharp and close to a first-order phase transition; for flexible chains, it is more smooth and is a second-order phase transition.

To discuss the obtained results, it is useful to recall the estimations for the virial coefficients B and C. Obviously, for the standard model of a polymer chain with spheric beads

$$B = b\tau \sim v\tau \ (b \sim v), \quad \tau = (T - \theta)/\theta, \quad C \sim v^2 \tag{21.7}$$

Accordingly, Eq. (21.4) for the transition point and Eq. (21.6) for the transition interval width may be rewritten in the form

$$\tau_{\text{tr}} \sim -(a^3/Nv)^{1/2}, \tag{21.8}$$

$$\Delta T \sim \theta(v/Na^3)^{1/2} \sim \theta|\tau_{\text{tr}}|(v/a^3). \tag{21.9}$$

The situation essentially depends on the characteristic dimensionless parameter v/a^3, in complete agreement with what could be expected on the basis of the qualitative considerations of subsection 13.3.

Recall that if the standard bead model is used for the description of a real polymer chain, then the characteristic parameter v/a^3 is determined by the stiffness of the polymer. Indeed,

$$v/a^3 \sim p^{-3/2}$$

[see Eq. (13.17)], where p is the ratio of the Kuhn segment to the thickness of the macromolecule, $p = l/d$.

Comparing Eqs. (21.8) and (21.9), we see that when $v/a^3 \ll 1$ (i.e., for stiff chains), $\Delta T/\theta \ll |\tau_{\text{tr}}|$. This means that the globule–coil transition in a stiff polymer chain proceeds in a relatively narrow temperature interval, clearly separated and substantially removed from the θ-point. Conversely, in a flexible chain where $\Delta T/\theta \sim |\tau_{\text{tr}}|$ because $v/a^3 \sim 1$, the globule–coil transition proceeds relatively smoothly, and the transition interval includes the θ-point. Both situations are shown in Figure 3.10.

As will be shown, this gives evidence for the difference in the order of the coil-globule phase transition in stiff and flexible chains. Indeed, the process of

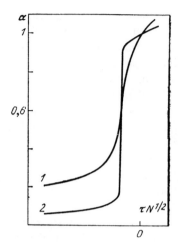

FIGURE 3.10. Dependence of the swelling parameter of a macromolecule on reduced temperature in the region of the globule–coil transition for very flexible chains, $v/a^3 = 0.2$ (*1*), and very stiff chains, $v/a^3 = 0.01$ (*2*)

gradual swelling of the globule and the temperature dependence $\mathscr{F} \sim -\tau^2$ associated with that process are distinctive for a second-order phase transition and occur only in the far vicinity of the transition point when already $|\tau| \ll 1$ but still $|\tau| \gg |\tau_{tr}|$. Of course, $\tau_{tr} \to 0$ for $N \to \infty$; therefore, the globule–coil transition is formally a second-order phase transition. This is because in the limit $N \to \infty$, it exhibits a discontinuity of the second derivative of the free energy ($\mathscr{F} \sim -\tau^2$ for $\tau < 0$ and $\mathscr{F} = 0$ for $\tau > 0$). In real chains, however, with finite values of N, the transition is observed for finite τ_{tr}. On approach to this point from the side of the globular phase, the globule near the temperature T_{tr} turns out to be (because of a pretransitional swelling) a rather loose system. Its density in terms of its dependence on N falls to that of a coil $\sim N^{-1/2}$:

$$n_0 \big|_{T=T_{tr}} \sim -(B/C)\big|_{T=T_{tr}} \sim -\tau_{tr}/v \sim (v/a^3)^{-3/2}(1/a^3 N^{1/2}). \quad (21.10)$$

If one approaches from the side of the coil phase, then the parameter z of volume interactions in the coil [*see* Eq. (13.12)] at $T = T_{tr}$ turns out to be

$$z\big|_{T=T_{tr}} \sim N^{1/2}(B/a^3)\big|_{T=T_{tr}} \sim N^{1/2} v \tau_{tr}/a^3 \sim -(v/a^3)^{1/2}. \quad (21.11)$$

Thus, in the immediate vicinity of the point T_{tr}, the situation strongly depends on the chain stiffness, that is, on the parameter v/a^3.

For a flexible chain $v/a^3 \sim 1$ and at the transition point, the volume interaction parameter (21.11) in the coil is of order unity, that is, the coil is noticeably non-ideal [appreciably compressed because $z < 0$; *see* Eq. (14.6)], whereas the globule density (21.10) is of the same order as the coil density. Consequently, the difference between the coil and globular states almost disappears in this case, that is, the transition essentially proceeds continuously and is naturally identified with a second-order phase transition.

Things are different for a stiff chain for $v/a^3 \ll 1$. Here, $z \ll 1$ [see Eq. (21.11)] at the transition point, that is, the coil is essentially ideal and the density of the globule (21.10) much lower than the density of the coil. Consequently, the coil and the globule are two essentially different states in this case. Specifically, the transition between them is accompanied by a substantial jump in the size or swelling parameter of the macromolecule:

$$R_{glob}/R_{coil} \sim \alpha_{glob}/\alpha_{coil} \sim (v/a^3)^{1/2}$$

because the link concentration inside the globule and the coil differ according to Eq. (21.10) by the factor $(v/a^3)^{3/2}$. It is easy to see that these two different states of the macromolecule provide two minima in the free energy; as the transition occurs, the probability of finding the system is "pumped" from one minimum to the other.

Theoretically, metastable states are also viable. Such a situation is typical for a first-order phase transition. Hence, the behavior of the stiff macromolecule in the immediate vicinity of a globule–coil transition is similar to that of the system near a first-order transition. This is true, as seen from Eq. (21.3), in such a vicinity of the transition point that $|\tau - \tau_{tr}| \ll |\tau_{tr}|$; in this region, $\mathscr{F} \sim \tau - \tau_{tr}$. Of course, however, it makes sense only outside the transition interval ΔT, that is, at $|\tau - \tau_{tr}|\theta = |T - T_{tr}| \gg \Delta T$. For a stiff chain at $v/a^3 \ll 1$, these inequalities are compatible, because according to Eq. (21.9), $|\Delta T/(\theta \tau_{tr})| \sim v/a^3$.

Thus, the globule–coil phase transition in the stiff macromolecule, or in the standard model of the polymer chain with $v/a^3 \ll 1$, is a first-order transition, while in the flexible macromolecule, or for $v/a^3 \sim 1$, it is a second-order transition. In both cases, the transition is preceded by a substantial swelling of the macromolecule in the globular state.

Finally, one general conclusion from the presented theory can be drawn. (This conclusion could also have been drawn earlier from the simple evaluations of subsection 13.3). In the globule–coil transition region, the swelling parameter of the macromolecule always depends on the temperature τ and the chain length N as

$$\alpha = \varphi(\tau N^{1/2}). \tag{21.12}$$

This expression defines a sort of law of corresponding states: the dependences $\alpha(\tau)$ for macromolecules of the same chemical nature and differing only by their chain lengths are distinguished only by the scale of the axis τ.

The expression of the function φ (21.12) is determined by the stiffness of the polymer chain. For all real macromolecules, however, as mentioned in subsection 13.9, the parameter v/a^3 is fairly small (<0.2), so the globule–coil transition proceeds almost always in a rather abrupt way (Fig. 3.10).

22. MORE COMPLEX GLOBULAR STRUCTURES AND RELATED PHASE TRANSITIONS

22.1. *A nucleus of a dense polymer globule may be analogous in its local structure not only to a liquid but to any other condensed phase.*

When discussing the globules in previous sections, we implied the condensed state of the standard bead model of a polymer chain. The volume interactions were assumed such that the system of disconnected links obeyed the simplest equation of state of the Van der Waals (15.11) or Flory-Huggins type (15.12), conforming to the dependence $\mu^*(n)$ similar to the one illustrated in Figure 3.9. In such a system of disconnected links (interacting point beads), in addition to the gas, only one condensed state is possible: liquid. Using either of the two equations of state (15.11) or (15.12), it is easy to demonstrate that the coil-globule transition temperature [*see* Eq. (21.5)], which is close to the θ-point for a long chain, is much higher than the critical temperature of the gas–liquid phase transition.[m] In other words, the merging of links into a chain favors condensation. In physical terms, this is quite natural, because the interlink bonds drastically reduce the freedom of independent link motion. Therefore, the entropy gain appearing on transition from the globule to the coil is incomparably less than that appearing on evaporation (or more exactly, on dissolving) of a condensed drop consisting of disconnected links and having the same volume. The energy loss, however, is the same by the order of magnitude for both indicated processes.

Actual volume interactions also may obey more sophisticated state equations for a system of disconnected links. Most significantly, apart from gas and liquid, other diverse condensed states may appear in the system. For example, one can picture a liquid-crystal globule (a so-called intramolecular liquid crystal), whose volume elements consist of chain sections with orientational ordering even though the globule as a whole may be isotropic. Also possible are globules having the structure of an ordinary or plastic crystal, quasiequilibrium globules with the structure of glass or amorphous solids, and so on. It also is obvious that phase transitions between different globular states (globule–globule transitions) as well as between any globular and coil states (globule–coil transitions) may exist.

Generally speaking, the study of each of the mentioned basic possibilities constitutes a self-contained, labor-consuming problem. In general form, the investigation can be performed only in the volume approximation in which the equilibrium density n_0 of the globule nucleus (as well as its volume $V=N/n_0$) is defined by Eq. (20.2) or, provided the plot $\mu^*(n)$ is known for the considered system of disconnected links, can be determined using the rule of "equal areas" illustrated in Figure 3.11. This figure shows the dependences $\mu^*(n)$ for the two

---

[m]To avoid misunderstanding, it should be noted that the system of disconnected links is assumed to be immersed in solvent. Accordingly, the dilute and more concentrated phases of solution are regarded as gaseous and liquid phases, respectively.

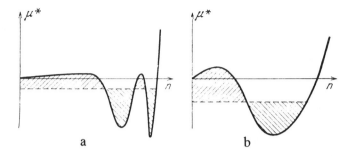

FIGURE 3.11. Concentration dependence of chemical potential of a link in a system undergoing a globule–globule phase transition (a) and a globule–coil phase transition with large concentration jump (b).

basic possible systems. Two different condensed phases can coexist in one of these systems (Fig. 3.11a). The condensed phase of the other system is thermodynamically favorable at a temperature above the θ-point (Fig. 3.11b; *see* subsection 13.11).

Recalling that the horizontal secant separating the equal areas in Figure 3.11 points out on the axis of ordinates the globule energy per link (*see* subsection 20.3), it becomes clear that a change in temperature or other conditions in the situation illustrated in Figure 3.11a may induce a globule–globule transition. (This is because the globular state, whose free energy is lower, is more stable thermodynamically). In the situation shown in Fig. 3.11b, the coil–globule transition proceeds in an unusual way (because the coil abruptly turns into a globule of finite density). Both transitions mentioned proceed as first-order phase transitions.

As an example of the realization of the situation illustrated in Figure 3.11a, we can cite a system in which a macromolecule exists in a multicomponent solvent. The globule–globule transition sharply changes the composition of the solvent penetrating the globule or in other words, the degree of absorption by the globule of low-molecular-weight ligand. The effect proves to be extremely strong when the solution is close to the critical point of component miscibility. An example of the realization of Figure 3.11b is provided by the macromolecule whose links contain mesogenic (forming liquid crystals) groups.

22.2. *In polymer globules, an intramolecular phase separation is possible, both a separation of the condensed nucleus from the gas-like fringe and a decomposition of the nucleus into different coexisting condensed phases.*

The simplest case of intramolecular phase separation can be examined in terms of the standard model of a polymer chain. If the temperature is below the critical gas–liquid transition point in the system of disconnected links, one can easily see that the dependence $v(\psi)$ (20.14) breaks into two branches (corresponding to gaseous and liquid phases). Therefore, Eq. (20.16) for $\psi(x)$ has not a continuous but a discrete solution. Accordingly, the spatial distribution of local link concentration $n(x)$ in the globule is discontinuous. In physical terms, this implies that the globule in the considered situation constitutes a two-phase system, including the dense nucleus surrounded by a gas-like fringe composed primarily of loop sections of the chain. The local link concentration undergoes a discontinuity at the phase boundary between the nucleus and the fringe. It can be shown that the chemical potential and pressure of the link system must remain smooth at the interglobular phase boundary.

In more complex polymer systems (where diverse condensed states can exist), the globular nucleus itself must separate into phases within the globule–globule transition interval. For example, in a heteropolymer globule, these could be the phases enriched by links of different types.

22.3. Condensation of a macromolecule of moderate length results in an emergence of the so-called small globule, whose size is compatible with the length of the Kuhn segment and whose structure strongly depends on the flexibility mechanism; for a persistent chain, a small globule takes the shape of a torus.
As noted in the Introduction, the large number N of links in a polymer chain is one of its fundamental properties. This does not mean, however, that all phenomena in polymer physics can be described adequately on the basis of the analysis of the asymptotic $N \to \infty$. For example, the globulization of the standard bead chain whose condensed phase density n_0 satisfies the inequality $n_0 a^3 \gg 1$ (i.e., each particle in the condensed phase occupies the average volume $1/n_0$, which is much less than a^3) leads in the case $N \gg n_0 a^3 \gg 1$ to a formation of the ordinary large globule that was discussed earlier. Apart from this case, however, the intermediate situation

$$n_0 a^3 \gg N \gg 1,$$

is possible (at least, in principle).

It is easy to see that for a sufficiently low temperature, a bead chain of such moderate length forms the globule whose nucleus has volume N/n_0 (i.e., its size is much less than a) and represents in fact a condensed "droplet" consisting of beads (links). Such a system is described fairly simply by the theoretic methods of subsection 7.4, where we discussed the globulization of the ideal chain subjected to the external field localized in a small (much smaller than a) region.

It is clear, however, that the standard bead model is not very suitable for the investigation of actual small globules, because the actual macromolecule has no immaterial bonds that could exist outside the globular nucleus without interfering with one another. (Generally, it should be noted that the standard model is adequate only if all dimensions in the system are much greater than a; *see* subsection 4.5). In addition, the structure of a small globule is much less universal than that of a large globule, and it is more dependent on the structure of the macromolecule.

As an example, here we refer only to the structure of a small globule formed by a homogeneous persistent macromolecule. Its fundamental feature, the torus-like shape (Fig. 7.12), can be clarified by observing that a persistent chain having no points of easy bending cannot fill the core of the globule where "a hole" appears.

CHAPTER 4

Polymer Solutions and Melts

Chapter 3 considered volume interactions inside a single macromolecule. This situation is realized experimentally in a dilute polymer solution in which the concentration is so small that the individual macromolecules do not entangle and seldom interact with one another (Fig. I.2a). This chapter investigates conformation properties of more concentrated polymer solutions in which the coils of different macromolecules become strongly overlapped with one another (Fig. I.2b). The ultimate case of a concentrated solution is the polymer melt containing no solvent.

23. BASIC PRELIMINARY DATA AND DEFINITIONS

23.1. *As the concentration of a polymer solution grows, it passes through three characteristic regimes: 1) the dilute solution in which individual coils do not overlap, 2) the semidilute solution in which coils are strongly overlapped but the volume fraction of the polymer in the solution is small (such an intermediate regime is intrinsic for polymers), and 3) the concentrated solution in which the volume fraction of the polymer is of order unity.*

A polymer solution can be described quantitatively by the concentration c, the number of links per unit volume of the solution, or by the volume fraction Φ of the polymer in the solution. Obviously, these quantities are proportional to one another: $\Phi = cv$, where v is the inherent volume of one link. Clearly, the value of Φ always lies between zero and unity; $\Phi \to 0$ corresponds to a dilute solution and $\Phi = 1$ to a melt.

Let us evaluate the boundary concentration c^* separating the concentration regimes of detached and overlapped macromolecules (Fig. I.2). It is evident that in order of magnitude, the value of c^* equals the link concentration inside an individual macromolecule, that is,

$$c^* \sim N/R^3, \tag{23.1}$$

where as usual, R is the characteristic size of the macromolecule. Indeed, the inequality $c \ll c^*$ implies that at any moment of time, the system has many "voids," where no macromolecules are present. This situation corresponds exactly to a dilute solution (Fig. I.2b). On the other hand, beginning from $c \gg c^*$,

148

the volumes occupied by the coils begin to overlap.

It is natural to begin by discussing the properties of a solution of coils (not globules), that is, a polymer solution in good, or θ, solvent (*see* subsection 13.5). Subsection 24.6 examines the poor-solvent situation associated with phase separation and precipitation.

We know the size R of an individual coil. It is defined by Eq. (3.2): $R \sim aN^{1/2}$ for the ideal coil or for the θ solvent, and by Eq. (16.1): $R \sim aN^{\nu}$ ($\nu \approx 3/5$) for the coil with excluded volume or good solvent. Accordingly, the boundary concentration of overlapping coils [*see* Eq. (23.1)] corresponds to the following volume fraction of polymer in the solution:

$$\Phi^* = c^* v \sim \begin{cases} N^{-1/2}, & \tau = 0, & (23.2) \\ N^{-3\nu+1} & (-3\nu+1 \approx -4/5), & \tau \sim 1, & (23.3) \end{cases}$$

where τ is the deviation from the θ point, $\tau = (T-\theta)/\theta$. For simplicity, we omitted in Eq. (23.3) a factor depending on the parameter v/a^3 (we return to it in subsection 26.4).

The value of Φ^* is seen to be very small for the polymer solution with $N \gg 1$, because it is proportional to N raised to a negative power. It is precisely because of this that there exists a broad region of concentrations $c^* \ll c \ll 1/v$ in which the coils are strongly overlapped ($\Phi \gg \Phi^*$) but the volume fraction of polymer in the solution is still small ($\Phi \ll 1$). The polymer solution in this concentration region is referred to as *semidilute*; the solution for which $\Phi \sim 1$ is referred to as *concentrated*.

23.2. *Semidilute polymer solutions in a good, or θ, solvent possess (just like dilute solutions) the property of universality and can be studied in terms of any polymer chain model (including the simplest of all, the standard bead model).*

Among the described concentration regimes, the most interesting in terms of theoretical physics is the semidilute regime of a polymer solution (apart from a dilute solution). It is easily seen that as long as the volume fraction of the polymer in such a solution is small, the binary link interactions predominate over the higher-order interactions, and near θ-conditions, where the contribution of binary collisions is compensated ($B=0$; *see* Sec. 13), the triple (three-particle) contacts prevail over others. Therefore, the behavior of the semidilute polymer solution remains universal (cf. subsection 13.5): the macroscopic equilibrium properties of the solution at $T \geqslant 0$ are universal functions of the parameters N, a, B, C, and c (or Φ) and are independent of the details of the volume link interaction. By the same reason, results to be obtained here for the semidilute solution of standard Gaussian chains can be applied to any other model of a polymer chain according to the rules described in subsection 13.8. Thus, an application of some particular polymer chain model (e.g., the standard bead model, lattice model, and so on) to studying semidilute solutions does not restrict generality.

In the region of concentrated polymer solutions ($\Phi \sim 1$), the universal behavior (in the simple sense of the word given earlier) of course disappears. We show later, however, that the dependence of conformational parameters on the chain length N still remains universal in a concentrated polymer solution, including the limiting case of polymer melt ($\Phi = 1$).

23.3. *The simplest experimentally observable characteristic of a polymer solution is osmotic pressure.*

From the theoretic viewpoint, the osmotic pressure p plays the same role in description of the solution that the pressure does in the case of conventional gas. The general idea of measuring the osmotic pressure is simple. Suppose that we have a vessel separated into two parts by a finely porous membrane. On one side of the membrane is a polymer solution; on the other side is a pure solvent. The pores in the membrane can be chosen or made so that solvent molecules could freely penetrate through the membrane while the polymer chains could not, that is, the membrane would be impenetrable for macromolecules (it is precisely such semipermeable membranes that are used in the special instruments called *osmometers*). Under equilibrium conditions, the solvent pressure therefore is the same on both sides of the membrane, while the macromolecules develop a certain pressure excess on one side. This excess is referred to as the *osmotic pressure*.

It is easily seen that if the free energy of the polymer solution is expressed as a function of the temperature T, the volume V, the number N_p of polymer chains, and the number of links in the chains N, then the osmotic pressure can be written as

$$\pi = -\frac{\partial \mathscr{F}(T,V,N_p,N)}{\partial V}\bigg|_{T,N_p,N=\text{const}} \tag{23.4}$$

Indeed, according to Eq. (23.4) π is the pressure associated with the volume change from an addition of solvent molecules provided that the number of polymer molecules is fixed (i.e., precisely the extra pressure acting on the semipermeable membrane).

Moving to the intensive quantities (the concentration $c = NN_p/V$ and the free energy density $\mathscr{F}/V$), Eq. (23.4) can be rewritten in a more convenient form:

$$\pi = -\frac{\mathscr{F}}{V} + c\frac{\partial}{\partial c}\frac{\mathscr{F}}{V}. \tag{23.5}$$

23.4. *In experiments on elastic scattering of light or neutrons, the statistical structure factor is measured, which is simply connected with both the pair correlation function of the solution and the local perturbation response function.*

Experiments on elastic (involving no frequency change) scattering of radiation provide a productive and widespread technique for experimental research on the structure of polymer systems. In these experiments, the intensity of the radiation scattered in the given direction k [*see* Eq. (5.12)] is measured. This intensity is proportional to the structure factor (or the form factor) of the scattering system, which is defined by Eq. (5.11) and can be written as

$$G(k) = (1/\mathcal{N})\langle|c_\Gamma(k)|^2\rangle = (1/\mathcal{N})\langle c_\Gamma(k)c_\Gamma(-k)\rangle, \qquad (23.6)$$

$$c_\Gamma(k) = \int c_\Gamma(x)\exp(ikx)d^3x = \sum_n \exp(ikx_n),$$

$$c_\Gamma(x) = \sum_n \delta(x - x_n), \qquad (23.7)$$

where x_n and $\mathcal{N}$ are the coordinates and the number of scattering centers, respectively (i.e., in our case, the links of all chains, or beads), and $c_\Gamma(x)$ the instantaneous (fluctuating) distribution of their concentration.

Subsections 5.5 and 19.6 discussed the structural factor of an individual polymer coil, a system of finite size. We saw that at $k=0$, it has a maximum, and that the width of that maximum equals $\sim 1/s$ according to Eq. (5.14), that is, is determined by the size of the system or, more accurately, by its radius of gyration s. For a macroscopic sample, however, the real situation conforms to the thermodynamic limit $\mathcal{N} \to \infty$, $s \to \infty$, and the previously mentioned maximum of $G(k)$ converts down to a delta function. From the definition (23.6), one can easily find that for an infinite homogeneous system

$$G(k) = (1/\mathcal{N})\int \langle\delta c_\Gamma(x)\delta c_\Gamma(x')\rangle\exp[ik(x-x')]d^3xd^3x' + c(2\pi)^3\delta(k), \qquad (23.8)$$

where $c = \langle c_\Gamma(x)\rangle$, $\delta c_\Gamma(x) = c_\Gamma(x) - c$.

Thus, at $k \neq 0$ (or more accurately, at $|k| \gg 1/s \sim V^{-1/3}$) the structural factor is expressed via the pair correlator of concentration fluctuations. Obviously, the pair correlator describes the interrelation of the fluctuations occurring in the system at points x and x'. Moreover, there is a general theorem of statistical physics stating that the pair correlator equals the so-called response function, namely, the concentration change at the point x provided that a unit action is applied at x'. Clearly, the response function (as well as the correlation function) describes the interrelation of the system elements at the distance $|x-x'|$. Consequently, measurements of $G(k)$ yield some definite information on the properties of the system investigated on length scales $1/|k|$.

A simple, general thermodynamic expression for $G(k)$ can be found in the limit $k \to 0$, $k \neq 0$, or rather, in the situation when the value of $1/|k|$ is much less than the size of the system ($k \neq 0$) but much greater than the radius of correlation of fluctuations ($k \to 0$). In this case, the second term in Eq. (23.8) can be disregarded (because $k \neq 0$), and $k=0$ can be assumed in the first term (because $k \to 0$). Subsequently, it becomes clear that $G(k \to 0)$ is determined by the rms fluctuation of the number of particles within a fixed volume. However, as is known from thermodynamics,[26] the last quantity is defined by the compressibility of the system (i.e., in our case, by the osmotic compressibility):

$$G(k \to 0) = T(\partial\pi/\partial c)^{-1}. \tag{23.9}$$

23.5. *Large-scale chain conformations are experimentally investigated by measuring the neutron scattering from a solution in which some chains contain deuterium nuclei incorporated to replace hydrogen nuclei.*

It is clearly apparent (and will be proved later) that the concentration growth is accompanied by a decrease in the correlation radius ξ, which eventually becomes less than the size of the chain. Accordingly, experiments on ordinary scattering provide no information about the conformation of the chain as a whole, because the structural factor tends to the universal limit (23.9) already on scales $1/|k| > \xi$.

The problem can be solved by labeling some chains (Fig. 4.1) to study only the radiation scattered by them. In this case, c_Γ in Eqs. (23.6) to (23.8) denotes the concentration of the labeled links. The difficulty is that the labeled links are possibly indistinguishable from the ordinary ones in terms of volume interactions, so the presence of the labeled atoms in some links would not disturb the structure of the solution under study. An optimal solution thus is to use isotopic but not chemical labels.

In fact, this problem is solved by introducing into a polymer solution or melt a small quantity of polymer of the same chemical structure but synthesized in heavy instead of ordinary water. In these chains, the hydrogen nuclei are replaced by deuterium nuclei, providing a substantially larger scattering amplitude of thermal neutrons. Measurement of neutron scattering by such a system allows examination of the conformation of the individual polymer chain in a solution or melt. To solve some other problems, more complicated labeling methods are also used, for example, only certain sections of the chains are labeled (the middle sections, the ends, the side branches, and so on).

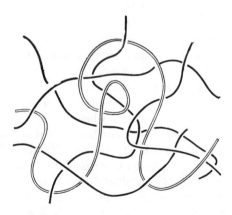

FIGURE 4.1. A labeled chain among ordinary ones.

24. THEORY OF POLYMER SOLUTIONS AND MELTS IN THE SELF-CONSISTENT FIELD APPROXIMATION

24.1. *Self-consistent field theory predicts that the osmotic pressure of a polymer solution is primarily determined by the contribution of volume interactions to the pressure in the disconnected link system, with the entropy contribution to the osmotic pressure being small because of the joining of the links into long chains.*

Let us set aside the problem of the applicability of the self-consistent field approximation to the description of a polymer solution. (This is discussed in Sec. 26.) For now, we only state that this approximation can be applied in a fairly broad region, and we consider results that can be obtained in this approximation.

The free energy of a homogeneous polymer solution can be expressed immediately in the self-consistent field approximation (*see* Sec. 15) as

$$\mathscr{F}/V = T(c/N)\ln(c/Ne) + f^*(c,T). \qquad (24.1)$$

Indeed,[a] the volume interaction contribution per unit volume equals f^* (15.7), while because of the homogeneity of the solution [$c(x) = \text{const} = c$], the conformational entropy (9.3) equals zero. Therefore, the entropy contribution is determined only by the independent motion of the chains [*see* Eq. (9.4)]. In correspondence with Eq. (24.1), one can also readily derive the osmotic pressure (23.5) of the solution:

$$\pi = T(c/N) + p^*(c,T). \qquad (24.2)$$

The result is physically obvious. If one recalls the definition of the quantity $p^* = p(c,T) - cT$ [*see* Eqs. (15.10) and (20.1)], it becomes clear that the difference between the osmotic pressure of the solution and the pressure $p(c,T)$ of the disconnected link system is reduced to the replacement of the ideal-gas pressure because of the independent motion of the links with concentration c by the ideal-gas pressure resulting from the independent motion of chains whose concentration equals c/N (i.e., is much less).

When the solution concentration is low, p^* can be calculated using the virial expansion (15.13):

$$\pi/T \cong c/N + Bc^2 + 2Cc^3. \qquad (24.3)$$

Using the expression for free energy (24.1), one can easily calculate not only the osmotic pressure but also other thermodynamic characteristics of the system, for example, chemical potentials, heat capacity, heat of mixing, and so on.

The first theory of molecular solutions and melts was developed in terms of a lattice model in the 1940s by P. J. Flory and P. Huggins. Later, it was realized that the Flory-Huggins theory is based on the self-consistent field approxima-

---

[a]Note that in this chapter, we traditionally denote the link concentration by c, whereas earlier [*see* Eqs. (9.3), (9.4), and (15.7)], this quantity was denoted by n (also by tradition, even though relating to a different context).

tion. Even though the Flory-Huggins theory is more complex, cumbersome, and less general than the conception given here, it nevertheless has been used frequently in the literature. Therefore, we give its brief outline in the Appendix to Sec. 24.

24.2. The conformation of an individual chain in a polymer melt or a concentrated or semidilute solution corresponds on large scales to a Gaussian coil (the Flory theorem).

A fundamental question of the conformation theory of polymer solutions and melts is: what is the conformation of a test (or labeled; *see* subsection 23.5) chain in a dense polymer system? Namely, does this conformation correspond to the state of a swollen coil, Gaussian coil, or globule?

A concentrated polymer solution (and more so a melt) is a rather dense system in which link interactions are very strong. Therefore, at first glance, it is impossible to assert any universal properties about the conformation of an individual macromolecule. In reality, however, this is not the case. There is a significant statement, first made by P. J. Flory in 1949 and called the Flory theorem, that the test chain in a system of entangled macromolecules has the conformation of an ideal Gaussian coil whose size R is proportional to $N^{1/2}$.

The Flory theorem is particularly simple for the polymer melt, where the size of the test macromolecule coincides in order of magnitude with that of the Gaussian coil that the given macromolecule would form in a dilute solution with no volume interactions (or under the θ conditions):

$$R \sim aN^{1/2} \text{ (in the melt).} \tag{24.4}$$

We show later that in a concentrated or semidilute polymer solution, the Flory theorem is also valid in the sense that the size R of the individual chain is proportional to $N^{1/2}$. The proportionality factor, however, depends on the solution concentration [*see* Eqs. (25.16) and (26.11)].

The proof of the Flory theorem is elementary in the self-consistent field approximation. In a homogeneous system, the self-consistent field potential (15.18) is the constant, coordinate-independent quantity $\mu^*(c,T)$. Consequently, the links experience no forces from the self-consistent field, and the chains neither swell nor shrink. We know that the coil swells if the self-consistent field has the shape of a potential hill (*see* subsection 17.2) and is compressed in the case of a potential well (*see* Sec. 20). Conversely, the chain remains Gaussian in a homogeneous system.

The physical meaning of these considerations is clarified by the following observation: in a homogeneous polymer melt, the surroundings of the test chain are the same regardless of whether the test chain swells or shrinks. (Even though the fraction of the links belonging to its "own" chain decreases among the spatial surroundings of a given link during swelling and increases during compression, this makes no difference when the links of all of the chains are identical). Hence, all chain conformations in the melt are equally probable, which complies with the Gaussian statistics.

Another comment should be made about the dynamic properties of the system. We know (*see* subsection 5.2) that a Gaussian coil is a strongly fluctuating system. It is obvious, however, that macromolecules in a solution or melt fluctuate only very slowly because of the exclusion of chains crossing. (*see* Sec. 11). This is why establishment of equilibrium Gaussian statistics in the test chain in a melt requires some considerable time, during which the ends of the test chain bring the macromolecule to equilibrium by diffusing through the loops and folds of the surrounding macromolecules (*see* Sec. 35).

It also follows from this that in the establishment of the equilibrium Gaussian conformation, a key role is played by the terminal links of the test chain. Therefore, the Flory theorem does not hold true, for example, for the solution or melt of ring macromolecules.

The proof of the Flory theorem given here is essentially correct, if somewhat formal. It is desirable to provide a deeper insight into this theorem: to make the test macromolecule ideal, it is necessary that no excluded volume interactions exist between its links; in other words, a certain mechanism of effective attraction between the links must exist in the solution or melt. This effective attraction must lead to a compensation or screening of the inevitable effect of excluded volume. The following two subsections are devoted to the investigation of the nature and properties of this effective attraction. The first step is calculation of the pair density correlator.

24.3. *The pair correlation function of concentrations in a polymer solution decays exponentially according to the Ornstein-Zernike law; the correlation radius is a diminishing function of solution concentration.*

Application of the self-consistent field method to the calculation of correlation functions is traditionally called the *random phase approximation*. There are several approaches to this method; here, we briefly describe the simplest one. It is based on the descriptive analogy with a simple mechanical system (i.e., a particle in a potential well). The equilibrium position corresponds to the minimum of the potential. To study the harmonic oscillations near equilibrium, one must find the quadratic terms in the expansion of the potential near the minimum. The situation in our case is exactly the same. In the self-consistent field approximation, the free energy minimum (15.2) corresponds to the equilibrium homogeneous state of the solution, and the quadratic terms of the expansion near the minimum must be found to analyze the fluctuations. Accordingly, let us write the free energy (15.5) containing the terms (9.3), (9.4), and (15.7):

$$F\{c(x)\} = \int \left[f^*(c(x)) - T(a^2/6)c^{1/2}(x)\Delta c^{1/2}(x) \right.$$

$$\left. + T(c(x)/N)\ln(c(x)/Ne) \right] d^3x; \qquad (24.5)$$

Then, after setting $c(x) = c + \delta c(x)$, where c is the mean (equilibrium) concentration independent of x and $\delta c(x)$ the small addition to c, we expand the free energy (24.5) into a power series of δc:

$$F\{c+\delta c(x)\}=\mathscr{F}(c)+\frac{Ta^2}{24c}\int\left[\frac{(\delta c)^2}{\xi^2}+(\nabla\delta c)^2\right]d^3x. \qquad (24.6)$$

Here, we designate the characteristic length ξ according to the relation:

$$\xi^2=\frac{a^2}{12}\left(\frac{1}{T}\frac{\partial p^*}{\partial c}+\frac{1}{N}\right)^{-1}. \qquad (24.7)$$

As expected, the minimum value of the free energy is reached at $\delta c(x)=0$, and it equals the free energy of the homogeneous solution (24.1). Because we are interested in the correlation function, however, the minimum of (24.6) must be found under certain additional conditions. For example, one may reason as follows. Suppose that one link is fixed at the origin $x=0$ and that we seek the concentration perturbation $\delta c(x)$ induced by this fact. Minimizing Eq. (24.6), we obtain the equation for δc:

$$\Delta\delta c-(1/\xi^2)\delta c=0 \quad (x\neq0); \qquad (24.8)$$

Its solution, which diminishes at infinity, takes the form

$$\delta c(x)=\text{const}\cdot(1/x)\exp(-x/\xi).$$

The constant factor in this formula can be found from the following physical consideration. At short distances from the fixed link, almost all concentration results from the close links of the same chain, and δc therefore must coincide with the local link concentration in the ordinary coil consisting of one chain with one fixed link. This quantity was found earlier [see Eq. (5.10)] to equal $3/(2\pi a^2 x)$. Hence,

$$\langle c_\Gamma(x)c_\Gamma(0)\rangle-c^2=c\delta c(x)=[3c/(2\pi a^2 x)]\exp(-x/\xi) \qquad (24.9)$$

[We recommend that the reader check that $\delta c(x)$ calculated by this method equals (up to the factor c) the pair correlation function, as written in Eq. (24.9).]

The exponential expression for the correlation function of the type (24.9) is usually referred to as the *Ornstein-Zernicke formula*, with the quantity ξ (24.7) being the correlation radius. One can easily see that the quantity ξ diminishes with the growth of the solution concentration, reaching in the melt a microscopic value (of the order of a link size). Hereafter, we need the expression for ξ (24.7) rewritten as a virial expansion:

$$\xi^2=(a^2/12)(2Bc+6Cc^2+1/N)^{-1}. \qquad (24.10)$$

Moving to a discussion of the correlation function (24.9), one can note that no correlations exist between chain sections separated by a distance exceeding ξ. This conclusion can be compared with that drawn when we discussed the flexibility of an individual polymer chain in subsections 2.3 and 3.2, because the orientational correlation also decays exponentially [see Eq. (2.2)] and disappears over lengths of the order of the persistent length. Following this analogy, one realizes that an individual chain in a polymer solution obeys Gaussian statistics. (A specific evaluation of its size, however, is to be postponed until subsection 25.8).

The expression (24.9) recalls a well-known fact from the physics of plasma or electrolytic solutions, where the Coulomb interaction of charged particles is screened at distances exceeding the Debye radius. On the basis of this analogy, one can state that the volume interactions in a polymer solution are screened at the distance of order ξ (S. Edwards, 1966). This result, however, has been obtained earlier as a purely formal one. Therefore in the next subsection, we try to give it a qualitative interpretation. It is desirable to answer the basic question: why is the screening effect only characteristic for a polymer solution?

To conclude this subsection, one other comment is necessary. The technique described for the calculation of the correlation functions is perfectly applicable to globules as well, while Eqs. (24.7) and (24.9) directly describe the correlations in the nucleus of a large globule in the volume approximation. [As will be clear from subsection 25.1, the term $1/N \to 0$ should be neglected in Eq. (24.7) in the globular regime.] More attention should also be placed on the equality of the value of ξ and the thickness of the surface layer of the globular nucleus a/ϑ_- [see Eq. (20.23)]. This is not an accidental coincidence; it implies that the influence of the globular surface reaches into the nucleus over a distance equal to the correlation radius.

*24.4. *The effective interaction of test particles through the medium of a polymer solution is attractive by its nature; it is strong because the high susceptibility of a polymer solution; and if the test particles are identical with the polymer links, then the effective attraction leads to a decrease of their virial coefficient by a factor of N, that is, to an almost total screening (compensation) of the excluded volume.*

Let us consider two particles immersed in a medium of polymer solution. Suppose that their interaction with the links is of a repulsive nature. Then, the so-called "correlation cavity" emerges around the first particle, the region where the solution concentration is somewhat lower because of the expelling action of the first particle. As long as the second particle is repelled by the links, the region of lower link concentration plays the role of a potential well for the second particle. Consequently, an effective attraction appears because of the squeezing of the polymer out of the space between the test particles. It should be emphasized that this effect proves to be particularly strong for the polymer medium. Indeed, as the entropy loss resulting from the redistribution of concentration depends on the number of independent particles (i.e., the chains), the susceptibility of the polymer system to any external influence must be particularly high. Specifically, the influence of the test particles is also strong just because the solution is polymeric. That is also why their effective attraction is essential.

It should be noted that if the interaction of test particles with links is attractive in nature, then the effective interaction is attractive all the same. The "correlation thickening" appears around one particle, which again acts as a potential well for the other.

Let us now try to formalize these qualitative considerations. For brevity, the calculations will only be outlined; the reader can perform them in detail. Suppose that the test particles are located at the points r_1 and r_2 of the polymer solution. For simplicity, we assume the test particles to be of the same nature as the links. The potential of their mutual interaction or their interaction with any other link is denoted by $u(x-x')$. Then, obviously, we must add to the free energy of the solution (24.6) the quantity [cf. Eq. (15.16)]

$$F_{ad} = \int (c+\delta c(x))u(x-r_1)d^3x + \int (c+\delta c(x))u(x-r_2)d^3x + u(r_1-r_2),$$

(24.11)

describing the interaction of the links with the test particles (the first and second terms) and the mutual interaction of the test particles (the third term). Minimizing the sum of contributions (24.6) and (24.4) according to the basic rule of the self-consistent field method (15.2), we obtain the equation for $\delta c(x)$:

$$\Delta \delta c - (1/\xi^2)\delta c(x) = (12c/Ta^2)[u(x-r_1)+u(x-r_2)]. \quad (24.12)$$

This equation is easily solved, because Eq. (24.9) is its Green's function. Substituting the solution into Eqs. (26.4) and (24.4), which is convenient to do in terms of k (i.e., after Fourier transformation), one can find the minimum value of the free energy. Obviously, this will depend (because of homogeneity) on r_1-r_2. To obtain the potential of effective interaction of the test particles, one only needs to subtract the constant limit, to which the indicated function tends for $|r_1-r_2| \to \infty$. This limit corresponds to the free energy of two independent test particles placed into the solution far from each other. Finally, the calculations yield the following expression:

$$F_{int}(r_1-r_2) = u(r_1-r_2) - \frac{24cT\xi^2}{(2\pi)^3a^2} \int \frac{\exp[ik(r_1-r_2)] \cdot u_k^2}{1+k^2\xi^2} d^3k.$$

(24.13)

Hence, we see that the additional interaction potential (as follows from our qualitative considerations) is attractive (negative) in nature. This attraction can be ascribed to a fluctuation exchange in the polymer medium. Further, the characteristic range of effective attraction is of the order of the correlation radius ξ. This can easily be seen when $u(r)$ is a short-range potential. Then, u_k in Eq. (24.13) is constant, and the effective potential turns out to be simply proportional to the correlation function (24.9).

Assuming the effect to be weak and using the expression ξ (24.10), one can use Eq. (12.4)[b] to calculate easily the virial coefficient of the effective interaction of the particles through the polymer medium:

$$B_{ef} = B/(2NcB+1). \qquad (24.14)$$

In particular, extrapolating this result to the melt $(cB=1)$, we obtain

$$B_{ef} = B/2N \ll B \text{ (in the melt)}. \qquad (24.15)$$

24.5. A test macromolecule in a melt of N-link chains behaves as a Gaussian coil if its length is shorter than N^2 and as a coil with excluded volume at longer lengths.

Let us return to the system illustrated in Figure 4.1 (a labeled chain in a polymer melt), but this time to consider the case when the length N_1 of the test macromolecule may differ from the length N of the chains of the melt. What would be the conformation of the chain of N_1 links? We know that at $N_1 = N$, the test chain has a Gaussian coil conformation (the Flory theorem; *see* subsection 24.2), while at $N=1$ or at $N_1 \to \infty$, we have a low-molecular-weight solvent for the test chain, which should then form a strongly swollen coil. The question follows: when does the transition from one regime to another take place on variation of N (or N_1)?

To answer this, consider the volume interaction parameter z_1 (13.12) in the coil of length N_1. Because the links of the test macromolecule interact through the medium of the melt of N-link chains, their virial coefficient should be expressed using Eq. (24.15):

$$z_1 \sim N_1^{1/2} B_{ef}/a^3 \sim N_1^{1/2}/N. \qquad (24.16)$$

If $N_1 = N$, then $z_1 \sim N^{-1/2} \ll 1$, which confirms that the "own" chain in the melt is ideal. If $N_1 \gtrsim N^2$, however, then $z_1 \gtrsim 1$ and the chain N_1 swells. Thus, transition from the ideal coil regime to the substantial swelling of the coil occurs at $N_1 \sim N^2$.

24.6. *On phase separation in the solution of N-link chains in poor solvent, the dilute phase concentration turns out to be very low ($\lesssim N^{-1/2}$) and may correspond to non-overlapping globules or slightly overlapping coils; the critical phase separation temperature is less than the θ point by a small value ($\sim N^{-1/2}$).*

Either lowering the temperature or another change of conditions, the polymer system can be placed into poor solvent, where an attraction prevails in the volume interaction of the links. A homogeneous state of the solution then becomes unstable, and the polymer becomes separated from the solvent, (i.e., precipitates). More precisely, the polymer solution separates into two phases; the concentrated (a precipitate), and the dilute.

---

[b]After having simplified it to the form $B = \int [u(r)/2T]d^3r$.

These circumstances can be classified graphically using the phase diagram of the solution. Figure 4.2 shows such a phase diagram plotted in the concentration–temperature plane. The state of the solution as a whole is shown by a point on the diagram. If the point is located in the unshaded area, then the equilibrium state of the solution is homogeneous. If at the given temperature the average concentration of the solution (i.e., the ratio of the total number of links to the total volume of the system) corresponds to the shaded area (e.g., the large point in Fig. 4.2), then the solution separates into two phases whose concentrations correspond to the edges of the separation region. In this case, the phase volumes are determined by the so-called "law of a lever": the total volume of the system is divided in proportion to the distances from the point depicting the state to the separation region boundaries.

This is true, of course, for any solutions, not only a polymer one. The specifics of polymer solution separation relate to the fact that one phase has a very low concentration. For a moderately sensitive experimental technique, the equilibrium dilute phase turns out to be indistinguishable from pure solvent. This circumstance can readily be explained qualitatively.

Indeed, if we start from the concentrated polymer system being in contact with pure solvent, establishing equilibrium requires the dissolving ("evaporation") of a certain amount of macromolecules, whose number is determined by a balance between the energy loss (because the contacts between the chains are more favorable than the contacts with the poor solvent) and the entropy gain. The energy loss, however, is proportional to the number of "evaporated" links, while the entropy gain is determined by the much lesser number of chains (cf. an analogous discussion in subsection 24.4). Therefore, it is clear that the concentration of the dilute phase c_{dil} must be very low indeed for $N \gg 1$. As for the concentrated phase, it can easily be described from physical considerations: from the local viewpoint, the concentrated phase can be treated as a nucleus of a "very large" globule, and the corresponding concentration c_{conc} therefore must be defined by the condition (20.2) or (20.27), that is, $c_{conc} = n_0$ or

$$c_{conc} = n_0 = -B/2C. \qquad (24.17)$$

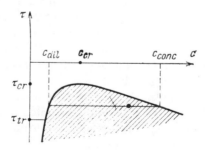

FIGURE 4.2. Diagram of phase separation of a polymer solution.

We now formalize these qualitative considerations. To determine the concentrations of coexisting phases in equilibrium, we can use the general thermodynamic rule, according to which the two unknown quantities c_{conc} and c_{dil} are determined by the two equations defining the conditions of equality of osmotic pressures and chemical potentials in the two phases.

The osmotic pressure of the concentrated phase is given by Eq. (24.2). Regarding the dilute phase as an ideal gas (or more accurately, a solution), we obtain

$$p^*(c_{conc}) + Tc_{conc}/N = Tc_{dil}/N. \tag{24.18}$$

Neglecting here the small terms $1/N$, we obtain the expected conditions $p^*(c_{conc}) = 0$ (20.2), that is, Eq. (24.17). Certainly, the correction $\sim 1/N$ to Eq. (24.17) can easily be found from Eq. (24.18). It should be noted that the result (24.17) is obtained in exact correspondence with the previous qualitative consideration, because the omitted ideal-gas terms ($\sim 1/N$) are small as the links are connected into chains.

Let us now turn to the condition of equality of the chemical potentials. This allows us to determine the dilute-phase concentration. As long as the phases cannot exchange links but only whole chains, the chemical potentials of the chains should be equal. The concentration of chains equals c/N, therefore, knowing the free energy $\mathscr{F}/V$, the chemical potential of the chain can be expressed as

$$\mu_p = \frac{\partial(\mathscr{F}/V)}{\partial(c/N)} = N\frac{\partial(\mathscr{F}/V)}{\partial c}.$$

Consequently, in the concentrated phase (24.1),

$$\mu_p = T\ln(c_{conc}/N) + N\mu^*(c_{conc}).$$

Regarding the dilute phase, the chemical potential of the ideal gas $T = \ln(c_{dil}/N)$ has to be increased by adding the free energy of the individual chain, which is written in different forms depending on whether the chains in the dilute solution are coils or globules:

$$N\mu^*(c_{conc}) + T\ln(c_{conc}/N) = T\ln(c_{dil}/N) + \begin{cases} 0, & T > T_{tr}; \\ N\mu^*(n_0) + \mathscr{F}_{surf}, & T < T_{tr}, \end{cases}$$

where T_{tr} is the globule–coil transition temperature and Eq. (20.18) used for the free energy of the globule.

First, we consider the case $T < T_{tr}$. Recalling that $c_{conc} = n_0$ [see Eq. (24.17)] and using expression (20.32) for the surface energy of the globule $\mathscr{F}_{surf}$, we find

$$c_{dil} \cong c_{conc}\exp(-\mathscr{F}_{surf}/T) \sim (-B/C)\exp(-\text{const} \cdot N^{2/3}|B|^{4/3}aC^{-5/6}). \tag{24.19}$$

If $T > T_{tr}$, then

$$c_{dil} \cong c_{conc} \exp(N\mu^*(c_{conc})) \sim (-B/C)\exp(-NB^2/4C). \qquad (24.20)$$

The results obtained are shown in Figure 4.2, where the phase diagram of the polymer solution is illustrated on the temperature-concentration plane. It should be noted that in the formulas obtained here, only the coefficient $B = b\tau = b(T-\theta)/\theta$ (12.5) depends on temperature. In Figure 4.2, the phase-separation region is shaded: if the solution parameters (τ, c) fall within this region (the large point in Fig. 4.2), then the equilibrium state of the solution is not homogeneous but rather phase separated.

The coordinates of the critical point are easily determined from Eq. (24.20), because near this point, $c_{dil} \sim c_{conc}$:

$$|B_{cr}| = b|\tau_{cr}| \sim C^{1/2}/N^{1/2}, \qquad (24.21)$$

$$c_{cr} \sim -B_{cr}/C \sim 1/(C^{1/2}N^{1/2}). \qquad (24.22)$$

Thus, the critical temperature is close to the θ-point, and the critical concentration is very low.

Attention should be drawn to the comparison of the critical temperature (29.21) with the globule–coil transition temperature (21.8):

$$|\tau_{cr}| \sim |\tau_{tr}|(C^{1/4}/a^{3/2}).$$

Because (*see* subsection 13.9) the parameter $C^{1/2}/a^3$ is known to be less than unity and for most polymers (especially rigid ones) is fairly small, then $|\tau_{cr}| < |\tau_{tr}|$, that is, the critical temperature is located closer to the θ-point. This is reflected in Figure 4.2: If we lower the temperature, then on emergence of phase separation, the dilute phase consists initially of coils and the globule–coil transition sets in only later.

As mentioned previously, the principal feature of polymer systems consists in an extremely low concentration of the dilute phase. Even near the critical point, the dilute phase concentration depends on N in the same way as the concentration c^* (23.1) does, even though it exceeds the value of c^* by the fairly large factor $a^3/C^{1/2}$. On lowering the temperature, the concentration of the dilute phase drops drastically. As to the dilute solution of globules, its concentration (24.19) is especially low, much lower than that (c_{conc}) at which the globules would begin to overlap.

It should also be mentioned that this last circumstance makes the experimental observation of the globule–coil transition extremely complicated. On lowering the temperature, the prevalent fraction of the chains does not pass into the globular state but rather precipitates.

In conclusion, we should note that the structure of the boundary between the concentrated and dilute phases of the polymer solution and its surface tension coincide with the structure of the boundary and surface tension of the large globule [i.e., are described by Eqs. (20.22), (20.26), and (20.30) to (20.32)]. The remarks made concerning the globule boundary in subsections 20.6, 22.2, and 24.3 also are fully applicable to the boundary between the concentrated and dilute phases.

APPENDIX TO SEC. 24. FLORY LATTICE THEORY OF POLYMER SOLUTIONS AND MELTS

Historically, the Flory theory (developed in the 1940s) has played a very significant role by providing a quantitative description of the conformational properties of polymer solutions. Many experimental results have been interpreted in the language of the Flory theory.

This theory is formulated in terms of the lattice model of a polymer solution (*see* subsections 18.5 and Fig. 2.2). Each pair of neighboring lattice sites occupied by off-neighbor links of the chain (i.e., the link-link contact) is assumed to possess the energy $T\chi_{pp}$. The energy $T\chi_{ps}$ is ascribed to the link-solvent contact (i.e., the pair consisting of an occupied and a free neighboring site) and the energy $T\chi_{ss}$ to the solvent-solvent contact (i.e., a pair of two free neighboring sites).

The free energy equals $\mathscr{F} = \widetilde{E} - T\widetilde{S}$. The internal energy $\widetilde{E}$ of the polymer solution is written in the Flory theory as

$$\widetilde{E}/(N_0 T) = \chi_{pp}\Phi^2/2 + \chi_{ps}\Phi(1-\Phi) + \chi_{ss}(1-\Phi)^2/2 = \text{const} + \text{const} \cdot \Phi - \chi\Phi^2, \tag{24.23}$$

$$\chi = \chi_{ps} - (\chi_{pp} + \chi_{ss})/2, \tag{24.24}$$

where N_0 is the total number of lattice sites and Φ the fraction of occupied sites (i.e., the volume fraction of the polymer in solution). The entropy $\widetilde{S}$ equals the logarithm of the number of different arrangements of N-link self-avoiding random walks with the volume fraction Φ over the lattice. This number is determined approximately in the Flory theory by combinatorial analysis. This is done [as in Eq. (24.23)] by assuming statistical independence of the links surrounding each chain, that is, by using the self-consistent field approach (*see* subsection 15.5). The result, which is correct within insignificant constants and terms linear in Φ, takes the form

$$\widetilde{S} = -(\Phi/N)\ln\Phi - (1-\Phi)\ln(1-\Phi). \tag{24.25}$$

The result of the Flory theory (24.23), Eq. (24.25) precisely conforms to the general formula (24.1) of the self-consistent field theory, provided that the interpolation (15.12) is used to obtain f^*.

Let us now explain the difference between E and $\widetilde{E}$, and also between S and $\widetilde{S}$. In the derivation by Flory, the mutual uncrossability of chains (the excluded volume effect) and their connectivity result in the formation of a set of geometric restrictions on possible link arrangements, thereby determining the entropy $\tilde{S}$. This is a real thermodynamic entropy; accordingly, $\tilde{E}$ is the internal energy. At the same time, S in Eq. (15.5) is only the part of the entropy that is associated with the chains (i.e., the conformational entropy). A contribution to the entropy provided by the local link arrangement is included in E (*see* subsection 15.2).

Thus, within the scope of the Flory theory, volume interactions are characterized by the single dimensionless quantity χ, called the *Flory-Huggins parameter*. The value $\chi = 0$ corresponds to the absence of energy interactions in the polymer solution when only the forces of steric repulsion (caused by excluded volume effects) act between the links. This is the case of the so-called athermal solution. Clearly, the athermal solvent is a good one. Because for θ-conditions, $\chi = 1/2$ [*see* Eq. (15.15)], then $\chi < 1/2$ conforms to the good-solvent region and $\chi > 1/2$ to the poor-solvent region.

25. SCALING THEORY OF POLYMER SOLUTIONS

25.1. *Predictions of the self-consistent field theory are unsatisfactory for a semidilute solution of a flexible-chain polymer in good solvent.*

In the previous section, the self-consistent field theory was shown to describe in simple and constructive form various properties of the polymer solution, such as an equation of state, correlations, test-chain conformations, phase separation, and so on. We know, however, that the self-consistent field method is based on the neglect of fluctuations (*see* subsection 15.1) or on the neglect of correlations between the links (*see* subsection 15.5). On the other hand, we know that in coils (both Gaussian and swollen), fluctuations are large and interlink correlations extend over the whole chain (*see* subsections 5.3, 5.5, and 18.6). Even though growth of the solution concentration suppresses fluctuations and leads to a decrease in the correlation radius [according to Eqs. (24.7) and (24.10) in the self-consistent field theory], the fluctuations apparently remain sufficiently large and the correlations long range in the semidilute solution regime. It follows from these general considerations that the results of the self-consistent field theory should not be used irresponsibly, at least when they are being applied to the semidilute solution.

This section develops the concepts of the consistent fluctuation theory of polymer solutions based on the fundamental (for a strongly fluctuating system) principle of scale invariance, or scaling (*see* Secs. 16 and 18). For simplicity, first consider the athermal (i.e., conforming to the ultimately good solvent) solution of flexible chains. According to Eq. (12.5), $B = v$ in the athermal limit, and for the flexible chains, $v \sim a^3$ [*see* Eq. (13.17)]. We return to the general case in the next section, where we analyze the effects of temperature and chain stiffness on the main characteristics of the solution.

*25.2. *A semidilute solution can be described formally in terms of the fluctuation field theory or by the polymer-magnetic analogy.*

The first fluctuation theory of the semidilute polymer solution of flexible chains in athermal solvent was developed by J. des Cloizeaux (1975) within the framework of the polymer-magnetic analogy. Having noticed that the multiple-chain system is analogous to a magnet in an external field and used the results of the fluctuation theory of magnets (known by that time), J. des Cloizeaux showed, for example, that the following expression is true for the osmotic pressure of a polymer solution:

$$\pi/T = a^{-d}N^{-vd}\varphi(\Phi N^{vd-1}), \qquad (25.1)$$

where d is the space dimensionality, v the critical exponent of the correlation radius, (see subsection 16.1: for $d=3$ we have $v \approx 3/5$; see also subsection 17.5), and φ a universal function of its argument with the asymptotic forms

$$\varphi(x) \cong \begin{cases} x, & x \ll 1, \\ \text{const} \cdot x^{vd/(vd-1)}, & x \gg 1. \end{cases} \qquad (25.2)$$

The formula (25.1) is called the *des Cloizeaux law*.

The primary drawback of this approach (based on the polymer-magnetic analogy) is the awkwardness of the necessary calculations and a complete lack of descriptiveness. As a result, the scaling method (see subsection 16.3) has gained wide recognition in the physics of polymer solutions. We used this method in Sec. 19. Let us now provide a consistent presentation of the scaling theory of polymer solutions.

25.3. *Scaling invariance (i.e., uniqueness of a characteristic size of a polymer coil) requires a single characteristic concentration of a polymer solution.*

We have already mentioned that the coil is characterized by only one macroscopic size R, and this significant conclusion has been proved by the renormalization group method (see subsection 18.6). Accordingly, the characteristic link concentration inside the coil can be determined in a unique way, $c^* \sim N/R^3$. It therefore is clear that only one characteristic concentration exists in the polymer solution. It has the order of c^*, and as mentioned in subsection 23.1, it corresponds to the transition from the dilute regime of individual chains to the semidilute regime of strongly overlapped coils.

The self-consistent field theory, however, is incompatible with the existence of a unique characteristic concentration. It is worth demonstrating this for further discussion. Considering the athermal solution of flexible chains ($B = v \sim a^3$), let us return to the expression for the osmotic pressure (24.3) and write out its asymptotic forms:

$$\pi/T \cong \begin{cases} c/N = \Phi/Na^3, & \Phi \ll 1/N, & (25.3) \\ c^2 B = \Phi^2/a^3, & 1/N \ll \Phi \ll 1. & (25.4) \end{cases}$$

Obviously, Eq. (25.3) corresponds to the dilute solution regime, because it yields the ideal-gas pressure of coils with density c/N. Eq. (25.4) corresponds to the semidilute regime, where the self-consistent field theory (see subsection 15.5) regards the pressure as a result of pair collisions of non-correlated particles.

Thus, in the self-consistent field theory, the transition concentration (or volume fraction) between the dilute and semidilute solutions is proportional to $1/N$. Meanwhile, from simple geometric considerations (*see* subsection 23.1), another result follows: $\Phi^* \sim N^{-4/5}$ [*see* Eq. (23.3)]. The presence of two characteristic concentrations conflicts (as we have shown) with the principle of scale invariance.

A quite similar situation appears in the analysis of Eq. (24.10) for the correlation radius. In this case, however, in addition to the incorrect estimate of the transition concentration, we also obtain the incorrect result $\xi \sim aN^{1/2}$ for the correlation radius. (The correct result is $\xi \sim R \sim aN^{3/5}$).

Thus, to describe fluctuation effects correctly, any theory has to be based on the uniqueness of the characteristic concentration c^*.

25.4. The osmotic pressure of a semidilute polymer solution depends on the concentration as $c^{9/4}$ (according to the fluctuation theory).

Consider the osmotic pressure as the first parameter to be calculated by the scaling method. We initially look at the dilute solution ($\Phi \ll N^{-4/5}$). The osmotic pressure π of such a solution obviously coincides with that of an ideal gas of coils:

$$\pi/T \cong c/N \quad \text{at} \quad c \ll c^*, \quad \text{i.e.,} \quad \Phi \ll N^{-4/5} \tag{25.5}$$

(c/N is the number of coils per unit volume), because for $\Phi \ll N^{-4/5}$ the polymer coils can be regarded as non-interacting in a first-order approximation. The conclusion (25.5) is trivial for the dilute regime, and of course, it coincides with the corresponding result (25.3) of the self-consistent field theory.

Now, we increase the polymer concentration c in the solution. According to the scaling concept of the uniqueness of the characteristic concentration c^*, the general expression for π, can be written as

$$\pi/T = (c/N)\varphi(c/c^*). \tag{25.6}$$

The dimensionless function $\varphi(x)$ in Eq. (25.6) must have the following asymptotic forms. For $x \ll 1$, we are in the dilute solution region; the relation (25.5) is satisfied if $\varphi(x) \cong 1$. In the semidilute solution region $x \gg 1$, ($c \gg c^*$), the function $\varphi(x)$ must possess a certain asymptotic of the power law type $\varphi(x) \sim x^m$ or, according to Eq. (23.3),

$$\pi/T \sim (c/N)(c/c^*)^m \sim (c/N)(ca^3 N^{3\nu-1})^m, \tag{25.7}$$

where m is some unknown power index. This index can be found from the physically obvious additional condition that the osmotic pressure of such a solution must be independent of the number N of links in the chain, because the coils in the semidilute solution are strongly overlapped. From Eq. (25.7), we find that this occurs when $m = 1/(3\nu-1)$, that is, for $c^* \ll c \ll a^{-3}$

$$\pi/T \sim c(ca^3)^{1/(3\nu-1)} \sim a^{-3}\Phi^{3\nu/(3\nu-1)} \tag{25.8}$$

or, because $\nu \approx 3/5$ (*see* subsection 17.5),

$$\pi a^3/T \sim \Phi^{9/4}. \tag{25.9}$$

Comparing Eq. (25.9) with the relation (25.4) that follows from the self-consistent field theory, one can see that they differ substantially. If according to the self-consistent field theory the osmotic pressure of the semidilute solution grows with concentration as c^2, then the scaling theory predicts the dependence $\pi \sim c^{2.25}$. Special experiments have confirmed the scaling relation (25.9) completely. As mentioned previously, the inaccuracy of the self-consistent field theory is caused by its failure to account for the appreciable correlated fluctuations in the polymer solution.

We described the scaling method for the calculation of the osmotic pressure of the semidilute polymer solution at great length, this is to show clearly the basic steps in reasoning of this type. As a rule, the expression for an unknown characteristic is initially written for the conditions of the regime at which this expression is trivial [in the given case, this regime corresponds to the dilute solution, Eq. (25.5)]. Then, the scaling assumption is made about the uniqueness of the characteristic parameter for the quantity (concentration, temperature, size) serving as an argument in the dependence that we want to find, and the unknown characteristic is written in a form analogous to Eq. (25.6). We are mainly interested in the power asymptotic form of the type of Eq. (25.6). The power index in this asymptotic expression must be determined using some additional physical considerations. On the whole, it should be noted that the determination of unknown quantities using the scaling method proves to be simpler than both the field-theoretic calculations of the fluctuation theory and those of the self-consistent field theory (similar to the Flory theory). To make sure that this is true, one can compare the derivation of Eq. (24.25) in the Flory theory with that of Eq. (25.9) in the scaling theory.

Even though we repeatedly used the scaling rationale (*see* Sec. 19), we shall dwell once again on the question: why was a power asymptotic expression of the function $\varphi(x)$ suggested to write Eq. (25.7)? Apart from the argument that such an asymptotic form is quite natural and compatible with what we know about the properties of other strongly fluctuating systems, it can be noted that no other asymptotic expression for this function allows us to ensure the independence of π and N in the semidilute regime. Finally, recall that in each specific case, the scaling rationale can be verified using the fluctuation theory technique.

To conclude this subsection, the three comments are appropriate. First, Eq. (25.6), found from the principle of scale invariance, corresponds exactly to the formally derived des Cloizeaux law (25.1).

Second, the virial coefficients of coil interaction can be evaluated from Eq. (25.6), because at low concentrations, there exists the virial expansion

$$\pi/T = c/N + A_2(c/N)^2 + \dots \tag{25.10}$$

[cf. Eq. (14.7)]. Comparing Eq. (25.10) with Eq. (25.6), we find $\varphi(x) = 1 + \text{const} \cdot x + \ldots$ at $x \ll 1$ and $A_2 \sim a^3 N^{3v}$ ($3v \approx 9/5$) in exact correspondence with Eq. (19.1).

Third, the osmotic pressure (25.9) coincides with the pressure exerted on the cavity walls by a globulized chain with excluded volume placed in the cavity (*see* subsection 19.4). This coincidence is not accidental, and we can now state that the globule considered in subsection 19.4 constitutes a semidilute solution from the local point of view.

25.5. *Because of the correlations, link collision probability in a semidilute solution is proportional not to the concentration* Φ *but rather to a smaller quantity* $\Phi^{5/4}$.

Here, we give an additional physical explanation of the reasons for the difference between the self-consistent result (25.4) and the scaling formula (25.9). We start with Eq. (23.5), expressing the osmotic pressure as a function of the free energy of the solution. The nontrivial part of the free energy $\mathscr{F}$ is obviously proportional to the product of the temperature by the number Q of contacts between links of different macromolecules. Q may be estimated as $Q \sim \mathscr{N}w$, where $\mathscr{N}$ is the total number of links in the solution and w the probability of contact for a given link. Thus, $\mathscr{F}/V \sim T\mathscr{N}w/V \sim T\Phi w/a^3$. Hence, $\pi/T \sim (\Phi^2/a^3)\partial w/\partial\Phi$, and the concentration dependence of π is associated with the dependence of the contact probability w on the volume fraction Φ of links in the solution. In the self-consistent field approximation (i.e., when no link correlation is assumed and the polymer coil pictured as a cloud of independent links), $w = \Phi$, whence follows Eq. (25.4). Conversely, the scaling result (25.9) is obtained when

$$w \sim \Phi^{1/(3v-1)} \sim (ca^3)^{1/(3v-1)}(1/(3v-1) \approx 5/4). \qquad (25.11)$$

Thus, in scaling theory, the probability of a given link contacting a link of another macromolecule does not equal the volume fraction Φ of links in solution, but it is substantially lower (because $\Phi \ll 1$). This decrease in the probability of contact is associated with the fact that in reality, individual links of a macromolecule are not independent but rather connected into a chain, each link surrounded by a cloud of neighboring links in the chain. Given the contact of two links, the surrounding clouds also come into contact, resulting in an additional effective repulsion, that is, decreasing the probability of the contact (cf. the similar reasoning in subsection 19.4).

25.6. *The correlation length of a semidilute polymer solution diminishes with the growth of concentration proportionally to* $c^{-3/4}$; *this length coincides with the size of a blob.*

An instantaneous conformation of a semidilute polymer solution can be depicted as a quasi-network with a characteristic size ξ. It is clear that the value of ξ has the meaning of the average spatial distance between two consecutive (along the chain) contacts with other chains. Let us estimate this size ξ. Taking into account Eq. (25.11), the average number of links g between two consecutive contacts along the chain in the semidilute solution is

$$g \sim w^{-1} \sim (ca^3)^{-1/(3v-1)}(-1/(3v-1) \approx -5/4). \qquad (25.12)$$

Next, because the chain of g links does not touch other chains between the two contacts, it can be treated as an isolated section of the ordinary chain with excluded volume,

$$\xi \sim ag^v \sim a(ca^3)^{-v/(3v-1)}(-v/(3v-1) \approx -3/4). \qquad (25.13)$$

The quantity ξ represents a very important characteristic of the semidilute polymer solution. On length scales $r < \xi$, the polymer chain behaves like an isolated one, whereas for $r > \xi$, the conformation of the macromolecule is perceptibly affected by the other chains. In particular, over the scales $r < \xi$, the polymer solution (like the isolated coil) is a strongly fluctuating system: the fluctuation of local link concentration is of the order of the concentration itself. At the same time, these fluctuations are suppressed for $r > \xi$, because the individual polymer coils behave independently. Thus, the quantity ξ defined by Eq. (25.13) is the correlation radius of concentration fluctuations or the so-called correlation length of the polymer solution. As mentioned in subsection 23.3, this quantity can be measured directly in experiments on elastic scattering of light or neutrons by the polymer solution. Such measurements confirm the validity of Eq. (25.13).

It is instructive to derive expression (25.13) for the correlation length of the semidilute polymer solution directly using the scaling reasoning. Indeed, Sec. 18 showed that in the dilute solution for $c \ll c^*$, the correlation length is of the order of the coil size (i.e., $\xi \sim R_{c \ll c^*} \sim aN^v$). As c^* is the only characteristic concentration, then

$$\xi = R_{c \ll c^*} \varphi(c/c^*). \qquad (25.14)$$

For $c \gg c^*$ (in the semidilute solution regime), the function $\varphi(x)$ must have the power asymptotic $\varphi(x) \sim x^l$. The power index l must be chosen from the condition that ξ is independent of N in the semidilute regime (Fig. 4.3). Therefore, we have $l = -v/(3v-1) \approx -3/4$, whence we obtain Eq. (25.13).

25.7. *A semidilute solution may be pictured as an almost close-packed system of blobs; at the same time, the number of contacts that a blob has with other chain sections is of order unity at every moment of time, with the free energy of the solution being of order T per blob.*

Figure 4.3 illustrates convenient visualization of a polymer chain conformation in a semidilute solution. The chains are represented as a sequence of so-called blobs of size ξ (cf. the blobs in various problems of Sec. 19). Inside the blob, the chain behaves as an isolated macromolecule with excluded volume, and different blobs are statistically independent of one another. This is because the blob size $\sim \xi$. The number g of links in the blob is specified by the condition $ag^v \sim \xi$, that is, $g \sim (ca^3)^{-1/(3v-1)}$ [see Eq. (25.12)]. Note that $g \sim c\xi^3$. Therefore, as Figure 4.3 shows, the semidilute solution is a close-packed system of blobs.

Blobs

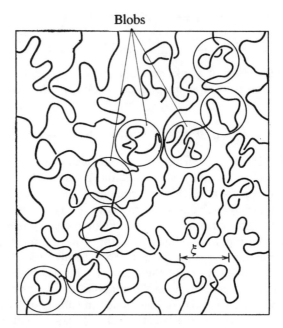

FIGURE 4.3. Instantaneous conformation of semidilute polymer solution as a "net" with mesh ξ or as a system of blobs.

It should also be noted that the osmotic pressure of a semidilute solution (25.8) can be written in the form $(\pi/T \sim c/g)$. From this viewpoint, the semidilute solution constitutes an ideal gas of blobs with the free energy $\sim T$ per blob (cf. subsection 19.3).

The specific property of a semidilute polymer solution consists in the fact that both indicated, seemingly opposing, concepts, (i.e., an ideal gas and a close-packed system) are valid with equal accuracy of order unity.

25.8. *Each macromolecule in a polymer solution is a Gaussian chain of blobs; in a semidilute solution, the size of each macromolecule diminishes with increasing concentration as $c^{-1/4}$.*

Let us consider another significant parameter of a semidilute solution: the size R of an individual macromolecule in such a solution. As mentioned in subsection 23.5, this can be measured by neutron scattering from the polymer solution with a small fraction of deuterated (labeled) chains.

The previous subsection showed that a semidilute polymer solution constitutes a close-packed system of blobs. If each macromolecule is treated as a chain of blobs, then the solution may be represented as a melt of chains of blobs. The Flory theorem (24.4) is valid for a melt. Consequently, in this case

$$R^2 \sim (N/g)\xi^2, \qquad (25.15)$$

because ξ is the size of the effective link (blob) and N/g the number of blobs in one macromolecule. The relation (23.15) also follows from the condition of statistical independence of the blobs. (Because there is no repulsive or attractive correlations between the blobs, the blob chain must behave as an ideal chain.) Inserting Eqs. (25.12) and (25.13) into Eq. (25.15), we obtain for $c^* \ll c \ll a^{-3}$

$$R^2 \sim Na^2(ca^3)^{-(2\nu-1)/(3\nu-1)}(-(2\nu-1)/(3\nu-1) \approx -1/4). \quad (25.16)$$

It can be seen that in the semidilute solution, the swelling coefficient of the macromolecule $\alpha^2 \equiv R^2/(Na^2)$ equals $\alpha^2 \sim (ca^3)^{-1/4}$. In concentrated solutions and melts, $ca^3 \sim 1$, so Eq. (25.16) yields $R^2 \sim Na^2$ and $\alpha^2 \sim 1$ (as it should be according to the Flory theorem). For $c \sim c^* \sim a^{-3}N^{-4/5}$, Eq. (25.16) yields $R \sim aN^{3/5}$, so that it gets smoothly transformed into the result of the dilute limit. Overall, it follows from Eq. (25.16) that in the semidilute solution, the swelling of macromolecules because of repulsive volume interactions diminishes with increasing concentration. This relates to the screening of volume interactions in concentrated polymer systems (*see* subsection 24.4).

We now derive the result (25.15) on the basis of scaling arguments. For $c \ll c^*$, we have $R^2 \sim a^2 N^{2\nu}$. By virtue of the main scaling assumption,

$$R^2 = R^2_{c \ll c^*} \varphi(c/c^*). \quad (25.17)$$

For $c \gg c^*$, $R^2 \sim a^2 N^{6/5}(c/c^*)^n$ asymptotically, where the exponent n is to be chosen from the condition that the quantity R^2 is proportional to N in the semidilute regime. This condition is a consequence of the correlation length ξ being much less than R in this regime; thus, individual chain sections are statistically independent on length scales $r > \xi$ and obey Gaussian statistics. Hence, we obtain $n = -(2\nu-1)/(3\nu-1)$, from which Eq. (25.16) follows.

25.9. Pair correlation functions of a labeled chain and of the total concentration of a semidilute solution diminish over characteristic distances of order of the coil size and the blob size, respectively.

As mentioned in Sec. 23, the experimental study of correlation properties of a polymer solution by the method of neutron (or light) scattering may be performed using either a solution of identical chains or a mixture of ordinary and labeled (deuterated) chains. In this case, the static structure is measured or in other words, the pair correlator (23.8) of the total link concentration $G(r)$ of the solution and of the link concentration of an individual labeled chain $G_s(r)$, respectively. Let us calculate both quantities.

To do this, we use the form of the correlation function that was discussed in Sec. 24 [*see* Eq. (24.9)]. Suppose that one link is fixed at the origin $r=0$, and find the perturbation induced at the point r. This will be $G(r)/c$ [or $G_s(r)/c$].

For distances $r < \xi$ from the fixed link, the concentration is totally caused by the links of the same chain. Therefore, $G(r) \cong G_s(r)$ on these scales, and both quantities coincide with the structure factor of an individual chain with excluded volume (19.13):

$$G(r) \cong G_s(r) \sim r^{-(3v-1)/v}a^{-1/v}, \quad -(3v-1)/v = -4/3, \quad r < \xi.$$
$$(25.18)$$

On scales exceeding the correlation radius ξ, fluctuations of the total concentration are suppressed. It becomes possible to describe them in terms of the self-consistent field approximation [or the random phase method; *see* subsection 24.3], and $G(r)$ is given by the Ornstein-Zernike formula $G(r) \cong (\text{const}/r)\exp(-r/\xi)$]. The quantity $G_s(r)$ on scales $r > \xi$ is determined by the fact that the labeled chain behaves on these scales as Gaussian [i.e. $G_s(r) = \text{const}/r$, as follows from Eq. (5.17)]. The constant factors in these expressions can be found from the condition of smooth joining with the asymptotic (25.18) for $r \sim \xi$. As a result, we obtain

$$G(r) \sim a^{-1/v}\xi^{-(2v-1)/v}r^{-1}\exp(-r/\xi), \quad \xi < r, \quad (25.19)$$

$$G_s(r) \sim a^{-1/v}\xi^{-(2v-1)/v}r^{-1}, \quad \xi < r < R. \quad (25.20)$$

Finally, for $r > R$, $G_s(r)$ also decays exponentially, even though this quantity remains [as is clear from Eqs. (25.19) and (25.20) for $r \sim R$] much greater than $G(r)$. This last circumstance is quite natural, because the fluctuations of one chain are of course greater than those of a chain system.

Thus, in the experiments with labeled chains, one can measure both the size R of an individual chain and the correlation length ξ. In experiments with ordinary solutions where the Fourier transform of the function $G(r)$ is measured, equal to

$$G(k) \sim \begin{cases} (ka)^{-1/v}, & 1/v \approx 5/3, \quad k\xi > 1; \\ \xi^{-(2v-1)/v}a^{-1/v}/[k^2+(1/\xi)^2], & (2v-1)/v \approx 1/3, \quad k\xi < 1, \end{cases}$$

ξ can be obtained. These experiments have been performed to find the value of ξ and to confirm the scaling formula (25.13) (M. Cotton *et al.*, 1976).

26. DIAGRAM OF STATES OF POLYMER SOLUTIONS

26.1. *In addition to the phase transition lines, the phase diagrams of polymer solutions also indicate the conditional lines of crossover separating the regions corresponding to the same phase but characterized by qualitatively different regimes of behavior of basic physical properties.*

Previous sections discussed many different properties of a polymer solution. We know that in a dilute solution, individual macromolecules can be swollen coils (*see* Secs. 17 to 19), Gaussian coils (*see* Sec. 4), or globules (*see* Secs. 20 to 22). We investigated the properties of concentrated and semidilute solutions in the self-consistent field approximation (*see* Sec. 24) but have not yet answered the question about its applicability. Finally, we also discussed the fluctuation theory of semidilute solutions of flexible chains in an athermal solvent (*see* Sec. 25).

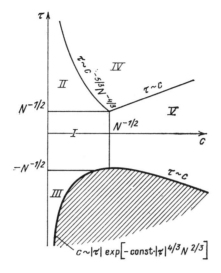

FIGURE 4.4. Diagram of states for a solution of flexible-polymer chains.

It thus is natural to undertake a classification of these data and to examine all possible states of a polymer solution. This is done using the temperature-concentration diagram of states of the solution.

Figures 4.4 and 4.5 illustrate the diagrams of states for solutions of flexible chains (M. Daoud, G. Jannink, 1976) and stiff macromolecules (D. Schaefer, J.

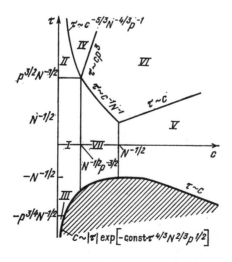

FIGURE 4.5. Diagram of states for a solution of stiff-polymer chains.

Joanny, P. Pincus, 1980; T. M. Birshtein, 1982). The shaded area in the diagrams depicts the phase-separation region; its boundary, shown by a thick line, was studied in subsection 24.6. The unshaded area corresponds to homogeneous phases. As shown in previous sections, however, different regimes of behavior are possible even in a homogeneous phase state. In the diagrams, such regimes are separated by thin lines. Certainly, these lines do not define any drastic, abrupt changes in the state of the solution (phase transitions), only regions of smooth transition from one regime of solution behavior to another.

While discussing various regimes, we shall calculate the three fundamental characteristics of a solution: 1) the osmotic pressure π, 2) the correlation length ξ, and 3) the size of an individual chain R. In this way, we obtain sufficiently complete information about the conformational properties of the polymer solution in all its regimes.

All results of this section are expressed via the critical exponent ν. In Figures 4.4 and 4.5 as well as in Table 4.1, the same results are given for the Flory value $\nu \approx 3/5$ (*see* subsection 17.5).

26.2. *The conformational properties of a polymer solution of moderate concentration are universal, that is, are completely determined by the four quantities: concentration, temperature deviation from the θ-point, and the length and stiffness of chain; they can be described using the standard bead model with correctly chosen parameters.*

We repeatedly discussed the universality of conformational properties of polymer solutions in subsections 13.6, 13.7, 20.8, 21.7, and 23.2. The phase

TABLE 4.1. *Basic characteristics of a polymer solution in various regimes**

Regime	No	$\pi d^\beta / T$	ξ / d	R / d
Dilute solution of Gaussian coils	I	Φ/N	$N^{1/2}p^{1/2}$	$N^{1/2}p^{1/2}$
Dilute solution of swollen coils	II	Φ/N	$N^{\beta/5}\tau^{1/5}p^{1/5}$	$N^{\beta/5}\tau^{1/5}p^{1/5}$
Dilute solution of globules	III	Φ/N	$p^{1/2}\lvert\tau\rvert^{-1}$	$N^{1/3}\lvert\tau\rvert^{-1/3}$
Semidilute strongly fluctuating solution	IV	$\Phi^{9/4}\tau^{3/4}p^{3/4}$	$\Phi^{-3/4}\tau^{-1/4}p^{-1/4}$	$N^{1/2}\Phi^{-1/8}\tau^{1/8}p^{1/8}$
Semidilute, weakly fluctuating solution in θ-solvent (with triple contacts dominating)	V	Φ^3	$\Phi^{-1}p^{1/2}$	$N^{1/2}p^{1/2}$
			For stiff chains	
Semidilute, weakly fluctuating solution in good solvent (with pair contacts dominating)	VI	$\tau\Phi^2$	$\Phi^{-1/2}\tau^{-1/2}p^{1/2}$	$N^{1/2}p^{1/2}$
Solution of overlapping but essentially non-interacting Gaussian coils	VII	Φ/N	$N^{1/2}p^{1/2}$	$N^{1/2}p^{1/2}$

*Boundaries between the regions are shown in Figure 4.5.
π—osmotic pressure; ξ—correlation length; R—rms end-to-end distance of a chain in solution; c—solution concentration; d—chain thickness; pd—effective segment length; Nd—contour length, τ relative deviation from the θ point, $\Phi = cd^\beta$.

diagrams of a polymer solution are naturally described in terms of the standard bead model of macromolecules. Just before passing to the direct construction of diagrams, we should recall how the parameters N, a, B, C, and c of the most convenient standard model are determined (*see* subsection 13.8).

In constructing diagrams of state, it is most natural to assume the link as consisting of a chain section whose length is of the order of the thickness d. In this case, both the number of links (beads) $N = L/d$, and their concentration $c = \Phi/d^3$ are associated with both the directly measured molecular mass of the chain and the weight concentration of the polymer in solution by the proportionality coefficients that are defined only by the properties of the monomer links but not the whole chain. According to subsection 13.8, in this case,

$$N = L/d, \quad a \sim dp^{1/2}, \quad B \sim d^3\tau, \quad C \sim d^6, \quad c \sim \Phi/d^3. \tag{26.1}$$

Because in the bead model $B \sim v\tau$, $C \sim v^2$, then

$$v/a^3 \sim p^{-3/2}, \quad p = l/d. \tag{26.2}$$

Among the parameters of the bead model, only B essentially depends on temperature, but the linear dependence $B \sim v\tau$ holds true only near the θ-point. Because $B \sim v$ far from the θ-point, however, one may (without restricting generality) always assume that $B \sim v\tau$ bearing in mind that in the athermic limit $\tau \sim 1$.

Let us now discuss the diagrams of state. We construct the diagram in Figure 4.5 for stiff chains ($p \gg 1$) and then return to Figure 4.4.

26.3. There are three regimes in the behavior of a dilute polymer solution; in these regimes, the individual chains form Gaussian coils, swollen coils, and globules, respectively.

At the θ-point as well as in its vicinity, volume interactions essentially do not disturb the structure of the Gaussian coil in a dilute solution. This region is shown in the diagrams of Figures 4.4 and 4.5 by area I. In regime I, $R \sim \xi \sim dN^{1/2}p^{1/2}$ (*see* subsections 3.2 and 5.1).

Even a small temperature increase above the θ-point, however, is sufficient for volume interaction parameter (13.12) in the coil $z \sim BN^{1/2}/a^3 \sim \tau N^{1/2}p^{-3/2}$ to exceed a value of order unity. For $z \sim 1$ or $\tau \sim N^{-1/2}p^{3/2}$, the macromolecule passes into regime II, in which the characteristics of the macromolecule vary substantially because of the excluded volume effect (*see* subsection 17.3). In this regime, according to Eqs. (17.1) and (26.2)

$$R \sim \xi \sim dN^\nu \tau^{2\nu-1}p^{2-3\nu}. \tag{26.3}$$

Thus, the boundary between regions I and II corresponds to $\tau \sim N^{-1/2}p^{3/2}$.

With a small temperature decrease below the θ-point, the behavior of the solution also varies: the coil turns into the globule that corresponds to regime III. The boundary between regions II and III is the globule–coil transition temperature, that is, according to Eq. (21.8), $\tau_{tr} \sim N^{-1/2}p^{3/4}$. In regime III, the size of the macromolecule (globule) is of order

$$R \sim dN^{1/3}|\tau|^{-1/3}, \tag{26.4}$$

and the correlation length in the globule is much less than R [the correlation length in a large globule is found from Eq. (24.10) for $N \to \infty$]:

$$\xi \sim dp^{1/2}|\tau|^{-1}. \tag{26.5}$$

The boundaries of the semidilute regimes from the side of high concentrations were discussed in subsections 23.1 and 24.6. For coil regimes I and II, this boundary corresponds to the coil overlapping concentration or (what is the same) the link concentration inside one coil $c^* \sim N/R^3$, that is,

$$c^* d^3 \sim N^{-1/2}p^{-3/2} \text{ (boundary I-VII)}, \tag{26.6}$$

$$c^* d^3 \sim N^{1-3\nu}\tau^{-3(2\nu-1)}p^{-3(2-3\nu)} \text{ (boundary II-IV)}. \tag{26.7}$$

In the case of globular regime III, the limiting factor is not a geometric overlapping of globules but rather their precipitation. According to Eqs. (24.19) and (26.1), this happens when

$$cd^3 \sim c_{\text{dil}}d^3 \sim |\tau|\exp(-\text{const} \cdot |\tau|^{4/3}N^{2/3}p^{1/2}). \tag{26.8}$$

26.4. *For a semidilute solution in good solvent, the decrease of the osmotic pressure $\sim \tau^{3/4}$, the size of an individual chain $\sim \tau^{1/4}$, and the increase in the correlation length $\sim \tau^{-1/4}$, when the temperature approaches the θ-point (with a decrease of τ) are easy to study by the scaling method; the chains in such a solution behave as swollen coils on small scales (less than correlation lengths) and as Gaussian coils on large scales.*

On crossing the polymer coil overlap concentration c^*, we move into the semidilute solution region. In the previous section, we calculated the values of π, ξ, and R in regime IV at $\tau \sim 1$ (athermic limit) and $p \sim 1$ (flexible chains). Let us now generalize the consideration carried out in Sec. 25, taking into account the dependences on τ and p.

In accordance with the basic principle of scale invariance and for arbitrary τ and p, the concentration c^* should be the only characteristic concentration associated with a crossover between the dilute and semidilute regimes. Accordingly, Eqs. (25.6), (25.14), and (25.17) should be valid as they were previously for the quantities π, ξ, and R. However, c^* is to be evaluated not as $c^* \sim a^{-3}N^{1-3\nu}$ (23.3) applicable only for $\tau \sim 1$, $p \sim 1$, but rather using the more general formula (26.7). As a result, we obtain for regime IV

$$\pi/T \sim (c/N)(c/c^*)^m \sim c(cd^3)^{1/(3\nu-1)}\tau^{3(2\nu-1)/(3\nu-1)}p^{3(2-3\nu)/(3\nu-1)}, \tag{26.9}$$

$$\xi/d \sim N^\nu \tau^{2\nu-1} p^{2-3\nu} (c/c^*)^l \sim (cd^3)^{-\nu/(3\nu-1)} \tau^{-(2\nu-1)/(3\nu-1)} p^{-(2-3\nu)/(3\nu-1)},$$
$$(26.10)$$

$$R/d \sim N^\nu \tau^{2\nu-1} p^{2-3\nu} (c/c^*)^n$$
$$\sim N^{1/2} (cd^3)^{-(2\nu-1)/2(3\nu-1)} \tau^{(2\nu-1)/2(3\nu-1)} p^{(2-3\nu)/2(3\nu-1)}. \quad (26.11)$$

Here, the exponents $m = 1/(3\nu-1)$, $l = -\nu/(3\nu-1)$, and $n = -(2\nu-1)/2(3\nu-1)$ are found from the same considerations and have the same values as in Sec. 25.

It is easy to verify (and the reader is advised to do so) that in regime IV discussed here, the interpretation based on the notion of a blob adopted in subsections 25.6 to 25.8 holds true, provided that the number g of links in the blob is formally defined so that the g-link chain would have the size ξ, that is, $\xi/d \sim g^\nu \tau^{2\nu-1} p^{2-3\nu}$ (26.3) or, according to Eq. (26.10),

$$g \sim (cd^3)^{-1/(3\nu-1)} \tau^{-3(2\nu-1)/(3\nu-1)} p^{-3(2-3\nu)/(3\nu-1)}. \quad (26.12)$$

Note that for $\tau \ll 1$, the blob thus defined does not correspond to a chain section between two contacts with other chains. The given definition, however, has definite merit in allowing an interpretation of the osmotic pressure (26.9) in terms of an ideal gas of blobs ($\pi/T \sim c/g$). Then, the blob system turns out to be close packed ($g \sim c\xi^3$), and the size of the macromolecule (26.11) can be defined as the size of a Gaussian coil from the chain of blobs $[R/d \sim \xi(N/g)^{1/2}]$.

Let us now turn to the problem of the boundary of regime IV, considered here from the side of high concentrations. Note that the size of the macromolecule (26.11) diminishes with increasing concentration. This is a manifestation of the screening of the excluded volume effect (see subsection 24.4) becoming more pronounced as c grows. This screening, however, cannot reduce the value of R to less than the size of an individual Gaussian coil on the order of $dN^{1/2}p^{1/2}$. One can see immediately that the mentioned quantities become equal at the concentration

$$c^{**}d^3 \sim \tau p^{-3} \quad \text{(boundary IV-VI)}. \quad (26.13)$$

Increasing the solution concentration c to higher than c^{**}, the solution behavior regime changes.

To make clear what happens at $c \sim c^{**}$, we calculate for regime IV the volume interaction parameter of a g-link chain section $g^{1/2} B/a^3$ constituting one blob. This quantity diminishes with increasing concentration. Using Eqs. (26.1) and (26.12), it is easy to verify that for $c \sim c^{**}$, it is of the order of unity. Consequently, for $c > c^{**}$, the chain statistics remains Gaussian not only on large scales but also on small ones (i.e., inside a blob).

26.5. *The regimes whose chain statistics are Gaussian on all length scales are described by the self-consistent field theory.*

The fact that for $c \gg c^*$ the chains become Gaussian implies from the physical viewpoint that link collisions are no longer affected by the surrounding "clouds of links" that are correlated with the initial links participating in the collision (*see* subsections 19.2, 19.4, and 25.5). One may say that concentration growth makes the links of other chains approach the given link so closely that they find themselves inside its "cloud." Consequently, the picture of volume interactions in a polymer solution at $c \gg c^{**}$ is identical to that in the cloud of non-correlated links, that is, conforms to the validity conditions of the self-consistent field method. This method is used to describe regions V to VII of the diagram in Figure 4.5.

It is seen from the formulas of the self-consistent field theory for the osmotic pressure (24.3) or the correlation radius (24.10) that for a semidilute solution there are two regimes, VI and V, where the binary and triple collisions, respectively, prevail. They can be called the regimes of good and θ-solvent. Using the previously mentioned formulas and the estimates (26.1), it is easy to find the boundary between the regimes

$$c^{***}d^3 \sim \tau \quad \text{(boundary V-VI)}, \tag{26.14}$$

as well as all solution parameters: for regime VI (predominantly pair collisions)

$$\pi/T \sim c(cd^3)\tau, \tag{26.15}$$

$$\xi/d \sim (cd^3)^{-1/2}\tau^{-1/2}p^{1/2}; \tag{26.16}$$

and for regime V (predominantly triple collisions)

$$\pi/T \sim c(cd^3)^2, \tag{26.17}$$

$$\xi/d \sim (cd^3)^{-1}p^{1/2}. \tag{26.18}$$

In both regimes VI and V, $R \sim dN^{1/2}p^{1/2}$.

Finally, regime VII is also feasible. It is realized in the region where Eqs. (26.16) and (26.18) yield a non-physical value of the correlation length ξ exceeding the coil size. From the condition $\xi \sim dN^{1/2}p^{1/2}$, it is easy to determine the appropriate boundaries:

$$cd^3 \sim N^{-1/2} \quad \text{(boundary V-VII)}, \tag{26.19}$$

$$cd^3 \sim \tau^{-1}N^{-1} \quad \text{(boundary VI-VII)}. \tag{26.20}$$

Note that boundary V-VII corresponds in order of magnitude to the same concentration as for the critical point of phase separation of solution (24.22). The solution parameters in regime VII are also simply determined from the general formulas (24.3) and (24.10): $\pi/T \sim c/N$, $\xi \sim R \sim dN^{1/2}p^{1/2}$. The osmotic pressure of a solution in regime VII thus equals the ideal-gas pressure of coils. This signifies that even though the coils strongly overlap in regime VII [a geometric overlap occurs on transition from I into VII, Eq. (26.6)], the interaction between them remains insignificant because of the proximity of the θ-point and low concentration.

26.6. *The region of regimes with developed fluctuations is large only for solutions of very flexible chains; the chain stiffness parameter p plays the role of a Ginzburg number, defining the applicability of the self-consistent field method.*

We have studied all regimes of a solution of stiff chains and wholly investigated the diagram of states in Figure 4.5. A simpler diagram in Figure 4.4 for flexible chains is described by the formulas of the present section at $p \sim 1$. This differs from Figure 4.5 by the absence of regimes VI and VII, which are as if "absorbed" by regions IV and I.

Note that increasing the chain stiffness p results in a drastic extension of application of the self-consistent field method and a narrowing of the fluctuation region, which shifts toward higher temperatures and lower concentrations. As $\tau \lesssim 1$, then for $p > N^{1/3}$ (or $N < p^3$) the fluctuation region is absent altogether. This remark is quite significant for many stiff-chain polymers (e.g., cellulose ethers, double-helix DNA, and so on) for which the case $N < p^3$ is typical. In phase-transition theory, the quantity determining the width of the fluctuation region is called the *Ginzburg number* **Gi**. Thus, for a polymer solution, **Gi** $\sim p^{-3/2} \sim v/a^3$ (26.2): the smaller v/a^3, the higher p (i.e., the higher the chain stiffness), the broader the region of applicability of the self-consistent field theory, and the less that of the fluctuation theory.

Note that even for the most flexible polymers, the number p equals a few units (approximately five; *see* subsection 13.9). Therefore, the fluctuation region is never excessively broad.

CHAPTER 5

Other Polymer Systems

Using the methods and concepts developed earlier, we consider in this chapter some special polymer systems that are of interest from the standpoint of both theory and practice, for example, mixtures of various polymers, solutions and melts of block copolymers or diphilic compounds, polymer liquid crystals, polyelectrolytes, and polymer networks. The contents of different sections of this chapter are not interrelated; therefore, Secs. 27 to 30 can be read in any order. In addition, this chapter is not needed for understanding Chapters 6 and 7.

27. POLYMER MIXTURES AND BLOCK COPOLYMERS

27.1. *As a rule, different polymers mix very poorly; a very slight repulsion between the links is sufficient to separate the mixture into virtually pure phases.*

Many common polymer systems consist of macromolecules of several different types. Functional requirements imposed on such a system may be very diverse. For example, in constructing some optical devices, a thorough homogeneity must be ensured to maintain the necessary transparency. On the other hand, in the production of various composite materials, required mechanical properties can be secured only with a certain segregation of the components in the material.

Here, we briefly consider only the simplest system of this type: a mixture of two different polymers, referred to as poly-A and poly-B. The first obvious question arising from the study of such a system pertains to the miscibility conditions or, conversely, to the conditions of the phase separation of the mixture. This is the problem tackled here.

Because this question is purely thermodynamic, one should begin by seeking an expression for the free energy of the mixture. Such an expression is easily found in the self-consistent field approximation. For a homogeneous system, we can immediately write

$$F/V = T(c_A/N_A)\ln c_A + T(c_B/N_B)\ln c_B + f^*(c_A, c_B, T), \qquad (27.1)$$

where the first two terms describe the entropy of translation (9.4) of A- and B-coils with chain lengths N_A and N_B and concentrations c_A and c_B, and where the third term is the volume interaction contribution. By analogy with Eq. (15.7), it is clear that $f^*(c_A, c_B)$ is the characteristic of the mixture of discon-

180

nected links, the contribution of the interactions to the free energy per unit volume. As for the analogous quantity $f^*(c)$ in a one-component system (see subsection 15.3), a virial expansion could be written for $f^*(c_A, c_B)$. Because, however, we deal most often with a dense system, concentrated solution, or melt, interpolation formulas are indispensable, and the most convenient is the Flory-Huggins formula. It is usually written in terms of the volume fractions of the components $\Phi_A \sim c_A$, $\Phi_B \sim c_B$, and the volume fraction of the solvent Φ_s:

$$f^*/T = \Phi_s \ln \Phi_s + \chi_{AB} \Phi_A \Phi_B + \chi_{As} \Phi_A \Phi_s + \chi_{Bs} \Phi_B \Phi_s. \qquad (27.2)$$

As a rule, the incompressibility condition

$$\Phi_s + \Phi_A + \Phi_B = 1. \qquad (27.3)$$

can additionally be assumed to be satisfied.

With Eqs. (27.1) to (27.3) in mind, investigation of the phase separation presents a purely algebraic task. We suggest the reader to solve this: only some results are cited here.

For a melt of two polymers (i.e., when $\Phi_s = 0$), the critical miscibility point is given by the following equalities:

$$\chi_{AB}^{(cr)} = (N_A^{1/2} + N_B^{1/2})^2 / (2 N_A N_B), \qquad (27.4)$$

$$\Phi_A^{(cr)} = N_B^{1/2} / (N_A^{1/2} + N_B^{1/2}), \quad \Phi_B^{(cr)} = N_A^{1/2} / (N_A^{1/2} + N_B^{1/2}). \qquad (27.5)$$

The formula (27.4) provides evidence of an extremely low miscibility of the polymers of comparatively long chains. If $N_A \sim N_B \sim N$, then $\chi_{AB}^{(cr)} \sim 1/N$, that is, a very slight repulsion of links is sufficient to segregate the mixture components. In fact, it is fairly difficult to find a couple of polymers for which the inequality $\chi_{AB} < 1/N$ would hold and intermixing be possible.

What is important is that the miscibility of the polymers with chains of different length is determined by short macromolecules. If, for example, $N_A \gg N_B$, then $\chi_{AB}^{(cr)} \sim 1/N_B$.

Two remarks to conclude this subsection. First, if a polymer mixture is in the homogeneous state, then the Flory theorem (see subsection 24.2) is valid for this mixture, and the chains of all of the components behave as Gaussian chains (on scales larger than the correlation length). Second, if a mixture separates, then a slight ($\sim N^{-1/2}$) excess of χ_{AB} over the critical value $\chi_{AB}^{(cr)}$ is sufficient for either phase to become an almost pure component (so that the concentration of each polymer in the "foreign" phase is negligible).

27.2. *If the chains of immiscible polymers are incorporated as blocks into copolymer macromolecules, then the solution or melt of such block copolymers cannot separate and forms a domain structure with domains of appropriate size and symmetry.*

Microscopic phase separation in a solution or melt of block copolymers (discussed briefly in this subsection) represents an interesting phenomenon.

Many problems associated with it are not solved yet, and the potential of its practical use is far from being exhausted.

Recall that a block copolymer is a chain consisting of consecutively joined blocks, each of which constitutes a long homopolymer chain. For example, the chemical structure of a two-block copolymer is A—...—A—B—...—B. The number of blocks in the molecule, just as the number of links in the block, can be arbitrary.

What happens when a sufficiently concentrated solution or melt is made from the chains of a block copolymer? From previous subsections, we know that in a typical case, the chains of poly-A and poly-B (or in our case, the blocks) are incompatible. A phase separation in such a system, however, is impossible because of the covalent bonding of the blocks into common chains.

As a result, the phase separation, which is impossible on the scale of the whole system, occurs on a certain limited length scale defined by the size of the blocks. The arising microdomain structure is schematically shown in Figure 5.1.

If the total amount of one of the components (e.g., A) is relatively small, then the corresponding phase enriched with A component (the A phase) occupies a small fraction of the total volume, and it constitutes a system of spherically shaped micelles scattered like "islets" in a "sea" of the phase enriched with B component (the B phase). On increasing the fraction of A links, the spherically shaped micelles become cylindric ones piercing the B phase like reinforcing wires. On further increasing the A fraction, a lamellar (or layer), structure appears, with A and B phases laid out in alternating planar layers. Finally, on still further increasing the fraction of A links, the so-called inverse phases emerge: first the cylindric phase (B phase cylinders piercing the A phase), then the spheric one (B "raisins" in the A "pudding").

To conclude, it should be noted that the microdomain (or micellar) structures are typical not only for block copolymers but also for systems consisting of the so-called diphilic molecules. One of their blocks has a low molecular weight, but because of its thermodynamic properties, it cannot mix with the other block. Examples include phospholipid molecules consisting of a hydrophilic "head" and a polymeric (usually not very long) "tail." The dissolution of such mole-

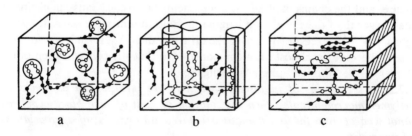

a   b   c

FIGURE 5.1. Microdomain structure in a melt of block copolymers. (a), Spheric A phase micelles in massive B phase. (b), Cylindric micelles. (c), Alternating planar lamellae.

cules in water leads to the formation of micellar structures with a hydrophobic (almost water-free) and hydrophilic (water-saturated) phases.

28. LIQUID-CRYSTAL POLYMERS

28.1. *In liquid crystals, the anisotropy of properties combines with the absence of long-range translational order; the liquid-crystal phases can be nematic (possessing a single direction of preferred orientation), cholesteric (nematic with orientational direction twisting along a helix), and smectic (with long-range translational order in one or two dimensions).*

The properties of matter in a liquid-crystalline state are intermediate between the liquid and the crystalline solid. As in liquids, long-range translational order is absent in liquid crystals. At the same time, the molecules in a liquid-crystalline phase retain long-range orientational order (i.e., liquid crystals are anisotropic like the crystalline solids).

Even though liquid crystals were discovered at the end of the 19th century, interest in their research increased in the 1960s, when some important practical applications of this new class of substances showed up (in particular, in the technology of indicating devices). The foundations of the physics of liquid crystals have been comprehensively described.[31–33]

Three basic modifications of liquid-crystalline phases are known: 1) nematic, 2) cholesteric, and 3) smectic (Fig. 5.2). In the nematic liquid crystal, there is a direction of preferred orientation, along which the molecules tend to orient themselves with their long axes (Fig. 5.2a); long-range translational order in the molecular pattern is totally absent. The cholesteric phase differs from the nematic by the direction of preferred orientation being twisted in a helix (Fig. 5.2b); such a phase may appear in a substance consisting of chiral molecules (i.e., molecules that cannot be superimposed on their mirror image). Smectic liquid crystals form layered structures (Fig. 5.2c); and in addition to local orientational order, they possess long-range translational order in one or two dimensions.

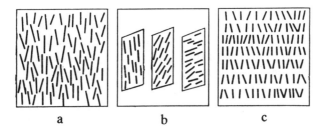

a   b   c

FIGURE 5.2. A sketch of nematic (a), cholesteric (b), and smectic (c) liquid crystals.

28.2. *Formation of a liquid-crystalline phase is typical for solutions and melts of stiff-chain macromolecules and for copolymers containing stiff- and flexible-chain fragments.*

Figure 5.2 makes clear that a trend toward formation of a liquid-crystalline phase should manifest itself most in substances whose molecules have a prolate form. In this case, anisotropy may arise even from purely steric effects (i.e., from the impossibility of arranging a sufficiently dense system of anisodiametric particles in an isotropic way).

From this viewpoint, it is clear that stiff-chain macromolecules, that is, macromolecules in which the length l of a Kuhn segment is much greater than the characteristic thickness d of the chain, should easily form a liquid-crystalline phase (primarily nematic or, if the links are chiral, cholesteric). This is indeed the case. Examples of macromolecules capable of forming liquid-crystalline phases of various types are provided by any helical macromolecules (α-helical polypeptides, DNA macromolecules, and so on), aromatic polyamides, some cellulose ethers, and so on. The asymmetry parameter of such macromolecules (i.e., the ratio l/d) may be very large; for the first two of the given examples, it may reach 20 to 50.

In this case, the nematic (or cholesteric) phase is formed, as a rule not only in a pure polymeric substance (in a polymer melt) but also in a more or less concentrated solution of such macromolecules. Liquid-crystalline polymer melts are frequently referred to as *thermotropic polymer liquid crystals* (because the liquid-crystalline transition in such substances can be induced in the most natural way by temperature change) and anisotropic polymer solutions as *lyotropic liquid crystals*.

In both thermotropic and lyotropic cases, the properties of the formed nematic (or cholesteric) phase in the system of highly anisodiametric polymer molecules ($l \gg d$) must, of course, differ considerably from the properties of low-molecular-weight liquid crystals, for which the asymmetry parameter l/d is usually not so large. Specifically, in theoretic studies of liquid-crystalline ordering in solutions and melts of stiff-chain polymers, the most significant problem is to find the asymptotic characteristics for $l/d \gg 1$. The existence of an additional large parameter leads to a simplification of the problem of liquid crystals consisting of stiff-chain polymers as opposed to the problem of low-molecular-weight liquid crystals.

Stiff-chain macromolecules do not constitute the only class of polymers capable of forming liquid-crystalline phases. Such a phase can appear in melts (less commonly, in concentrated solutions) of copolymers containing both flexible and stiff sections of a chain. As a rule, the specifics of the equilibrium properties of such liquid crystals manifest themselves to a much lesser degree than in the case of stiff-chain macromolecules, because the asymmetry parameter of a stiff section of a chain is usually not so large. In terms of dynamic properties, however, these systems remain very peculiar objects, combining the properties of both polymers and liquid crystals. While for the solutions (melts) of stiff-chain macromolecules the formation of the nematic and cholesteric phases is typical, copolymers often exhibit a smectic ordering as well. The layered smectic structure is formed in this case because of microscopic phase separation similar to that considered in subsection 27.2. As a result, the strata enriched with stiff-chain component are formed and interlaid with flexible-chain sections (similar to Fig. 5.1c).

This section presents the foundations of the theory of nematic liquid-crystalline ordering in dilute solutions of stiff-chain macromolecules, which is the most advanced at present. Other aspects of the statistical physics of liquid-crystalline polymers are discussed elsewhere.[34]

We begin by considering a solution of long, stiff rods. This is one of the simplest systems in which the nematic phase can appear.

28.3. *The Onsager approximation in the theory of nematic ordering corresponds to the second-order virial approximation for a system of asymmetric particles; this approximation is asymptotically correct for the limiting case of a solution of long, thin rods.*

The first molecular theory of nematic ordering was proposed by L. Onsager in 1949 for a solution of long, cylindric, stiff rods of length L and diameter d ($L \gg d$). In terms of polymer theory, such a system constitutes a model of a solution of ultimately stiff-chain macromolecules whose flexibility is so negligible that it fails to manifest itself over the length L.

Suppose that N rods are located in the volume V so that their concentration is $c = N/V$ and the volume fraction of the rods in the solution $\Phi = \pi c L d^2/4$. Assume that the solution is athermal, that is, only repulsive forces act between the rods (because of the mutual impermeability of the rods). The liquid-crystal ordering then occurs, resulting from purely steric causes. Let us now introduce the function of rod distribution over orientations $f(u)$ (cf. subsection 9.4); $cf(u)d\Omega_u$ is the number of rods in a unit volume oriented within the small solid angle $d\Omega_u$ around the direction defined by the unit vector u. In the isotropic state, $f(u) = \text{const} = 1/4\pi$ according to Eq. (9.5); in the nematic phase, $f(u)$ is a function with a maximum around the anisotropy axis.

In the Onsager approximation, the free energy of the solution of rods is written as a functional of $f(u)$ as

$$F = NT \ln \Phi + NT \int f(u)\ln[4\pi f(u)]d\Omega_u$$

$$+ NTc \iint f(u_1)f(u_2)B(\gamma)d\Omega_{u_1}d\Omega_{u_2}. \tag{28.1}$$

Here, the first term represents the free energy of translational motion of the rods [cf. Eq. (9.4)]. The second term describes the orientational entropy losses because of the nematic ordering of macromolecules; the general method to calculate these losses can be found in subsection 9.4. In the considered case of stiff rods, these losses are defined by the same formula (9.8) as for chains of freely jointed stiff segments.

The third term in Eq. (28.1) is the free energy of rod interaction in the second-order virial approximation. In this last term, $B(\gamma)$ is the second virial coefficient of interaction of the rods whose long axes defined by unit vectors u_1 and u_2 make an angle γ with each other.

As long as we examine the case when the interaction of the rods is reduced to their mutual impermeability, the value of $B(\gamma)$ equals half the volume excluded by one rod for a motion of the other [see Eqs. (12.4) and (14.8)]. Calculation of this excluded volume is illustrated in Figure 5.3. The result then takes the form

$$B(\gamma) = L^2 d |\sin \gamma| \qquad (28.2)$$

to agree with the estimate (13.18).

Thus, the fundamental approximation of the Onsager method lies in the fact that rod interaction is accounted for in the second-order virial approximation. Consequently, this method is only applicable for sufficiently low concentrations of the solution of rods. To characterize quantitatively the range of applicability of the Onsager method, let us now estimate the value of third-virial coefficient for the solution of rods.

In the case of the steric interaction the third virial coefficient, C is proportional to the volume of domain in the phase space of three particles within which each of these particles overlaps with the two others.[35] If in the system of three rods one rod is fixed, then the volume $\sim L^2 d$ will be excluded for the motion of the second rod and the same volume for the motion of the third rod. To ensure the contacts of each rod with the others, however (because only such configurations contribute to C), all three rods must lie approximately in one plane. This requirement implies that the additional factor d/L should be introduced in the estimate for C. Therefore, we can evaluate the coefficient C as

$$C \sim dL^2 \cdot dL^2 \cdot d/L \sim d^3 L^3. \qquad (28.3)$$

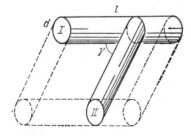

FIGURE 5.3. Calculation of the second-virial coefficient of interaction between rod-like macromolecules. The figure shown in broken lines is close to a prism of thickness d, its base is a parallelogram with area $l^2 |\sin\gamma|$, and its volume is close to $l^2 d |\sin\gamma|$. The volume $2l^2 d |\sin\gamma|$ (on the both sides of rod I) is unavailable for rod II.

[cf. Eq. (13.18)]. A more accurate calculation yields $C \sim d^3 L^3 \ln(L/d)$, that is, confirms the estimate (28.3) to be correct up to a logarithmic factor.

The second-order virial approximation for a solution of rods with concentration c is valid under the condition $Bc \gg Cc^2$. Inserting into this inequality the estimate (28.3) for C and the estimate for B following from Eq. (28.2) $B \sim L^2 d$, we obtain the condition of applicability of the Onsager approximation in the form

$$c \ll 1/(Ld^2) \quad \text{or} \quad \Phi \ll 1. \tag{28.4}$$

Subsection 28.4 shows that in the limiting case of long rods $(L \gg d)$, a liquid-crystal transition occurs for $\Phi \ll 1$. Therefore, the Onsager method is asymptotically correct when used to study this transition and the properties of the anisotropic phase emerging in the limit $L \gg d$ (which is the most interesting in the research of stiff-chain macromolecules).

28.4. *In the athermal solution of rods of length L and diameter d, the liquid-crystalline ordering is a first-order phase transition proceeding when the volume fraction of the rods in the solution* $\Phi \sim d/L \ll 1$; *the nematic phase appearing at the transition point is highly ordered.*

To find the equilibrium distribution function, the expression (28.1) must be minimized with respect to $f(u)$. Straightworward minimization leads to an integral equation that can only be solved numerically. Onsager used an approximate variational method with the test function

$$f(u) = \frac{\alpha}{4\pi} \frac{\text{ch}(\alpha \cos \vartheta)}{\text{sh } \alpha}, \quad \int f(u) d\Omega_u = 1, \tag{28.5}$$

where ϑ is the angle the vector u makes with the direction of the anisotropy axis and α the variational parameter. The test function (28.5) is inserted into the expression (28.1), which is then minimized with respect to α. The minima found corresponds to the possible phases (isotropic and nematic). Transitions between these phases can be studied by a conventional method, equating the pressures $\pi = -F/V + c\partial(F/V)/\partial c$ and chemical potentials $\mu = \partial(F/V)/\partial c$ of the two phases.

As a result of the conceptually trivial but mathematically awkward calculations, the following conclusions are obtained.

1. The orientational ordering in a solution of long, stiff rods is a first-order phase transition occurring at low concentrations of rods in the solution $(\Phi \sim d/L \ll 1)$ when the second-order virial approximation still holds.

2. For $\Phi < \Phi_i$, the solution is isotropic. For $\Phi > \Phi_a$, it is anisotropic, and for $\Phi_i < \Phi < \Phi_a$, the solution separates into the isotropic and anisotropic phases and

$$\Phi_i = 3.34 d/L, \quad \Phi_a = 4.49 d/L, \quad w \equiv \Phi_a/\Phi_i - 1 = 0.34. \tag{28.6}$$

The degree of ordering of the appearing nematic phase can be characterized by the "order parameter"

$$s = \langle 3\cos^2\vartheta - 1 \rangle / 2 = (1/2) \int (3\cos^2\vartheta - 1) f(u) d\Omega_u. \qquad (28.7)$$

In the isotropic phase, $s=0$; In the totally oriented one, $s=1$. In the intermediate cases, the value of s varies from 0 to 1, assuming larger values as the degree of ordering of the solution increases. Calculations using the formula (28.7) show that the order parameter in an athermal solution of stiff rods at the point of emergence of the nematic phases (i.e., at $\Phi = \Phi_a$) equals

$$s_0 = 0.84, \qquad (28.8)$$

that is, the anisotropic phase is sufficiently highly oriented.

It should be noted that the only fundamental physical limitation of the Onsager method is connected with the second-order virial approximation (i.e., with the condition $\Phi \ll 1$). Using the variational procedure only makes the calculations simpler. The integral equation arising from accurate minimization of the functional (28.1) can be solved numerically with a high degree of accuracy. As a result, the following values of Φ_i, Φ_a, and s_0 are obtained

$$\Phi_i = 3.290 d/L, \quad \Phi_a = 4.223 d/L, \quad s_0 = 0.796. \qquad (28.9)$$

This proves that the variational technique ensures a very small uncertainty (approximately 5%) of the determinations of liquid-crystal transition parameters.

28.5. *In the solutions of semiflexible chains, for which the Kuhn segment length (even though exceeding considerably the chain thickness) is much less than the total contour length of the chain, nematic ordering can occur; the parameters of the corresponding phase transition depend on the flexibility mechanism of the polymer chain.*

Real stiff-chain macromolecules always possess a certain finite flexibility. Therefore, the effects of chain flexibility on the properties of the liquid-crystal transition should be analyzed to apply the results obtained earlier to real polymer solutions.

Depending on the ratio of the total contour length L of a macromolecule to the effective Kuhn segment length l, the stiff-chain macromolecules may be subdivided into the following three classes:

1. If the Kuhn segment length is so large that $l \gg L \gg d$, then the flexibility of the polymer chain can be ignored, and we have the case of the ultimately stiff-chain macromolecules (stiff rods) considered earlier.

2. If $L \gg l \gg d$, then the stiff-chain macromolecule comprises many Kuhn segments and from the global viewpoint resides in the state of a statistical coil. (Such macromolecules are called *semiflexible*.)

3. The possible intermediate cases when the contour length of the macromolecule and the Kuhn segment length are of the same order of magnitude, $L \sim l$, are also possible.

In all of these three cases, the molecules composing the solution are characterized by a pronounced asymmetry ($L \gg d$, $l \gg d$); therefore, the nematic phase can be expected to appear already in a dilute solution. Now, we illustrate how the chain flexibility affects the liquid-crystalline ordering by the example of a solution of semiflexible macromolecules ($L \gg l \gg d$). One can read about the transition from an isotropic to a nematic phase for macromolecules with $L \sim l$ elsewhere.[34]

Semiflexible macromolecules differ according to the flexibility mechanism of the polymer chain. We consider two well-studied flexibility mechanisms: 1) freely jointed (Fig. 1.1; the chain is pictured as a sequence of long, freely jointed stiff rods of length l and diameter d, $l \gg d$), and 2) persistent one (Fig. 1.2; the chain is a uniform, cylindric, elastic filament with Kuhn segment length l and diameter d, $l \gg d$).

Do the characteristics of the transition from an isotropic to nematic phase or only on the values of the parameters l and d depend on the flexibility mechanism of the semiflexible chain? To answer this, we write the expression for the free energy of an athermal solution of semiflexible chains, analogous to the expression (28.1) for the solution of rods.

In this case, the first term corresponds to the entropy of the translational motion of macromolecules as a whole relative to one another. For long, semiflexible chains ($L \gg l$), this contribution to the free energy is quite insignificant (cf. the analogous conclusion in subsection 24.6). Thus, it does not produce any substantial effects on the properties of nematic ordering, and it therefore can be neglected. Consequently, the free energy of the solution of semiflexible macromolecules in the Onsager approximation must contain the contribution $F_{conf} = -TS\{f(u)\}$ associated with the entropy loss $S\{f(u)\}$ because of orientational ordering and the free energy F_{ster} of the steric interaction of macromolecules in the second-order virial approximation:

$$F = F_{conf} + F_{ster} = -TS\{f(u)\} + F_{ster}. \qquad (28.10)$$

The method of calculation of the entropy contribution $F_{conf} = -TS\{f(u)\}$ was described in subsection 9.4; the expressions (9.7) and (9.8) were derived for the entropy $S\{f(u)\}$ of solutions of long persistent and freely jointed macromolecules. These expressions are seen to differ conspicuously in their form. This is sufficient to conclude that the characteristics of the liquid-crystal transition in a solution of semiflexible chains depends not only on the values of l and d but also on the flexibility mechanism of a polymer chain.

Before writing the expression for F_{ster}, it should be noted that as soon as $l \gg d$ for semiflexible macromolecules, the polymer chains can always be subdivided into sections of length λ, such that $d \ll \lambda \ll l$, and called *elementary links*. Elementary links defined in this way are, in fact, long, stiff rods, and the second-virial coefficient of interaction of two links with orientations u_1 and u_2 equals $B_\lambda(\gamma)$ $=\lambda^2 d |\sin \gamma|$ [cf. Eq. (28.2)]. Taking this into account, the expression for F_{ster} can be written in the form [cf. the third term in Eq. (28.1)]

$$F_{ster}=N\frac{L}{\lambda}T\frac{4}{\pi}\frac{\Phi}{\lambda d^2}\iint f(u_1)f(u_2)\cdot\lambda^2 d|\sin \gamma|\,d\Omega_{u_1}d\Omega_{u_2} \qquad (28.11)$$

[N is the total number of macromolecules in the solution and $f(u)$ the distribution function over the orientations of the vectors u tangential to the chain], because L/λ is the number of elementary links in the macromolecule and $4\Phi/(\pi\lambda d^2)$ their concentration in the solution. It is obvious that the quantity λ in Eq. (28.11) will be canceled, and we finally obtain

$$F_{ster}=N\frac{4}{\pi}T\frac{L\Phi}{d}\iint f(u_1)f(u_2)|\sin \gamma|\,d\Omega_{u_1}d\Omega_{u_2}. \qquad (28.12)$$

28.6. In solutions of semiflexible persistent macromolecules, nematic ordering occurs at high solution concentrations, and the resulting anisotropic phase is less ordered than in the case of freely jointed macromolecules with the same Kuhn segment length and the same chain diameter.

Equations (9.7), (9.8), and (28.12) totally define the free energy (28.10) of the solution of semiflexible persistent or freely jointed macromolecules as a functional of the distribution function $f(u)$. The minimization of F with respect to $f(u)$ and all subsequent calculations can be carried out completely in analogy with subsection 28.4. As a result, we reach the following conclusions:

1. For both flexibility mechanisms, the orientational ordering of the athermal solution is a first-order phase transition proceeding at low polymer concentrations in the solution.

2. For $\Phi < \Phi_i$, the solution is homogeneous and isotropic. For $\Phi > \Phi_a$, the solution homogeneous and anisotropic, and $\Phi_i < \Phi < \Phi_a$ is separated into isotropic and nematic phases, with $\Phi_i \sim \Phi_a \sim d/l \ll 1$.

3. For an athermal solution of freely jointed semiflexible chains, the characteristics of the liquid-crystal transition are

$$\Phi_i=3.25d/l, \quad \Phi_a=4.86d/l, \quad w=0.50, \quad s_0=0.87. \qquad (28.13)$$

Comparing the result (28.13) with (28.6) and (28.8), we conclude that the fact that the rods are freely jointed into long chains affects only slightly the parameters of the transition from an isotropic to nematic phase: the phase separation region broadens somewhat, and the order parameter of the emerging orientation-ordered phase grows slightly.

For an athermal solution of persistent semiflexible chains, we have

$$\Phi_i=10.48d/l, \quad \Phi_a=11.39d/l, \quad w=0.09, \quad s_0=0.49. \qquad (28.14)$$

It is easy to see that the orientational ordering in the solution of persistent chains proceeds at substantially higher concentrations than in the solution of freely jointed macromolecules (with the same values of the parameters d and l). In addition, the relative jumps in the polymer concentration in the solution during the transition and in the order parameter on emergence of the liquid-crystalline phase prove to be appreciably smaller.

These differences result from the different form of the expressions (9.7) and (9.8) for entropy losses because of orientational ordering of persistent and freely jointed chains [because Eq. (28.12) for F_{ster} is the same in both cases]. The analysis of Eqs. (9.7) and (9.8), which we recommend that the reader perform, shows that when the orientational ordering is high [e.g., for large values of α in Eq. (28.5)], the entropy losses for the persistent flexibility mechanism are much higher than for the freely jointed chains. The physical meaning of this fact is simple: to orient the segment of a freely jointed chain within the solid angle $\Delta\Omega$, it is sufficient to direct it only once according to the given orientation. The same procedure requires multiple corrections of the chain direction for an effective segment of the persistent chain at $\Delta\Omega \ll 1$. This is why persistent chains are more difficult to orient: nematic ordering proceeds at high solution concentrations, and the order parameter is smaller than for freely jointed chains.

28.7. *As the orientational ordering grows in a solution of persistent chains, so does the mean size of the chain along the ordering axis (i.e., the macromolecules straighten); in a solution of freely jointed chains, the orientation does not force the macromolecules to straighten.*

The difference in character of nematic ordering in solutions of semiflexible macromolecules with different flexibility mechanisms manifests itself not only in the thermodynamic characteristics of the phase transition itself but also in the conformations of polymer chains in the liquid-crystalline phase. Let us consider, for example, the mean square end-to-end distance $\langle R_z^2 \rangle$ of a polymer chain along the ordering axis for the freely jointed and persistent flexibility mechanisms.

Clearly, in these two cases, the chain orientation in the nematic phase affects the value of $\langle R_z^2 \rangle$ differently. In the case of the ultimately oriented solution ($s = 1$), each segment of the freely jointed chain may assume two opposing directions so that $\langle R_z^2 \rangle = Ll$, whereas the persistent chain must straighten completely (i.e., $\langle R_z^2 \rangle = L^2$). Therefore, it is natural to expect that a similar difference would also be observed for finite degrees of ordering ($s < 1$). The orientation in the solution of freely jointed chains does not result in a straightening of the chains along the ordering axis ($\langle R_z^2 \rangle < Ll$) because of the equivalence of opposite directions and the possibility of a sudden change of chain orientation at the points of the free joint. At the same time, the orientation leads to appreciable straightening of the chains in the solution of persistent chains. Indeed, because of the continual character of the persistent chain, the transition to the opposite direction involves a realization of all intermediate chain orientations. This is very unfavorable, because the mean field emerging in such a solution tends to orient all macromolecules along the ordering axis. The value of

$\langle R_z^2 \rangle$ therefore must grow substantially as the order parameter approaches unity: $s \to 1$. Numeric calculation confirms these quantitative considerations, showing exponential growth of the value of $\langle R_z^2 \rangle$ for the persistent model as $s \to 1$.

28.8. *In the presence of strong attraction of the segments, the phase separation region corresponding to nematic ordering broadens substantially; two different nematic phases can also coexist in this case.*

Until now, we have dealt solely with athermal polymer solutions in which nematic ordering proceeds at low concentrations of a stiff-chain polymer. When attempting to account for the effects of attraction between macromolecules on the properties of the liquid-crystal transition, the problem associated with the fact that the anisotropic phase at the transition point may be so highly concentrated as to render the second-order virial Onsager approximation inapplicable to describe of the phase immediately arises. Consequently, to analyze the nematic ordering with allowance for the attraction of segments, the theory presented earlier must be generalized to the case of higher concentrations of a polymer in the solution. One such generalization is described in Ref. 34; we show only the general form of the obtained phase diagrams.

Figure 5.4 illustrates the phase diagrams for the liquid-crystal transition in a solution of stiff rods for the variables θ/T, Φ, and several values of the asymmetry parameter L/d. θ is the temperature at which the osmotic second-virial coefficient becomes zero. For larger values of L/d, the phase diagram has a peculiar form, shown in Fig. 5.4a. In the region of moderately high temperatures, there is a narrow corridor of phase separation into isotropic and anisotropic phases lying in the dilute solution region, conversely, at low temperatures, the phase separation region is very broad, with the concentrated strongly anisotropic phase coexisting with the virtually totally diluted isotropic one. Between these two regions is an interval between the triple-point temperature T_t and the critical temperature T_{cr} ($T_{cr} > T > T_t$). In this interval, there are two regions of phase separation: 1) between the isotropic and nematic phases, and 2) between two nematic phases with different degree of anisotropy. The temperatures T_t and T_{cr} substantially exceed the θ-temperature.

As the ratio L/d diminishes, the interval between T_{cr} and T_t becomes narrower, and it disappears at $(L/d)_{cr1} \approx 15$. For $L/d < 15$, the phase diagram has neither critical nor triple points (Fig. 5.4b), and one may note only the crossover temperature T_{cros} between the narrow high-temperature corridor of phase separation and the very broad low-temperature phase separation region. The temperature T_{cros} diminishes with the diminishing of L/d. At $(L/d)_{cr2} \approx 3.5$ (when this temperature drops below the θ-point), the situation changes qualitatively once again: the triple and critical points now correspond to an additional phase transition between the two isotropic phases (Fig. 5.4c). The physical meaning of this result is obvious: at small values of the asymmetry parameter, the properties of the solution of rods should come closer to those of the solution of isotropic particles. Specifically, at sufficiently low temperatures,

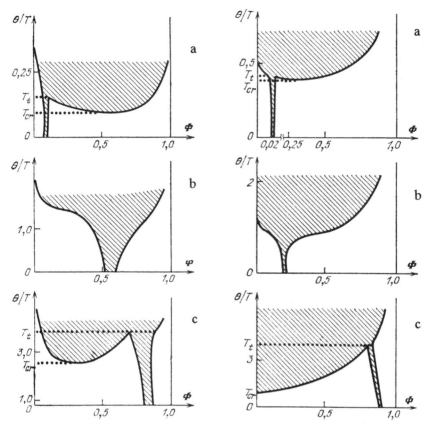

FIGURE 5.4. State diagrams of liquid–crystalline solution of rod-like macromolecules of different lengths.

FIGURE 5.5. State diagrams of liquid–crystalline solution of persistent macromolecules of different stiffness.

there must exist the region of conventional isotropic-liquid phase separation (*see* subsection 24.6).

Moving from the solution of stiff rods to the solution of semiflexible freely jointed macromolecules, the phase diagrams undergo (with the decrease of the ratio l/d) the same sequence of changes as in Figure 5.4, with

$$(l/d)_{cr1} \approx 20, \quad (l/d)_{cr2} \approx 6.8.$$

The situation is the same for the solution of semiflexible chains with persistent flexibility mechanism (Fig. 5.5). In this case, the asymmetry parameter takes the following critical values: $(l/d)_{cr1} \approx 125$; $(l/d)_{cr2} \approx 50$. This means that in the solution of persistent chains, the conventional isotropic-liquid phase separation must occur, even for the case of macromolecules with a rather high degree of stiffness ($l/d \lesssim 50$). Moreover, as seen from comparison of Figures 5.4 and 5.5, all of the characteristic temperatures (T_{cr}, T_t) for the solution of persistent chains are substantially lower than for the solution of rods with the same asymmetry parameter. Hence, in the presence of attractive forces, the orientationally ordered phase is still much more difficult to obtain in the solution of persistent chains than in the corresponding solution of stiff rods or freely jointed chains.

29. HIGH ELASTICITY OF POLYMER NETWORKS

29.1. *Network polymers may possess the property of high elasticity, that is, the capacity for very large deformations retaining their elastic (reversible) nature even in a strongly nonlinear region.*

As the Introduction noted, the polymer network (or gel) constitutes in the simplest case a number of chain macromolecules joined together by chemical covalent bonds and, because of these bonds, forming a common spatial frame. The most spectacular feature of polymer gels is the property of high elasticity possessed by all networks that are not cross-linked too densely. A highly elastic body can endure very large (nonlinear) deformations and still remain elastic (i.e., it is capable of recovering the initial non-deformed size and shape on release of an external load). This property is well known by the example of a common vulcanized rubber, in which the limit of elasticity is often reached only at deformations of approximately tens or hundreds of percent.

It should be stressed that highly elastic polymers differ substantially from conventional (low-molecular-weight) solids. The primary difference lies in the fact that they remain elastic in the range where the deformation–stress relationship is not linear, whereas in ordinary solids, the linearity (i.e., the Hooke law) ceases approximately in the same region where the deformations are no longer elastic.

Finally, we note that as a rule, the adiabatic deformation of a highly elastic polymer leads to an increase in its temperature. This circumstance suggests that entropy factors play a primary role in the nature of high elasticity. Later, we show that the elasticity of polymer networks is indeed determined by the conformational entropy of the chains.

29.2. *The classical theory of high elasticity is based on the assumption that the chains forming the network are Gaussian and phantom.*

High elasticity is one of the well-known properties of polymer materials that are frequently used in practice. Therefore, the development of a molecular theory of nonlinear high elasticity of polymer networks was continually one of the fundamental problems of the statistical physics of macromolecules.

The classical theory of high elasticity was independently developed in the 1940s by several authors (H. M. James and E. Guth, F. T. Wall, L. R. Treloar, P. J. Flory); A detailed review of the results of the classical theory is given in Ref. 36. The main simplification of the classical theory lies in disregarding the topologic constraints on network chain conformations. Each subchain[a] of the

---

[a]We use the term *subchain* to denote a polymer chain connecting two branching points (or depending

network, even though strongly entangled with surrounding chains, is supposed to be capable of assuming any conformation compatible with the given end-to-end distance (i.e., the distance between cross-linking or branching points). This corresponds to the assumption that the subchains are phantom (*see* subsection 11.1).

An additional simplification consists in the phantom subchains being assumed to be Gaussian. This assumption can partly be justified by the polymer networks being usually rather concentrated systems. As we know from the Flory theory (*see* subsection 24.2), chains indeed obey Gaussian statistics in a concentrated polymer solution.

In view of the simplicity of the classical theory, it is convenient to present its positive content first. Then, we discuss the conditions of applicability of this theory.

29.3. *In the classical theory, the elasticity of a polymer network is explained by the entropy losses occurring during the extension of subchains; the property of high elasticity of a not-too-dense network depends on the low modulus and high limit of elasticity of long Gaussian subchains.*

Provided that the network consists of phantom Gaussian subchains, its elasticity is obviously of entropic nature and appears because the network deformations lead to a shifting of the branching points (i.e., to the change of end-to-end distances of the subchains). This results in a depletion of the set of conformations that the system of subchains may assume and in a loss of conformational entropy. The entropic elasticity of an individual Gaussian chain was discussed in subsection 8.1; an N-link chain ($N \gg 1$) was shown to possess a low elastic modulus ($\sim 1/N$) and a very high limit of elasticity (because the end-to-end distance of a free ideal coil is $\sim N^{1/2}$ and of a stretched chain $\sim N \gg N^{1/2}$). Obviously, these facts qualitatively explain the previously mentioned basic features of the highly elastic behavior. They also make clear that only the network in which the concentration of branching and cross-linking points is not high (i.e., the subchains are long) can be highly elastic.

29.4. *The classical theory of high elasticity yields a quite definite nonlinear stress–strain relation for each type of strain (extension or compression, shear, and so on), with the result being universal: the stress is independent of any molecular parameters of the chain and is defined only by the temperature and cross-linking density.*

Let us now calculate the free energy of the elastic strain of a polymer network, taking into account the classical theory assumptions made earlier. Suppose that a sample of the polymer network experienced a strain that changed the sample size in the directions of the axes x, y, z by the factors of λ_x, λ_y, λ_z. The simplest presumption is that the system of branching (or cross-linking) points is deformed affinely with the network sample. This means that if the end-to-end vector of some subchain was equal before the deformation to R_0, that is, had the

---

on the method of network formation, two points of cross-linking; *see* Appendix to this section). The subchain is not supposed to involve any other branching points, except those that limit the subchain.

components $(R_0)_x$, $(R_0)_y$, $(R_0)_z$, then after the deformation, the components of the corresponding vector R are determined by the relations $(R)_i = \lambda_i (R_0)_i$, $i = x$, y, z. In this formulation, the assumption of affinity implies the neglecting of fluctuations of the branching point positions and thus cannot be satisfied accurately. The resulting errors, however, have no effect on the final expression for the free energy of the strained network.[36]

According to the assumption made, each subchain is Gaussian (*see* subsection 29.2), and the change of the subchain's free energy on deformation of the polymer network is thus given by Eq. (8.1), that is,

$$\Delta F(R) = F(R) - F(R_0)$$

$$= (3T/2Ll) \sum_i [(R)_i^2 - (R_0)_i^2]$$

$$= (3T/2Ll) \sum_i (\lambda_i^2 - 1)(R_0)_i^2, \tag{29.1}$$

where L is the contour length of the subchain, l the Kuhn segment length, and the summation taken over all Cartesian components $i = x$, y, z. To obtain the change of the free energy of the entire polymer network ΔF, Eq. (29.1) should be summed over all subchains of the network, that is, averaged over all possible values of R_0 and then multiplied by the total number of subchains in the network νV, where V is the sample volume and ν the number of subchains per unit volume:

$$\Delta F = (3T\nu V/2l) \sum_i (\lambda_i^2 - 1)\langle (R_0)_i^2/L \rangle. \tag{29.2}$$

Taking into account that for Gaussian statistics

$$\langle R_0^2/L \rangle = \sum_i \langle (R_0)_i^2/L \rangle = l \tag{29.3}$$

and that all three coordinate directions are equivalent in the nonstrained sample, we obtain $\langle (R_0)_i^2/L \rangle = l/3$ and

$$\Delta F = T\nu V \sum_i (\lambda_i^2 - 1)/2 = T\nu V(\lambda_x^2 + \lambda_y^2 + \lambda_z^2 - 3)/2. \tag{29.4}$$

It should be noted that the relation (29.4), as well as all of the conclusions of the classical theory of high elasticity, prove to be universal, and they are independent of the structural details of the subchains or of the Kuhn segment length. Analyzing the calculations presented, it is easy to conclude that this universal result is a direct consequence of the two initial assumptions about the subchains being Gaussian and phantom.

Now let us apply Eq. (29.4) for determining the stress–strain relation for uniaxial extension or compression of the polymer network. As a rule, ordinary polymer substances are relatively weakly susceptible to overall compression. This signifies that the uniaxial extension does not practically vary the network volume: the extension (compression) along the axis x by a factor of λ results in the compression (extension) along the axes y and z by a factor of $\lambda^{1/2}$. In this case, therefore, $\lambda_x = \lambda$, $\lambda_y = \lambda_z = \lambda^{-1/2}$, and from Eq. (29.4), we obtain

$$\Delta F = T v V (\lambda^2 + 2/\lambda - 3)/2. \tag{29.5}$$

Hence, we find the stress τ, that is, the elastic force per unit area of the unstrained cross-section of the sample:

$$\tau = \frac{1}{A} \frac{\partial(\Delta F)}{\partial D} = \frac{1}{V} \frac{\partial(\Delta F)}{\partial \lambda} = T v \left(\lambda - \frac{1}{\lambda^2} \right) \tag{29.6}$$

(A is the cross-sectional area of the unstrained sample and D the length of the strained sample.) As it should be, the strain is proportional to the temperature, because the elasticity in the considered model is of purely entropic nature.

The formula (29.6) is one of the basic results of the classical theory of high elasticity of polymer networks. Importantly, this formula predicts not only the elastic modulus in the region of linear stress–strain relation[b] but also the nonlinear properties.

The formula (29.6), as well as similar formulas of the classical theory of high elasticity for other types of deformation (twisting, biaxial extension, shear, and so on) agree quite well with experimental results in many cases. Numeric values of elastic moduli, their temperature dependence, and the form of the nonlinear stress–strain relation are all in many cases confirmed with satisfactory accuracy. It is easy to recognize that the relative success of the simple classical theory is associated with the mentioned universality of Eq. (29.4), that is, essentially with the fact that the high elasticity of networks is caused by large-scale properties of entire chains and not small-scale features of the chemical structure and interactions of individual atomic groups and links.

On the whole, however, the classical theory must be regarded as only the first approximation, because for most networks, it needs certain (sometimes quite substantial) corrections. Discussion of the limits of applicability of the classical theory, deviations from the theory, and relevant refinements of theoretic approaches is convenient to begin by considering the general question of the relationship between the properties of the network and the method of its formation.

29.5. *The structure and properties of the polymer network depend substantially on the conditions and technique of its formation.*

---

[b]The relative extension of the sample apparently equals $\lambda - 1$. If $(\lambda - 1) \ll 1$, then Eq. (29.6) yields the linear "Hooke law" $\tau \cong 3 T v (\lambda - 1)$, that is, the coefficient of linear elasticity equals $3 T v$.

Section 26 showed that the equilibrium properties of a solution of linear polymer chains are determined by such parameters as the concentration, thermodynamic quality of the solvent (temperature), and length and flexibility of chains, but they are independent of the conditions and methods of the synthesis of macromolecules. For polymer networks, the situation is quite the opposite: properties of the network not only depend on the conditions at which it exists at the given moment, but also on the conditions at which it was produced as well as the preparation technique.

The reason for this is simple. First, the covalent cross-linking fixes the points of cross-links or branching points on all of the network chains; in other words, the lengths and number of subchains connecting branching points and subchains connected to the basic framework of the network by its one end are fixed. In addition, in the process of synthesis, the topologic structure of the network is fixed: because the polymer chains cannot cross one another without breaking (*see* subsection 11.1), they fix the topology of their arrangement relative to one another by joining into a common spatial framework. The topologic structure remains constant in all processes involving the network provided that these processes do not break the covalent bonds.

It thus may be stated that the polymer network retains the memory of the conditions and technique of its synthesis. Accordingly, we now discuss the properties of networks synthesized by various methods.

29.6. *Gaussian statistics of subchains is typical for a dry (without solvent) network produced by cross-linking the melt chains; the classical theory of high elasticity is valid for compression and weak extension of such a network; for other methods of network synthesis, the statistics of subchains may not necessarily be Gaussian.*

Let us consider the simplest technique of polymer network synthesis. Suppose that we performed a fast cross-linking of macromolecules in a conventional polymer melt (by chemical agents or ionizing radiation). According to the Flory theorem (*see* subsection 24.2), the conformations of the solution macromolecules before cross-linking correspond to a Gaussian coil. The fast cross-linking fixes the structure, and naturally, the ensemble of network subchains, still obeys Gaussian statistics. This is precisely what was used in Eqs. (29.2) and (29.3) when averaging over all subchains of the network. Apparently, the subchains can be approximated by Gaussian coils even in the case of a dry network synthesized from a monomeric melt in the presence of multifunctional links.

In other methods of network synthesis, by no means must the statistics of the subchains necessarily be Gaussian. Let us assume, for example, that the network is obtained by fast cross-linking of a semidilute polymer solution with subsequent removal of the solvent. Because both the degree of chain entanglement in the semidilute solution is substantially less than in the melt and this degree of entanglement becomes fixed during cross-linking, the network subchains after removal of the solvent must assume "partially segregated" conformations (Fig. 5.6) to provide the lesser degree of entanglement (i.e., the lesser order of topologic linkage of each subchain with surrounding ones) than is typical for Gaussian conformations of subchains.

If the polymer network swells in a solvent, the statistical properties of the constituent subchains naturally vary with the change in quality of the solvent. For example, when the network is synthesized by fast cross-linking of the melt chains (so that the initial statistics of the subchains is Gaussian), then a good solvent is introduced into the system, the excluded volume effect appears in the conformations of the subchains, and the statistics conform not to Gaussian but to swollen coils. Conversely, on addition of the θ-solvent to the "partially segregated" network of the type shown in Figure 5.6, the swelling leads to Gaussian statistics.

The list of similar examples could be continued. It is clear, however, that the Gaussian statistics of subchains assumed in the classical theory is far from typical for all types of polymer networks.

29.7. *The classical stress–strain relation can be used for the description of a relatively fast compression or extension of swollen polymer networks or of networks synthesized in the presence of solvent; this is accomplished by introduction into the classical dependence $\tau(\lambda)$ of the phenomenologic coefficient, the front factor.*

Suppose that the polymer network is produced by cross-linking melt macromolecules or some other method resulting in Gaussian statistics for the subchains in the dry state. When such a network swells in good solvent, the statistical properties of the subchains change (*see* subsection 29.6); therefore, the classical theory of high elasticity (*see* subsection 29.4) cannot be used directly for the description of the stress–strain relation of the swollen network. Let us now discuss the modifications that should be introduced into the classical theory in this case.

Generally, the deformation of the swollen network is accompanied not only by a change in the entropy of its subchains, but also by a change of link concen-

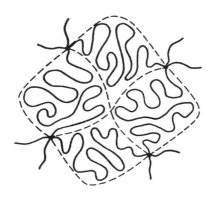

FIGURE 5.6. Structure of a polymer network prepared by cross-linking in a semidilute solution with subsequent extraction of the solvent.

tration in the network (i.e., the energy of link interaction). The link concentration varies only very slowly, however, which is compatible with the slow establishment of equilibrium between the solvent molecules residing both within the network and outside. In typical cases, the corresponding characteristic time may amount to many hours.[36] Accordingly, if experiments on elastic deformation of swollen chains are carried out fast enough (as assumed in this subsection), then only the entropy effect can be taken into account in their analysis.

Suppose that as a result of contact with solvent, the size of the network increased by a factor of α with respect to that of the dry network. Under the affinity assumption (*see* subsection 29.4), we should suppose that the end-to-end vector R of the subchains increases by the same factor. The distribution function of R in the case of the swollen networks then must take the form [cf. Eq. (4.2)]

$$P(R) = (3\alpha^2/2\pi Ll)^{3/2} \exp(-3\alpha^2 R^2/2Ll). \tag{29.7}$$

Writing an equation analogous to Eq. (8.1) for the given function $P(R)$ and using the same rationale as in subsection 29.4, we obtain the following expression for the change in free energy of the swollen network under deformation, replacing Eq. (29.4):

$$\Delta F = TvV\alpha^2(\lambda_x^2 + \lambda_y^2 + \lambda_z^2 - 3)/2. \tag{29.8}$$

Hence, the stress τ in the uniaxial extension-compression [cf. Eq. (29.6)]:

$$\tau = Tv\alpha^2(\lambda - \lambda^{-2}). \tag{29.9}$$

The factor α^2 appearing in this relation is called the *front factor*. It defines the increase in the stress for swollen networks for the same values of λ and network concentrations v). As to the form of the dependence $\tau(\lambda)$ itself, it is identical for both swollen and dry networks [cf. Eqs. (29.6) and (29.9)].

Earlier, we discussed networks obtained in the dry state (by cross-linking or by synthesis in the presence of multifunctional monomers). Subsection 29.6 has already mentioned that in the general case of synthesis under arbitrary conditions (e.g., for networks produced in the presence of a solvent), statistics of subchains in the dry state cannot be assumed to be Gaussian any longer, so Eq. (29.1) is, strictly speaking, not correct. A rigorous account of the dependence of the statistical properties of the subchains on synthesis conditions, however, would exceed the accuracy of the considered classical theory, which (as pointed out earlier), is based on the notion of phantom chains. This is why the elastic properties of the networks are analyzed in the general case by the following approximate method.

Let us choose a state (generally speaking, swollen) of the polymer network in which the statistics of the subchains is closest to being Gaussian, and term it the *reference state*. For the networks considered earlier, the reference state is that of a dry network. At the same time, if the network is produced by fast cross-linking in the θ-solvent, then the initial state of the swollen network is naturally to be regarded as a reference state, and in the dry state, the subchain must assume the essentially non-Gaussian conformation shown in Figure 5.5. In the general case, given network synthesis in the presence of solvent, the reference state corresponds to the swollen network, with the degree of swelling being higher for a higher volume fraction of solvent under the conditions of synthesis. At the same time, it should be stressed that the volume fraction of solvent under the conditions of synthesis, generally speaking, does not equal that in the reference state.

Assuming the subchain statistics in the reference state to be approximately Gaussian, we can repeat all of the reasoning of subsection 29.4 for this state and obtain Eq. (29.6) for the dependence $\tau(\lambda)$ under uniaxial extension-compression. If, on the other hand, the network is swollen or compressed relative to the reference state (by a factor of α), then having performed the calculations analogous to those described earlier [Eqs. (29.7) and (29.8)], we obtain Eq. (29.9).

Hence, in the framework of the discussed approximation of the classical theory, Eq. (29.9) also remains valid in the general case, with the front factor α^2 acquiring the meaning of the square of the swelling (compression) coefficient for the linear dimensions of the deformed polymer network before deformation relative to the dimensions of the reference state.

29.8. *Deviations of the stress–strain relation for a network from the classical law in the case of extension are described by the phenomenologic Mooney-Rivlin formula; for typical polymer networks, the Mooney-Rivlin corrections stem from topologic constraints imposed on subchain conformations.*

Thus, by introducing the front factor, one can take account approximately of the dependence of network properties on the conditions of their synthesis. In this case, the theory predicts that the stress–strain relation for the uniaxial extension-compression remains classical: $\tau \sim \lambda - \lambda^{-2}$. How faithfully, however, does this prediction come true?

For most networks (even those whose subchain statistics is Gaussian in the initial non-deformed state), the classical dependence $\tau(\lambda)$ agrees with experimental data with an accuracy of better than 10% only for $0.4 < \lambda < 1.2$; for $\lambda > 1.2$, Eq. (29.9) yields highly excessive values of τ (Fig. 5.7). M. Mooney in 1940 and R. Rivlin in 1948 suggested an empiric formula that as a rule makes it possible to describe accurately the dependence $\tau(\lambda)$ under uniaxial extension:

$$\tau = 2(\lambda - \lambda^{-2})(C_1 + C_2\lambda^{-1}) \text{ at } \lambda > 1, \tag{29.10}$$

where C_1 and C_2 are called the *Mooney-Rivlin constants*.

According to the current concept, Mooney-Rivlin corrections are caused primarily by topologic constraints on conformations of network subchains, and in the densely cross-linked networks also by anisotropy of interaction between the subchains orienting under deformation. The detailed microscopic theory of these effects, however, has not been developed yet.

29.9. *The size of a polymer network sample in a solvent is determined by the balance between entropic elasticity and volume interaction energy; on crossing the*

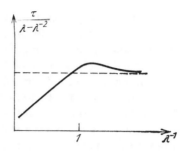

FIGURE 5.7. Typical dependence $\tau(\lambda)$ observed in experiments on uniaxial extension-compression of polymer networks in so-called Mooney-Rivlin coordinates: $\tau/(\lambda-\lambda^{-2})$ and λ^{-1}. In these coordinates, the result of classical theory corresponds to a broken straight line parallel to the abscissa.

θ-region, as the solvent quality deteriorates, the network dimensions decrease drastically (i.e., the network collapses).

In addition to deformation properties, a significant parameter of the polymer network is its equilibrium size under free swelling in solvents of different quality. This size is determined by the balance between the free energy F_{el} associated with entropic elasticity of network subchains and the free energy F_{int} of the volume interactions of network links for the given degree of swelling (cf. subsection 13.2). To determine F_{el}, it is convenient to choose the reference state discussed earlier as an initial state. The free energy change ΔF_{el} because of network sample swelling by a factor of α ($\alpha < 1$ corresponds to compression) relative to this state can be obtained in complete analogy with the method in subsection 29.4. The final results also coincide if one sets $\lambda_x = \lambda_y = \lambda_z = \alpha$ in Eq. (29.4):

$$\Delta F_{el} = (3/2)T\nu V(\alpha^2 - 1). \qquad (29.11)$$

To find F_{int}, we assume for simplicity that network swelling is so substantial that the volume fraction of polymer in the network is much less than unity. Then, we may use the virial expansion and write [cf. Eq. (12.3)]

$$F_{int} = T\nu VN(Bn + Cn^2), \qquad (29.12)$$

where N is the number of links in the subchain, B and C the second- and third-virial coefficients of link interaction and n the link concentration in the network. Let us denote the link concentration in the reference state by n_0. Then, $n = n_0/\alpha^3$, and the change of F_{int} resulting from the network swelling by a factor of α relative to the reference state equals

$$\Delta F_{\text{int}} = T\nu VN[\,B(n-n_0) + C(n^2 - n_0^2)\,]$$

$$= T\nu VN[\,Bn_0(\alpha^{-3} - 1) + Cn_0^2(\alpha^{-6} - 1)\,]. \qquad (29.13)$$

The equilibrium swelling coefficient α can be found by minimizing the sum $\Delta F = \Delta F_{\text{el}} + \Delta F_{\text{int}}$ with respect to α (*see* subsection 15.1). The resulting equation takes the form

$$\alpha^5 - y/\alpha^3 = x, \qquad (29.14)$$

where $y = 2Cn_0^2N$, $x = Bn_0N$ [cf. Eq. (13.5), defining the swelling parameter of an individual macromolecule]. The relation (29.14) implicitly defines the function $\alpha = \alpha(x,y)$. The dependence of the swelling coefficient α on the parameter x determining the solvent quality ($x > 0$ conforms to a good solvent, $x < 0$ to a poor solvent, and $x = 0$ to the θ solvent), for the typical value $y = 0.1$ is shown in Figure 5.8.

In the limit of very good solvent, we have $\alpha \sim x^{1/5} \sim (Bn_0N)^{1/5}$, that is, $n = n_0/\alpha^3 \sim n_0^{2/5} B^{-3/5} N^{-3/5}$. Thus, the size of a network sample and its link concentration are determined by both the number N of links in subchains and the conditions of the network synthesis expressed via the link concentration n_0 in the reference state.

With a decrease of x (i.e., with deterioration of solution quality), the parameter α diminishes according to Eq. (29.14), that is, the size of the network sample diminishes. At large (by modulus) negative x Eq. (29.14) yields $\alpha^{-3} \cong -x/y$ and $n = -B/2C$, that is, the link concentration in the network in poor solvent coincides, according to Eq. (20.27), with the link concentration in a globule. Certainly, this means that deterioration of the quality of the solvent, in which the network swells (e.g., by lowering the temperature), on passing the θ-point leads to the globule–coil transition in each of the network subchains. In

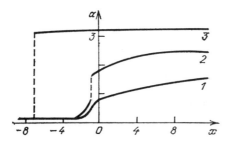

FIGURE 5.8. (*1*), Dependence $\alpha(x)$ (29.14) for $y = 0.1$. (*2*) and (*3*), Corresponding dependences for charged networks, defined by (30.43), for $y = 0.1$, $\lambda = 4$, and $s = 6$ (*curve 2*) and $s = 12$ (*curve 3*).

this case, network size diminishes very drastically (Fig. 5.8). Such a phenomenon is called the *network collapse*; we discuss it in more detail in subsection 30.10.

APPENDIX TO SEC. 29. METHODS OF POLYMER NETWORK SYNTHESIS

Most widespread methods of polymer network preparation are based either on a synthesis of macromolecules with participation of multifunctional monomers or on vulcanization (i.e., cross-linking of linear polymer chains).

In the former case, a chemical linking of sections of linear macromolecules is accomplished directly during the polymer synthesis. For this purpose, molecules with three or more functional groups are introduced into the initial mixture of monomers in addition to the bifunctional particles joining to form the linear chains. If such a mixture has a low concentration, then a polydisperse (i.e., containing macromolecules of different mass) system of branched macromolecules is formed. As the concentration of the initial monomer mixture grows, the size of the appearing macromolecules increases, but not in a smooth way: if the concentration exceeds a certain finite value (called a *gel formation threshold*), then together with the branched macromolecules of finite dimensions, an infinite (i.e., macroscopic) cluster emerges in the system, which is in fact a macroscopic network. In the immediate vicinity of the gel formation threshold, the network has a very large (infinite at the point of actual threshold) intrinsic cell size. On further concentration growth, the cell size diminishes, (i.e., the cross-linking density rises). From a theoretic viewpoint, gel formation is analogous and closely associated with so-called percolation.[29,37] In this book, the gel formation theory is not discussed. The classical foundations of this theory are presented in Ref. 4 and the current approaches in Ref. 8.

A polymer gel is also produced from a melt or a concentrated or semidilute solution of linear macromolecules by cross-linking with chemical bonds either by exposure to ionizing radiation, or by the introduction of bivalent "linking" molecules into the system. The first highly elastic substance, rubber, was obtained precisely in this way by sulphur treatment (with sulphur atoms acting as cross-links) of the concentrated solution of chains of natural rubber (vulcanization). Vulcanization boosts formation of the microscopic polymer network (but not the system of finite branched macromolecules) provided that the concentration of cross-links is high enough.

Apart from polymer networks, there are also so-called physical gels, in which the linear macromolecules are linked into a network not by covalent but weaker bonds (e.g., hydrogen, ionic, dipole-dipole, and so on). In principle, some peculiar gels, comprising topologically interlocked ring macromolecules, are also feasible. Properties of such gels are quite interesting, but they have been insufficiently investigated to be discussed here.

30. POLYELECTROLYTES

30.1. *The conformations of charged macromolecules of polyelectrolytes in solution depend on the fraction of charged links, concentration of polymer and low-molecular-weight salt, and for weakly charged polyelectrolytes, also on non-Coulomb link interactions.*[49]

Macromolecules containing charged links are called *polyelectrolytes.* As a rule, they dissociate in solution to form charged links and low-molecular-weight counter ions. The number of charged links equals the number of counter ions, and the polymer solution as a whole is electrically neutral (as it should be).

Polyelectrolytes are classified into strongly and weakly charged ones. In strongly charged polyelectrolytes, every link (or considerable fraction of links) carries a charge; therefore, Coulomb link interaction predominates over the nonelectrostatic (e.g., Van der Waals) interaction. An important example of a very strongly charged polyelectrolyte is the DNA double helix (*see* subsection 37.4). A small fraction of charged links is typical for weakly charged polyelectrolytes; and the non-Coulomb link interaction may play an appreciable part in them.

Let e denote the link charge and ε the dielectric constant of the solution. Then, the potential energy of the Coulomb interaction between the charged links i and j separated by the distance r_{ij} equals

$$V(r_{ij}) = (e^2/\varepsilon r_{ij})\exp(-r_{ij}/r_{\mathrm{D}}), \qquad (30.1)$$

where r_{D} is the so-called Debye radius determined by the screening of the electrostatic interaction by the ions in the solution.[26] As was noted above, the low-molecular-weight counter ions are necessarily present in the solution, thus effecting its electric neutrality. Very often, however, the solution also contains ions of added low-molecular-weight salt. If the total concentration of all low-molecular-weight ions (or more correctly, ionic strength) in the solution equals n, then[c] (*see* Ref. 26):

$$r_{\mathrm{D}} = (\varepsilon T/4\pi n e^2)^{1/2}. \qquad (30.2)$$

30.2. *The size of polyelectrolyte macromolecules in a dilute salt-free solution is proportional to the number N of charged links (i.e., the charged macromolecule proves to be fully stretched in terms of the dependence on N).*

Consider an individual macromolecule of polyelectrolyte, a macro-ion, in an infinite solution containing no low-molecular-weight salt. This situation corresponds to the dilute limit salt-free solution. Because each macromolecule of finite length N occupies a very large volume in such a solution, the counter ions cannot stay near the macroion and recede from it over long distances to acquire a great entropy gain. Accordingly, the concentration of counter ions is very low near the

---

[c]In the general case, when the i-th ions carry the charges ez_i, the quantity n should be replaced in Eq. (30.2) with the so-called ionic strength of the solution, $n_{ef} = \Sigma_i n_i z_i^2$.

macroion and therefore $r_D \to \infty$ [see Eq. (30.2)] and the charges interact via a nonscreened Coulomb potential.

In this situation, the charged macromolecule (because of the repulsion of charged links) must become fully extended in the sense that its mean square end-to-end distance is proportional to its length, that is, $\langle R^2 \rangle \sim N^2$ (in contrast to a Gaussian coil, for which $\langle R^2 \rangle \sim N$). Indeed, the energy of the Coulomb interaction of links is of order $E_C \sim N^2 e^2 / (\varepsilon \langle R^2 \rangle^{1/2})$ (because the number of link pairs $\sim N^2$). When the chain is not extended, for example, $\langle R^2 \rangle \sim N^{2-\delta}$, then $E_C \sim N^{1+\delta/2} \gg N$. In other words, at large N, it is the repulsive Coulomb interaction that contributes the most to the free energy forcing the macromolecule to extend. Therefore, only the extended conformations with $\langle R^2 \rangle \sim N^2$ may correspond to equilibrium.

30.3. *A strongly charged polyelectrolyte macromolecule is almost fully extended in the limit of a very dilute salt-free solution; in this case, the conformation of a molecule of weakly charged polyelectrolyte can be visualized as an extended chain of blobs.*

The result obtained in the previous subsection does not mean that the molecule of polyelectrolyte in the dilute salt-free solution is fully extended. Consider the simple model of a charged macromolecule. Suppose we have a standard Gaussian chain of $N\sigma$ links ($\sigma \gtrsim 1$) with rms distance a between neighboring links in the chain. Let all of the links interact by means of ordinary non-Coulomb forces of the Van der Waals type. Moreover, let each first link in the sequence of σ links of the chain possess the charge e. The case $\sigma \gg 1$ corresponds to a weakly charged polyelectrolyte and $\sigma = 1$ to a strongly charged one.

First, let us assume that we are at the θ-point with respect to non-Coulomb interactions, where the influence of these interactions on the conformation is inessential. Then, it is convenient to picture the conformation that the considered macromolecule acquires only because of Coulomb interaction as a sequence of blobs (Fig. 5.9; cf. Sec. 19). Each blob comprises g consecutive charges of the chain (i.e., $g\sigma$ links). The blob size D is found from the condition that the energy of electrostatic repulsion of two neighboring blobs in the chain is of the order of the thermal energy T:

FIGURE 5.9. Conformation of a chain of weakly charged polyelectrolyte in an extremely dilute salt-free solution.

$$g^2 e^2/(\varepsilon D) \sim T. \tag{30.3}$$

Then, on the one hand, the polymer chain inside the blob (being weakly disturbed by Coulomb interactions) remains Gaussian, that is,

$$D \sim a(g\sigma)^{1/2}, \tag{30.4}$$

and on the other hand, the blob system forms an extended chain whose longitudinal and transverse dimensions are of order

$$R_{\parallel} \sim (N/g) D, \tag{30.5}$$

$$R_{\perp} \sim (N/g)^{1/2} D; \tag{30.6}$$

N/g is the number of blobs in the chain (cf. Sec. 19). From Eqs. (30.3) to (30.6), one can evaluate the parameters of the chain of blobs

$$g \sim \sigma^{1/3} u^{-2/3}, \quad D \sim a\sigma^{2/3} u^{-1/3}, \quad R_{\parallel} \sim Na(u\sigma)^{1/3}, \quad R_{\perp} \sim N^{1/2} a\sigma^{1/2}, \tag{30.7}$$

where the designation

$$u \equiv e^2/(\varepsilon aT) \tag{30.8}$$

is used for a characteristic dimensionless parameter of the problem.[d] Under normal conditions (e is the electron charge, $a \sim 1$ nm, $T \sim 300$ K, $\varepsilon \approx 80$, water solution), the parameter u is of order unity; the estimate $u \sim 1$ will often be used here.

From the relation (30.7), one can see that for a weakly charged polyelectrolyte ($\sigma \gg 1$), the number g of charges in the blob is much greater than unity (taking into account the estimate $u \sim 1$), and $R_{\parallel} \ll N\sigma a$, that is, the chain is far from forming a fully extended conformation. This is precisely the case when the macromolecule is to be treated as a chain of blobs (Fig. 5.9). However, if $\sigma \sim 1$ (a strongly charged polyelectrolyte), then the conformation of the charged macromolecule in the dilute solution is almost fully extended ($R_{\parallel} \sim Na$, $g \sim 1$), and there is no sense in introducing the notion of a blob. The last conclusion about a substantial stiffening of the chain of strongly charged links is valid regardless of the presence of non-Coulomb link interaction, because in this case, its influence is insignificant against the background of the strong electrostatic repulsion of the links.

For weakly charged polyelectrolytes, however, the non-Coulomb interactions must be taken into account. Let us assume, first, that in terms of these interactions exists the good solvent regime. Then, the conformation of a weakly charged polyelectrolyte macromolecule in the dilute solution can be described as before

---

[d]The quantity $l_B = e^2/(\varepsilon T)$ has the dimensionality of length and is called the *Bierrum length*; in water at room temperature, $l_B = 0.7$ nm. The dimensionless parameter u (30.8) equals the ratio of the Bierrum length to the link size, $u = l_B/a$.

using the concept of a chain of blobs (Fig. 5.9). The only difference is that instead of Eq. (30.4), the following relation should be written in this case [cf. Eq. (19.2)]:

$$D \sim (g\sigma)^{3/5} a^{2/5} B^{1/5}, \tag{30.9}$$

where B is the second-virial coefficient of non-Coulomb link interaction. We suggest that the reader obtain the basic parameters of the chain of blobs for this case using Eqs. (30.3), (30.5), (30.6), and (30.9).

Consider now the conformation of a weakly charged polyelectrolyte in poor solvent when the attraction predominates in the non-Coulomb link interaction. Since the relation $\langle R^2 \rangle \sim N^2$ (see subsection 30.2) is valid for the very dilute solution regardless of the presence of non-Coulomb interaction, the macromolecule forms a chain of blobs in a poor solvent as well, although each of the blobs is in the globular state. To form the conformation in the uncharged globules shown in Fig. 5.9, the free energy per blob should obviously equal $\mathscr{F}_{\text{surf}}$, the surface free energy of the globular blob of $g\sigma$ links [see Eq. (20.32)]. Therefore, in this case Eq. (30.3) is replaced by

$$g^2 e^2 / (\varepsilon D) \sim \mathscr{F}_{\text{surf}}. \tag{30.10}$$

The blob size D can be written [to replace Eq. (30.4)] as

$$D \sim (g\sigma / n_0)^{1/3}, \tag{30.11}$$

where n_0 is the average link concentration in the globular blob unperturbed by electrostatic interaction.

The quantities n_0 and $\mathscr{F}_{\text{surf}}$ for the globular state were derived in Sec. 20. Suppose that the solution temperature is only slightly below the θ temperature for the non-Coulomb interaction: $|\tau| \equiv |T - \theta| / \theta \ll 1$. Then, according to Eqs. (20.27) and (20.32)

$$n_0 \sim |\tau| / v, \tag{30.12}$$

$$\mathscr{F}_{\text{surf}} \sim T (g\sigma)^{2/3} |\tau|^{4/3} (a^3 / v)^{1/3}; \tag{30.13}$$

where v is the volume of the monomer link.

Taking into account Eqs. (30.10) to (30.13), we obtain the following relations for the isolated macromolecule of weakly charged polyelectrolyte in poor solvent:

$$g \sim \sigma |\tau| / u, \quad D \sim v^{1/3} \sigma^{2/3} / u^{1/3},$$

$$R_{\parallel} \sim N v^{1/3} u^{2/3} / (\sigma^{1/3} |\tau|), \quad R_{\perp} \sim N^{1/2} v^{1/3} (u\sigma)^{1/6} / |\tau|^{1/2}. \tag{30.14}$$

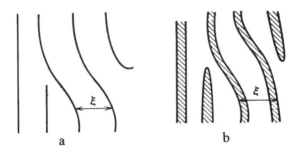

FIGURE 5.10. Structure of semidilute salt-free solution of polyelectrolytes. (a), Strongly charged chains. (b), Weakly charged chains.

30.4. *In a polyelectrolyte solution of finite concentration, the Coulomb interaction is screened by counter ions so that the chains become coiled on large length scales; the distance between neighboring chains in such a solution is of the order of the Debye screening radius.*

Macromolecules of polyelectrolyte thus take in the very dilute salt-free solution highly extended conformations: $R_\parallel \sim N$. It also follows from this, however, that the situation of a dilute solution is difficult to realize in this case, because it is necessary that $c \ll c^* \sim N/R_\parallel^3 \sim N^{-2}$ [cf. Eq. (23.1)]. More interesting is the concentration region $c \gg c^*$, in which spheres circumscribed around the macromolecules overlap strongly. The structure of the salt-free polyelectrolyte solution in this region is illustrated by Figure 5.10; the lines in the figure depict the polymer chains in the case of strongly charged polyelectrolytes and blob chains in weakly charged polyelectrolytes. In contrast to the dilute limit where the counter ions recede to "infinity", in the situation of in Figure 5.10 they stay near the polymer chains, their concentration n is finite, and their effect on the chain conformation is substantial.

This is connected with the fact that in the solution of finite concentration of low-molecular-weight ions, there is a definite screening radius of Coulomb interactions defined by Eq. (30.2). Consequently, the conclusion that $\langle R^2 \rangle \sim N^2$ is no longer true in this case, because the Coulomb repulsion of the links is screened on large scales and the chain acquires a coil conformation. At the same time (on sufficiently small scales, at least smaller than the Debye radius), the screening is ineffective, and the electrostatic interaction stiffens (or extends) the polyelectrolyte chain in accordance with the discussion in subsection 30.3 (Fig. 5.10).[e]

A significant intrinsic dimension of a polyelectrolyte solution is associated with the distance between neighboring chains (Fig. 5.10). Let us now determine this quantity for a salt-free, strongly charged polyelectrolyte, whose chains repre-

---

[e]Note that with the presence of a low-molecular-weight salt in a solution of finite concentration, the latter statements are also valid for extremely low chain concentrations. In this case, the radius r_D is finite even in the very dilute solution, so the chains form coils on large length scales no matter what the polymer concentration in solution is.

sent stiff filaments each containing N charges e and with the distance between neighboring charges being equal to a. From geometric considerations, it is obvious that

$$Na\xi^2 \sim N/c, \quad \text{i.e. } \xi \sim (c/a)^{-1/2}, \tag{30.15}$$

where c is the concentration of charged links in the solution. Taking into account that $c=n$ in a salt-free solution and comparing Eqs. (30.15) and (30.2) for this case:

$$r_D \sim (\varepsilon T/ne^2)^{1/2} \sim (na)^{-1/2} u^{1/2} \sim (ca)^{-1/2} u^{1/2}, \tag{30.16}$$

we find that in the most typical case, $u \sim 1$ [see Eq. (30.8)]

$$\xi \sim r_D. \tag{30.17}$$

The evaluation (30.17) can also be derived for salt-free solutions of weakly charged polyelectrolytes using more complicated calculations. (One should take into account that Coulomb interactions are screened not only by low-molecular-weight counter ions but also by the polymer chains themselves). Hence, for most cases, the characteristic parameters ξ and r_D are of the same order of magnitude. This circumstance simplifies the formation of model concepts on the structure of polyelectrolyte solutions of finite concentration.

30.5. *In a solution of sufficiently strongly charged polyelectrolytes, a fraction of counter ions stays in the immediate vicinity of the polymer chains, effectively neutralizing their charge; this phenomenon is referred to as counter ion condensation.*

The role of counter ions in a polyelectrolyte solution is not always reduced to a simple Debye screening of the Coulomb interaction of links. In some cases, it becomes favorable for a fraction of counter ions to stay in the immediate vicinity of the polymer chain (for weakly charged polyelectrolytes, inside the blobs), effectively diminishing its charge. This phenomenon was predicted by L. Onsager in 1947 and is usually called counter ion condensation.

An explanation of the condensation phenomenon is associated with the well-known fact of the logarithmic distribution of electric potential around a charged cylinder. (Recall that the potential at a point separated by a distance r from an infinite, charged straight line is proportional to $\ln r$.) Certainly, the comprehensive self-consistent determination of the potential distribution, with allowance made for the screening effect of counter ions, non-zero thickness, and finite length of macromolecules, is a very difficult problem. Some approaches to its solution are discussed in subsection 30.7; however, it is useful to discuss preliminarily the physical meaning of the phenomenon of counter ion condensation using the simplest assumptions.

Consider a salt-free solution of strongly charged polyelectrolytes comprising stiff filaments with charges e located a distance a from one another. The linear charge density of the filaments equals $\rho_0 = e/a$. Now select "region 1" in the solution, corresponding to the molecular vicinity of the polymer chains (*see* the shaded area in Fig. 5.10b); the remaining solution space will be called "region 2." We assume that the counter ions in region 1 are bound to the polymer chain (i.e., are in the "condensed" state), whereas the counter ions in region 2 are "free" in the sense that they can move throughout the whole solution. It should be noted that when speaking here of the bound or condensed state, we do not imply any chemical bonds or real condensation but only indicate that the counter ions stay near the polymer chains (in region 1). Let us clarify under what conditions the binding of a finite fraction of counter ions in the condensed state (in our sense) is thermodynamically favorable.

To do this, the following simplifying assumption (the so-called two-phase approximation) is made: The electrostatic potentials inside regions 1 and 2 are constant and equal ψ_1 and ψ_2, $\delta\psi = \psi_1 - \psi_2 \neq 0$. This is, of course, a very rough approximation, but it allows one to analyze the basic qualitative properties of counter ion condensation (*see* subsection 30.7). In the two-phase approximation, the concentrations of counter ions in regions 1 and 2 are constant and differ by the Boltzmann factor:

$$c_1 = c_2 \exp(-e\delta\psi/T), \qquad (30.18)$$

where $-e$ is the charge of a counter ion; $-e\delta\psi > 0$. Let β denote the fraction of counter ions in region 2 (i.e., of "free" counter ions) and φ the volume fraction of region 1 in the solution. Then, Eq. (30.18) is transformed to

$$\ln[(1-\beta)/\beta] = \ln[\varphi/(1-\varphi)] - e\delta\psi/T. \qquad (30.19)$$

In the situation shown in Figure 5.10, $\delta\psi$ can naturally be expressed as a potential difference in the field of a charged cylinder (because the sections of region 1 represent the cylinders on not-too-large scales; Fig. 5.10b) at points removed from the cylinder axis by the distances r_1 and r_2, where r_1 is the radius of the sections of region 1 and $r_2 = \xi/2$:

$$\delta\psi = -2(\rho/\varepsilon)\ln(r_2/r_1) = -(\rho/\varepsilon)\ln(1/\varphi), \qquad (30.20)$$

where ρ is the linear charge density on the cylindric sections of region 1. The last equality is written taking into account that $r_2^2/r_1^2 = 1/\varphi$ (Fig. 5.10b). The Debye screening is disregarded in Eq. (30.20), because as shown in subsection 30.4, $r_D \sim \xi$, and the possible corrections are consequently less than the terms taken into account in Eq. (30.20).

One should not insert the ratio $e/a = \rho_0$ as the quantity ρ into Eq. (30.20), but rather the effective linear charge density calculated with allowance made for the fraction $1 - \beta$ of counter ions in the "condensed" state in region 1 and partially neutralizing the chain charge. Therefore,

$$\rho = e\beta/a. \qquad (30.21)$$

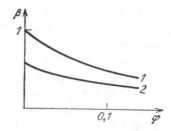

FIGURE 5.11. Dependence $\beta(\varphi)$ (30.22) for $u=0.5$ (*1*) and $u=1.5$ (*2*).

It should also be noted that $\varphi \ll 1$ for the moderately concentrated polyelectrolyte solution. (This corresponds to the most important and interesting cases.) Then, Eq. (30.19) takes the form

$$\ln[(1-\beta)/\beta] = (1-u\beta)\ln \varphi, \tag{30.22}$$

where $u=e^2/(\varepsilon aT)$. The dependence $\beta(\varphi)$ implicitly specified by this equation is shown in Figure 5.11 for different values of u.

Both from Figure 5.11 and the structure of Eq. (30.22), it is seen that depending on the value of u, two different regimes of behavior of the function $\beta(\varphi)$ can be distinguished for $\varphi \ll 1$. If $u<1$, then for $\varphi \to 0$ $\beta \to 1$; if $u>1$, then $\beta \to 1/u$ for $\varphi \to 0$, that is, the final fraction $1-\beta$ of counter ions remains in the "condensed" state even in a very dilute solution. This corresponds to the counter ion condensation.

Figure 5.12 shows how the effective linear charge density ρ, with the partial counter ion condensation (30.21) taken into account, depends on the genuine linear charge density of the polyelectrolyte chain $\rho_0=e/a$ at $\varphi \ll 1$. It is clearly seen that up to the value $\rho_0=\varepsilon T/e$ corresponding to $u=\rho_0 e/(\varepsilon T)=1$, both the "effective" and "genuine" values of ρ coincide (i.e., counter ion condensation does not occur). In the region $\rho_0 > \varepsilon T/e$, a further growth of the value of ρ stops: an increase in the linear charge density of the polymer chain is totally compensated by a corresponding counter ion condensation.

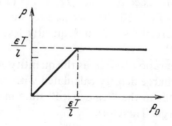

FIGURE 5.12. Dependence $\rho(\rho_0)$ for a polyelectrolyte chain.

The physical meaning of this result can be interpreted additionally as follows. Suppose the counter ion is retained at a mean distance r from the polymer chain with the linear charge density ρ_0. Then, the energy of electrostatic attraction of the counter ion to the macromolecule is of the order of $2\rho_0 e \ln r/\varepsilon$. On the other hand, entropy loss in the free energy because of a restriction of the region in which the counter ion may move is of the order of $T \ln r^2$. Both contributions to free energy are seen to be proportional to $\ln r$. Therefore, depending on the coefficient of $\ln r$, one or the other contribution prevails at any r: when $\rho_0 < \varepsilon T/e$, it is favorable for the counter ions to move away from the polymer chain and when $\rho_0 > \varepsilon T/e$ to condense on the macromolecule.

This consideration was carried out for a salt-free solution; however, it is easy to realize that in the presence of a low-molecular-weight salt in the solution, the same approach can also be applied. It is only necessary that the Debye radius r_D be much greater than the distance a between the charges along the chain (i.e., that the sections of region 1 be cylindric on small length scales). In this case, the Debye radius r_D should be taken as the quantity r_2 in Eq. (30.20); in other respects, the considerations of this subsection remain unchanged.

30.6. *For weakly charged polyelectrolytes, a pronounced counter ion condensation only occurs in a poor solvent (where the blobs are globular) and in this case constitutes an avalanche-like process resulting in a nearly total condensation of the counter ions on the macromolecules.*

The theory presented in subsection 30.5 may also be applied to study solutions of weakly charged polyelectrolytes, because in this case, the counter ion condensation may occur not on the polymer chain itself but rather on the chain of blobs. Therefore, region 1 for weakly charged polyelectrolytes corresponds to the space inside the blobs. Equations (30.18) to (30.20) hold true for this case. The linear charge density ρ_0 of the chain of blobs (without allowance made for the presence of a fraction of counter ions in region 1, that is, their "condensation" on the chain of blobs) equals $\rho_0 = ge/D$, where the expressions for g and D are defined, depending on the type of non-Coulomb interaction, by either Eqs. (30.7) and (30.14) or the corresponding relation for a good solvent [*see* Eq. (30.9)].

To determine the quantity ρ in Eq. (30.20), the partial neutralization of the charges on the chain by the counter ions must be taken into account. It should also be noted that in this case, not only the effective number N of charges on each macromolecule but the parameters g and D themselves change as a result of counter ion condensation. Understandably, these parameters can be found by substituting $\sigma \to \sigma/\beta$ in all relations of subsection 30.3. Indeed, because only the fraction β of charges remains uncompensated, the number σ of links between two uncompensated charges increases by the factor $1/\beta$. Hence, we obtain for the θ-temperature with respect to the non-Coulomb interaction [*see* Eq. (30.7)]

$$g \sim \sigma^{1/3} u^{-2/3} \beta^{-1/3}, \quad D \sim a\sigma^{2/} u^{-1/3} \beta^{-2/3}, \quad \rho = ge/D \sim e\beta^{1/3}/[a(u\sigma)^{1/3}].$$
$$(30.23)$$

Inserting Eq. (30.23) into Eqs. (30.20) and (30.19), we obtain instead of Eq. (30.22)

$$\ln[(1-\beta)/\beta] = (1 - Q\beta^{1/3})\ln \varphi, \qquad (30.24)$$

where $Q \sim u^{2/3}/\sigma^{1/3}$. In weakly charged polyelectrolytes, $(\sigma \gg 1)$, $Q \ll 1$. This means that for the function $\beta(\varphi)$ defined by Eq. (30.24), $\beta \to 1$ for $\varphi \to 0$, that is, counter ion condensation does not occur: for $\varphi \ll 1$, the fraction of counter ions inside region 1 is small. A similar conclusion can also be made for the counter ion condensation in a chain of blobs in good solvent.

Now consider a chain of globular blobs in poor solvent. Substituting $\sigma \to \sigma/\beta$ in Eq. (30.14), we obtain

$$g \sim \sigma|\tau|/u\beta, \quad D \sim v^{1/3}\sigma^{2/3}/(u^{1/3}\beta^{2/3});$$

$$\rho = ge/D \sim e|\tau|\sigma^{1/3}/(u^{2/3}v^{1/3}\beta^{1/3}). \qquad (30.25)$$

Consequently, for $\varphi \ll 1$ and with allowance made for Eq. (30.20), Eq. (30.19) takes the form

$$\ln[(1-\beta)/\beta] = (1 - Q'\beta^{-1/3})\ln \varphi, \qquad (30.26)$$

where $Q' \sim (u\sigma a^3/v)^{1/3}|\tau|$. Because $\sigma \gg 1$, $u \sim 1$, and $a^3 \gtrsim v$, the value of Q' remains greater than unity even for a slight decrease in temperature below the θ-point with respect to the non-Coulomb interaction ($|\tau| \ll 1$). On the other hand, it can easily be seen that for $Q' > 1$, Eq. (30.26) has no physically reasonable solution for $\varphi \ll 1$. The physical meaning of this fact can be seen as follows.

In accordance with Eq. (30.25), the quantity ρ for the chain of globular blobs is proportional to $\beta^{-1/3}$. This implies that a decrease in β (i.e., an increase of the fraction of condensed counter ions in region 1) leads not to a decrease but rather a growth of the linear charge density because of the "collapse" of the chain of blobs as the charge becomes neutralized. In its turn, this circumstance induces an additional influx of counter ions into region 1, and so on. An avalanche-like process sets in, called the *avalanche-like counter ion condensation*. This process stops only when almost all of the counter ions get into region 1, the charge of this region becomes totally compensated, and the chain conformation can no longer be represented as a sequence of blobs. In other words, a globule is formed whose shape differs insignificantly from a sphere and that keeps almost all of its counter ions inside.

Thus, for weakly charged polyelectrolytes, the counter ion condensation is significant only in poor solvent, where the blobs are in the globular state. The avalanche-like process in this case leads to a cardinal change in the polyelectrolyte chain conformation.

30.7. *The behavior of a system of counter ions near the polyelectrolyte chain can be investigated more consistently by using the nonlinear Poisson-Boltzmann equation.*

The two-phase approximation used in subsections 30.5 and 30.6 is, of course, very crude. In fact, the electrostatic potential near a charged polymer chain is distributed in a complicated way, and it cannot be approximated by the two constants ψ_1 and ψ_2. It can, however, be determined from the solution of the Poisson equation

$$\Delta\psi = -4\pi\tilde{\rho}/\varepsilon, \tag{30.27}$$

where $\tilde{\rho}$ is the total charge density, including the charges on the polymer chains and the pertinent counter ions as well as the charges of other ions (existing in the solution in the presence of salts). In the self-consistent field approximation, the concentration distribution of mobile ions is in turn defined by the potential $\psi(x)$ according to the Boltzmann distribution

$$c_i(\mathbf{x}) = c_i^{(0)} \exp[-ez_i\psi(\mathbf{x})/T], \tag{30.28}$$

where the index i enumerates the types of ions, ez_i is the charge of the ion of type i, $c_i^{(0)}$ are constants.

Equations (30.27) and (30.28) form a closed system, because the total charge density of all mobile ions featured in Eq. (30.27) equals $\Sigma_i ez_i c_i(x)$. For example, suppose we deal with a solution comprising the polyelectrolyte chains, inherent counter ions of infinitely low concentration and some added univalent salt (like common NaCl). In this case, there are only two types of mobile ions with opposite charges and equal concentrations, that is, $z_1 = 1$, $z_2 = -1$, $c_1^{(0)} = c_2^{(0)} = n/2$, where n is the total concentration of mobile ions. Combining Eqs. (30.27) and (30.28), we find that outside the polymer chains, the potential satisfies the following nonlinear equation:

$$\Delta\tilde{\psi} = r_D^{-2} \operatorname{sh} \tilde{\psi},$$

where $\tilde{\psi} = e\psi/T$ and r_D is the Debye radius (30.2). This equation, combined from the Poisson (30.27) and Boltzmann (30.28) equations, is referred to in literature as the *Poisson-Boltzmann equation*.[f] Certainly, it should be supplemented by the electrostatic boundary conditions at the surface of the macromolecules.

If $\tilde{\psi} \ll 1$, then the Poisson-Boltzmann equation reduces to the well-known linear Debye-Hückel equation.[g] $\Delta\tilde{\psi} = r_D^{-2}\tilde{\psi}$. Its spherically symmetric solution has the form $(\mathrm{const}/r) \cdot \exp(-r/r_D)$. Analysis of this solution substantiates the

---

[f]The same name is applied to the equations based on the other forms of nonlinearity obtained for different combinations of the concentrations c_i and the charges z_i of counter ions and salts. We recommend that the reader derive by himself the Poisson-Boltzmann equation for arbitrary c_i and z_i.
[g]The Debye-Hückel equation can be derived by linearizing the exponential (20.28) for arbitrary c_i and z_i as well.

usual concept of the screened Coulomb interaction, which proved to be very productive in the physics of electrolytes or plasma (*see*, e.g., Sec. 78 in Ref. 22). For polyelectrolytes, however, a linear approximation is often inapplicable, because the dimensionless potential $\tilde{\psi}$ is not small, especially near the chain of polyions. Specifically, the counter ion condensation is, of course, a purely nonlinear phenomenon.

In its complete nonlinear form, the Poisson-Boltzmann equation is only numerically tractable. The numeric solution obtained for the strongly charged polyelectrolyte (when the polymer chain is pictured as an infinite uniformly charged cylindric surface) allows a deeper insight into the conditional notion of condensed counter ions and, correspondingly, the condensation effect.

As expected, the macromolecule with linear charge density ρ_0 above the critical one is surrounded by a diffuse cloud of counter ions, which stays close to a polyion even in a system with a zero concentration of chains (where the entropy gain because of "evaporation" of the cloud is the highest) and a charge just sufficient to compensate the excess ρ_0 over the critical density.

It should be noted that the Poisson-Boltzmann equation, even in the totally nonlinear form, is far from being accurate. It is a result of the self-consistent field approximation, because it is supposed in Eq. (30.28) that the ions interact only via the self-consistent potential ψ. Analysis, however, shows that in most real situations, the accuracy of the given approximation is quite satisfactory.

30.8. *The Coulomb interaction stiffens the chain of strongly charged polyelectrolytes, that is, leads to an increase in the persistent length l of the chain; the corresponding contribution into $\tilde{l}$ is called the electrostatic persistent length.*

In a salt-free polyelectrolyte solution of finite concentration (as well as in the presence of a low-molecular-weight salt), the Coulomb interaction between the links is screened. The radius of screening r_D is defined by Eq. (30.2). As subsection 30.4 mentioned, the polyelectrolyte chain on large scales acquires the conformation of a coil; at the same time, the electrostatic interaction leads to a substantial "stiffening" (extension) of the polymer chain. Such "stiffening" conforms to an effective increase in persistent length as well as the Kuhn segment length of the polymer chain. Let us now assess qualitatively the electrostatic contribution to the persistent length.

Suppose that the polyelectrolyte macromolecule constitutes in the uncharged state a persistent chain with Kuhn segment length l_0 [the persistent length $\tilde{l}_0 = l_0/2$; see Eq. (3.8)]. Next, suppose that the charges e separated by distance a appear on this chain (e.g., after immersing the chain into solvent, some fraction of links dissociates). Let r_D denote the Debye screening radius in this solvent. Because of the electrostatic interaction, the persistent length of the polyelectrolyte chain will increase and become equal to $\tilde{l} = \tilde{l}_0 + \tilde{l}_e$ (or $l = l_0 + l_e$ for the Kuhn segments). The contribution $\tilde{l}_e$ is referred to as the *electrostatic persistent length*.

That the Coulomb interaction really leads to an effective renormalization of the persistent length can be illustrated using Figure 5.13, in which the considered

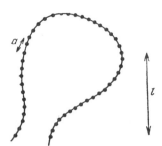

FIGURE 5.13. Explaining the concept of electrostatic persistent length.

chain is shown for the case $a \ll r_D \ll l = l_0 + l_e$ (a strongly charged polyelectrolyte; moderate concentration of a low-molecular-weight salt in the solution). In this case two types of Coulomb interaction are possible:

1. Between the charges separated by a distance $\lesssim r_D$ along the chain (i.e., short-range interaction tending to increase the persistent length).

2. Between the charges separated by a distance $\gtrsim l$ along the chain (i.e., such charges approach one another closer than the distance r_D as a result of random bending of the chain; such an interaction should naturally be classified with volume interaction).

This subdivision into short-range and volume interactions is quite unambiguous for $r_D \ll l$, because the links separated by the distance exceeding r_D but less than l can neither interact directly because of the Debye interaction nor draw together as a result of chain bending.

Thus, for $a \ll r_D \ll l$, the Coulomb short-range interaction brings about an increase in the persistent length. In the next subsection, we demonstrate that for strongly charged polyelectrolytes, these inequalities as a rule hold true. As for weakly charged polyelectrolytes, their Kuhn segment contains less than one charge. Therefore, the two classes of electrostatic interaction cannot be identified in this case, and the notion of a persistent length cannot be introduced. Nevertheless, in this case, one can define an electrostatic persistent length of the chain of blobs.

*30.9. *The electrostatic persistent length in strongly charged polyelectrolyte chains substantially exceeds the Debye screening radius.*

Let us now determine the quantity $\tilde{l}_e$ for the chain of strongly charged polyelectrolyte shown in Figure 5.13 for $a \ll r_D \ll l$. Consider a short section of such a chain of length s, much less than l ($s \ll l$) but greater than r_D ($s \gg r_D$), and suppose that this section bends slightly with a constant curvature radius so that the directions of the ends make the angle $\theta \ll 1$. Were the chains uncharged, the energy ΔE_0 of such a bending would be determined by Eqs. (2.7) and (2.9) to give

$$\Delta E_0 = (1/2) T \tilde{l}_0 \theta^2/s. \tag{30.29}$$

In the charged chain, an additional repulsion of the links (30.1) occurs so that the bending energy ΔE increases:

$$\Delta E = \Delta E_0 + \Delta E_e = (1/2) T \tilde{l}_0 \theta^2/s + (1/2) T \tilde{l}_e \theta^2/s. \tag{30.30}$$

This last equality, written with allowance for the electrostatic bending energy ΔE_e being proportional at small bendings to the product of the curvature square $(\theta/s)^2$ and the length s of the section, defines the electrostatic persistent length $\tilde{l}_e$.

Charges on the considered chain section, however, interact with the energy (30.1). Therefore,

$$\Delta E_e = \frac{e^2}{\varepsilon} \sum_{i=1}^{M} \sum_{j=i+1}^{M} \left[\frac{\exp(-r_{ij}/r_D)}{r_{ij}} - \frac{\exp(-|j-i|a/r_D)}{|j-i|a} \right], \tag{30.31}$$

where the contributions of the Coulomb interaction energy from all link pairs of the chain section are summed, M is the total number of links in the given chain section, and r_{ij} is the distance between the links i and j in the conformation bent with curvature θ/s. To calculate ΔE_e using formula (30.31), we substitute the variables n, m for i, j $[n = ia/s, \ m = ja/s \ (0 \leqslant n < m \leqslant 1)]$ and change from a summation over i and j to an integration with respect to n and m:

$$\Delta E_e = \frac{e^2 s^2}{\varepsilon a^2} \int_0^1 dn \int_n^1 dm \left[\frac{\exp(-r_{mn}/r_D)}{r_{mn}} - \frac{\exp(-|m-n|s/r_D)}{|m-n|s} \right]. \tag{30.32}$$

The distance r_{mn} is to be calculated to the accuracy of square terms in θ^2. The result takes the form

$$r_{mn} = s(m-n)[1-(\theta^2/24)(m-n)^2]. \tag{30.33}$$

Substituting Eq. (30.33) into Eq. (30.32) and extracting the basic term proportional to θ^2 in Eq. (30.32), we evaluate the integrals to obtain

$$\Delta E_e \cong (1/24)(e^2 s/\varepsilon a^2) h(s/r_D) \theta^2, \tag{30.34}$$

where the function $h(x)$ is expressed as

$$h(x) = 3x^{-2} - 8x^{-3} + \exp(-x) \cdot (x^{-1} + 5x^{-2} + 8x^{-3}). \tag{30.35}$$

For the considered section of length $s \gg r_D$, it is necessary to use the asymptotic form of the function $h(x)$ for $x \gg 1$: $h(x) = 3x^{-2}$. Then,

$$\Delta E_e \cong (1/8) [e^2 r_D^2/(\varepsilon a^2)] (\theta^2/s). \tag{30.36}$$

Comparing Eqs. (30.36) and (30.30), we obtain the expression for the electro-static persistent length $\tilde{l}_e$

$$\tilde{l}_e = (1/4)e^2 r_D^2/(\varepsilon a^2 T) = (u/4)(r_D^2/a), \tag{30.37}$$

where $u \equiv e^2/(\varepsilon a T)$. Taking into account that for typical cases, $u \sim 1$ (see subsection 30.3) and $r_D \gg a$ (if the salt concentration is not overly high), we reach the conclusion that $\tilde{l}_e \gg r_D$, that is, the stiffening of the polymer chain because of electrostatic interaction occurs on length scales substantially larger than the Debye radius r_D (despite the fact that this interaction has the action radius r_D).

The total persistent length of the macromolecule is $\tilde{l} = \tilde{l}_0 + \tilde{l}_e$. In many cases, provided that the corresponding uncharged chains are not too stiff, we obtain $\tilde{l}_e \gg \tilde{l}_0$ (i.e., the electrostatic contribution to the persistent length prevails).

Note also that the inequalities $a \ll r_D \ll l = l_0 + l_e$ that we adopted initially are fulfilled automatically because of the result (30.37). Equation (30.37) is obtained without allowing for possible counter ion condensation, that is, it is valid for the case $\rho_0 = e/a < \varepsilon T/e$ or $u < 1$ (see subsection 30.5). For $u > 1$, a fraction of counter ions will condense on the polymer chain, thus neutralizing effectively its charge. Taking into account the results of subsection 30.5, one can easily generalize this reasoning to this case. The electrostatic persistent length takes the form

$$\tilde{l}_e = (1/4)\varepsilon T r_D^2/e^2 \quad \text{when } u > 1. \tag{30.38}$$

Note that in this regime, the quantity $\tilde{l}_e$ is independent of the linear charge density of the polymer chain (i.e., of the quantity a), as it should be, because the charge density exceeding the value ρ_0 is compensated by the counter ions precipitating on the chain. Note also that for $u > 1$, the inequality $\tilde{l}_e \gg r_D$ also holds true for all physically reasonable situations.

Regarding the picture of the salt-free polyelectrolyte solution shown in Figure 5.10, one can understand why the chains are shown essentially straight on these scales. According to Eq. (30.17), $\xi \sim r_D$. On the other hand, $\tilde{l}_e \gg r_D$; hence, $l = l_0 + l_e \gg \xi$. Thus, the stiffening of a polyelectrolyte chain in the salt-free solution occurs on length scales substantially exceeding the distance between two neighboring chains.

Finally, note that Eq. (30.37) can also be used for weakly charged polyelectrolytes if it is applied to the chain of blobs. The reasoning given previously remains valid if the substitutions $e \to ge$, $a \to D$ are made, where the quantities g and D for the non-Coulomb interaction of different types are determined in subsection 30.3. In particular, for the θ-point relative to the non-Coulomb interaction we have, with allowance made for Eq. (30.7),

$$\tilde{l}_e \sim r_D^2/D. \tag{30.39}$$

For $r_D \gg D$ (according to subsection 30.8, only in this case can one introduce the electrostatic persistent length), we have $\tilde{l}_e \gg r_D$ (i.e., the stiff section of the chain of blobs exceeds considerably the Debye radius).

30.10. *In the presence of a small fraction of charged links on a polymer network, its collapse proceeds (as far as solvent quality deteriorates) as a discrete first-order phase transition; the abrupt change of the size is associated with additional osmotic pressure of the gas of counter ions in the charged network.*

Let us now pass from polyelectrolyte solutions to a discussion of the effects of charged links on properties of other polymer systems. Subsection 29.9 considered the theory of free swelling of polymer networks in solvents of different quality. The presence of even a small fraction of charged links on the network chains proves to affect substantially the network swelling process with variation of the solvent quality. Now, we consider this problem in more detail.

Returning to the designations of subsection 29.9, we suppose that in addition to ordinary uncharged links, the network subchains contain N/σ charges (σ thus is the mean number of neutral links between the two consecutive charged links in the chain; cf. subsection 30.3).[h] Assume that the fraction of charged links is small ($\sigma \gg 1$), that is, the networks are weakly charged. Because the network sample as a whole is electrically neutral, there must be N/σ counter ions per subchain inside the sample. Suppose also that the solution in which the network swells is salt-free, so counter ions are the only low-molecular-weight ions in the considered system.

The free energy of the charged polymer network will comprise, (apart from the contributions F_{el} and F_{int} allowed for in subsection 29.9) two additional contributions F_0 and F_C:

$$F = F_{el} + F_{int} + F_0 + F_C. \qquad (30.40)$$

The term F_0 is the ideal part (without taking into account the electrostatic interaction) of the free energy of a gas of counter ions. Because the total number of counter ions inside the network equals NVv/σ and their concentration n/σ, then

$$F_0 = (TNVv/\sigma)\ln(n/\sigma). \qquad (30.41)$$

The presence of the free energy F_0 leads to a general increase in the equilibrium size of the network because of the osmotic pressure of the gas of counter ions extending the network.

The term F_C is associated with the proper Coulomb interaction of charged links and counter ions in the network. In the simplest approximation, this contribution may be identified with the free energy of Coulomb interaction in an electrically neutral plasma with an average link concentration per unit volume of n/σ. It is known[24] that in the Debye-Hückel approximation, which is definitely

---

[h]As distinguished from subsection 30.3, hereafter N will denote the number of links in the chain (subchain), not the number of charges.

valid for the case of a weakly charged network ($\sigma \gg 1$),

$$F_C = -TNVvu^{3/2}\sigma^{-3/2}(na^3)^{1/2}, \tag{30.42}$$

where $u \equiv e^2/(\varepsilon aT)$. The fact that $F_C < 0$ implies that this term corresponds to the effective link attraction.

Taking into account the relation $n = n_0/\alpha^3$ (*see* subsection 29.9; n_0 is the link concentration in the reference state of the network), one can express the contributions F_0 and F_C as a function of the parameter α. The following minimization of the free energy (30.40) with respect to α (cf. subsection 29.9) brings about the equation

$$\alpha^5 - s\alpha^3 + \lambda\alpha^{3/2} - y/\alpha^3 = x, \tag{30.43}$$

where the designations x and y are defined in subsection 29.9 and $s \equiv N/\sigma$ is the number of charges on the subchain, $\lambda \equiv (N/2)(u/\sigma)^{3/2}(n_0a^3)^{1/2}$. The distinction of the dependence $\alpha(x)$, given by the relation (30.43) for typical values of the parameters y and λ and for various values of s from the corresponding dependence for the neutral network (Fig. 5.8) is associated with the term $-s\alpha^3$ (30.43), that is, with the osmotic pressure of the gas of counter ions. (For physically reasonable values of the parameters, the contribution of the term $\lambda\alpha^{3/2}$ is insignificant.) It is seen from Figure 5.8 that for the conditions of very poor solvent, the size of the network, which in this case is in the globular (collapsed) state, is independent of the degree of its charging. On increasing x (or improvement in solvent quality) starting from a certain critical value x_{cr}, however the attraction of the links of the charged network becomes too weak to counteract the osmotic pressure of the gas of counter ion and to confine the network in the collapsed state. As a result, at $x = x_{cr}$, a sharp increase in the equilibrium size of the charged network (i.e., a first-order phase transition) occurs. For $x > x_{cr}$, the network size varies weakly with x, being determined by the balance of the elastic energy and the energy connected with the osmotic pressure exerted by counter ions [i.e., by the balance of the first and second terms on the left-hand side of Eq. (30.43)]. The described features of swelling and collapse of charged networks become more pronounced for larger values of the parameter s, that is, at larger number of charges per subchain (Fig. 5.8).

The phenomenon of abrupt collapse of charged polymer networks was discovered in 1978 by T. Tanaka, and it has been intensely investigated since that time. Interest in this phenomenon is connected with large, abrupt changes in the volume (by a factor of a few thousands) of a polymer network that can be triggered by a very small variation of external conditions. It has already been mentioned that from the theoretic viewpoints, the collapse of the polymer network is a macroscopic manifestation of the globule–coil transition occurring in each subchain. Additional sharpness of this transition is associated in charged networks with the thrusting pressure of counter ions.

30.11. *The compatibility of a mixture of two polymers improves substantially after a weak charging of one of the components.*

The results obtained in the previous subsection can be interpreted as a consequence of the condition of macroscopic electric neutrality of a polymer system. Because of this condition, a gas of counter ions cannot leave the space inside the polymer network and thus exerts an additional and very substantial osmotic pressure on the network. The electric neutrality condition specifies the behavior of polyelectrolyte systems in many other instances. Here is another example of this kind.

Subsection 27.1 showed that different polymers mix poorly in most cases: a minor repulsion of links is sufficient for the mixture to separate into essentially pure phases. We now show that compatibility can be substantially improved if one of the components is made weakly charged.

Let us return to the notation adopted in subsection 27.1. Assume that the polymers of type A include a small number N_A/σ of charged links ($\sigma \gg 1$). Because of the condition of electric neutrality, the mixture must contain the same number of counter ions per chain A. As in the case of the charged network [cf. Eq. (30.40)], the presence of charges will bring about two additional contributions to the free energy, F_0 and F_C, associated with the translational entropy of the gas of counter ions [cf. Eq. (31.41)] and with the Coulomb interaction [cf. Eq. (30.42)]. Analysis shows that for $\sigma \gg 1$, the contribution F_C is insignificant, as in the case of the networks. Therefore, we write here only the expression for F_0

$$F_0 = (TVc_A/\sigma)\ln(c_A/\sigma), \qquad (30.44)$$

where c_A is the concentration of the links A.

It can easily be seen that the appearance of the additional contribution (30.44) in the expression (27.1) results in a renormalization of the number of links in the chain A: the expression for the free energy of the mixture remains the same provided that the substitution $1/N \to 1/N_A + 1/\sigma$ is made. Consequently, the coordinates of the critical point of phase separation of the mixture can immediately be written in the considered case as [cf. Eqs. (27.4) and (27.5)]

$$\chi_{AB}^{(cr)} = \frac{1}{2}\left[\left(\frac{1}{N_A}+\frac{1}{\sigma}\right)^{1/2}+\left(\frac{1}{N_B}\right)^{1/2}\right]^2,$$

$$\Phi_A^{(cr)} = \left[1+\left(\frac{1}{N_A}+\frac{1}{\sigma}\right)^{-1/2}\left(\frac{1}{N_B}\right)^{1/2}\right]^{-1}. \qquad (30.45)$$

For $N_A \sim N_B \gg \sigma$, that is, when there are many charged links per chain A, $\chi_{AB}^{(cr)} \sim 1/\sigma$. Comparing this result with the estimate $\chi_{AB}^{(cr)} \sim 1/N$, obtained in subsection 27.1 for a mixture of uncharged polymers with $N_A \sim N_B \sim N$, we conclude that in this case, the phase separation starts at essentially larger values of χ. In this way, the phase separation region substantially decreases, which corresponds to an improvement of compatibility of the polymer mixture.

In physical terms, this result signifies that because of electric neutrality, the phase separation of polymer A is accompanied with a separation of counter ions. This leads to a much more substantial loss of translational entropy in comparison with the separation of only the polymer chains. This is because each counter ion possesses three independent translational degrees of freedom (just as a whole polymer chain), but the number of counter ions per chain A is much greater than unity. As a result, the separation into two macroscopic phases proves to be less favorable thermodynamically than in the absence of charged links.

CHAPTER 6

Dynamical Properties of
Polymer Solutions and Melts

Until now, we have considered only the equilibrium properties of polymer systems. Such systems are also known to possess remarkable dynamic properties. For example, polymeric liquids (solutions and melts) are usually very viscous, they "keep memory" of their previous flow history, and they often provide a qualitatively different response to weak and strong action. A fundamental property of polymeric liquids is viscoelasticity: when exposed to sufficiently rapidly changing actions, such liquids behave as elastic rubber-like materials, whereas under slowly varying forces, a flow typical for a viscous liquid sets in. This chapter presents the basic theoretical approaches, allowing one to describe the dynamic properties of polymeric liquids on the basis of molecular concepts. As in studies of equilibrium properties, we start with the analysis of the dynamics of an individual polymer chain (i.e., with a consideration of a dilute solution of non-overlapping coils), the first three sections of this chapter are devoted to these problems.

31. THE ROUSE MODEL: A PHANTOM CHAIN IN
IMMOBILE SOLVENT

31.1. *The simplest theory of polymer chain dynamics, formulated for a standard Gaussian model, assumes the chain to be ideal, phantom, and the solvent immobile.*

Earlier, we introduced several polymer chain models, of which the standard Gaussian bead model (*see* Sec. 4) proved to be the most convenient for a theoretical analysis of equilibrium properties. Therefore, it is natural to begin investigating polymer coil dynamics within the framework of this model as well.

Consider a standard Gaussian chain of N links. Let us try to simplify as much as possible the analysis of the dynamic properties of such a chain. First, we neglect for the present volume interactions of the links, that is, we assume the chain is ideal. Second, we do not take into account the motion of the solvent, that is, assume the solvent to be an immobile viscous medium, in which the moving chain links (beads) experience friction but which is not carried along with their motion. Third, the polymer chain we assume to be phantom, that is, we neglect topological constraints on possible chain motions (*see* subsection 11.1 and Fig. 1.16) or allow chain sections to pass freely through one another.

The dynamic behavior of a polymer chain under these assumptions was first examined by V. A. Kargin and G. L. Slonimskii in 1948 and by P. Rouse in 1953. The given model of the coil dynamics is known in the literature as the *Rouse model*.

31.2. *The mathematical description of the Rouse model is based on equations of motion of the links, allowing for random forces acting on them (the Langevin equation).*

In the Rouse model, each link is subjected first to the forces f^{ch} from neighboring links in the chain; second to the force f^{fr} of friction against the solvent; and third to the random force f, which appears when the given link collides with solvent molecules. Therefore, the equation of motion for the n-th link of the polymer chain can be written as

$$m \frac{\partial^2 x_n}{\partial t^2} = f_n^{ch} + f_n^{fr} + f_n, \tag{31.1}$$

where x_n is the radius vector of the position of the n-th link, $\partial^2 x_n/\partial t^2$ the acceleration of the n-th link, and m its mass. For conventional motion of the link in dense solvent, the inertial term in Eq. (31.1) is quite insignificant, and the equation of motion takes the form

$$f_n^{ch} + f_n^{fr} + f_n = 0. \tag{31.2}$$

Consider now each force in Eq. (31.2) separately.

The force exerted by the neighboring links in the chain is caused by chain connectivity. In the standard Gaussian model, link bonding is specified by the correlation functions (4.14), which can be treated according to Eq. (6.1) as energy terms if one introduces the energy of interaction of the neighboring links $U_{n,n+1} = -T \ln g(x_{n+1}-x_n)$. The total interaction energy appearing because of chain connectivity equals

$$U = \sum_{n=1}^{N-1} U_{n,n+1} = \text{const} + \sum_{n=1}^{N-1} \frac{3T}{2a^2} (x_{n+1}-x_n)^2, \tag{31.3}$$

where const is a constant independent of the conformation of the macromolecule. The force f_n^{ch} exerted on the n-th link by neighboring links in the chain is found by differentiating the expression (31.3) with respect to x_n:

$$f_n^{ch} = -\frac{\partial U}{\partial x_n} = \frac{3T}{a^2} (x_{n+1}-2x_n+x_{n-1}) \quad (n \neq 1, N). \tag{31.4}$$

The forces of friction against the solvent, f_n^{fr}, naturally are assumed to be proportional to the velocity, because the links move with thermal velocities in a viscous solvent that is regarded as immobile. Therefore,

$$f_n^{fr} = -\xi \frac{\partial x_n}{\partial t}, \tag{31.5}$$

where ξ is the friction coefficient.

The friction force is, in fact, a regular constituent of the total force exerted by the solvent on a particle moving within it. Because of the discrete molecular structure of the solvent, the indicated total force contains a non-regular term caused (roughly speaking) by the impacts of individual molecules and leading to conventional Brownian motion. As previously mentioned, this force is a random function of time, $f_n(t)$, with zero mean value

$$\langle f_n^r(t) \rangle = 0, \tag{31.6}$$

because the regular part of the force (the friction) is extracted.

Statistical properties of the random force are investigated in the theory of Brownian motion in full detail (see, e.g., Ref. 38), where that force is shown to be Gaussian (i.e., its value is distributed according to the Gaussian law) and delta correlated. The Gaussian character stems from the fact that the number of collisions with solvent molecules is large, and f_n is formed as a sum of a very great number of random contributions. The delta correlation means that

$$\langle f_{n\alpha}^r(t) f_{m\beta}^r(t') \rangle = 2\xi T \delta_{nm} \delta_{\alpha\beta} \delta(t-t'), \tag{31.7}$$

where α and β enumerate the Cartesian components, and δ_{nm} and $\delta_{\alpha\beta}$ are Kronecker's symbols ($\delta_{pq} = 0$ at $p \neq q$, and $\delta_{pq} = 1$ at $p = q$).

The physical meaning of Eq. (31.7) is simple. First, it reflects the fact that random forces have no preferred direction. Second, forces acting on different links or on one link at different moments do not correlate with one another at all (i.e., are statistically independent). The last circumstance (the zero correlation time) is, of course, an idealization. It causes a divergence in the equality (31.7) at $t = t'$, which is insignificant in calculations of any observable characteristics.

Because the random quantity $f_n(t)$ is Gaussian, its probability distribution is wholly determined by the first two moments [i.e., by Eqs. (31.6) and (31.7)], which can be used to calculate any necessary average values. To illustrate this circumstance and to check the correctness of the choice of proportionality factor $2\xi T$ in Eq. (31.7), we show that the free link (i.e., not integrated in the chain) performs ordinary Brownian motion under the action of the described random forces. According to Eqs. (31.2) and (31.5), the equation of motion can be written in this case as

$$\xi \frac{\partial x}{\partial t} = f^f(t). \tag{31.8}$$

Suppose that at the initial moment, the link resided at the origin. Then, integrating Eq. (31.8), we obtain

$$x(t) = \frac{1}{\xi} \int_0^t f(t') dt'. \tag{31.9}$$

Hence,

$$\langle x^2(t) \rangle = \frac{1}{\xi^2} \left\langle \left(\int_0^t dt' f(t') \right) \left(\int_0^t dt'' f(t'') \right) \right\rangle$$

$$= \frac{1}{\xi^2} \int_0^t dt' \int_0^t dt'' \langle f(t') f(t'') \rangle$$

$$= \frac{6T}{\xi} \int_0^t dt' \int_0^t dt'' \delta(t' - t'')$$

$$= (6T/\xi) t$$

$$= 6Dt, \tag{31.10}$$

where the last equality is written with regard to the well-known Einstein equation[38]

$$D = T/\xi, \tag{31.11}$$

relating the coefficient of translational diffusion of a Brownian particle, D, and the friction coefficient ξ. From Eq. (31.6), one can see that the free link subjected to the random forces with moments (31.6) and (31.7) actually performs Brownian motion with the diffusion coefficient D.

Returning to dynamics of the polymer chain in the Rouse model, we write the equation of motion for the n-th link (31.2), with allowance made for Eqs. (31.4) and (31.5), in the form

$$\xi \frac{\partial x_n}{\partial t} = \frac{3T}{a^2} (x_{n+1} - 2x_n + x_{n-1}) + f_n(t). \tag{31.12}$$

Equations of motion of the type (31.12), containing the random forces, are called in statistical physics the *Langevin equations*.

31.3. In the continuous limit, equations of motion for links in the Rouse model are reduced to a diffusion equation; the state of the chain ends defines the boundary conditions of this equation.

Let us formally regard the index n in Eq. (31.12) as a continuous variable and move to the continuous limit of the function $x(t,n)$ on the right-hand side of Eq. (31.12). This passage to the continuous limit is possible for a long polymer chain, because the value of x slowly varies with the argument. (Substantial changes appear only on length scales of the order of the whole polymer coil.) In the continuous limit, we obviously have

$$x_{n+1}-x_n=\Delta x/\Delta n \to \partial x/\partial n, \tag{31.13}$$

$$x_{n+1}-2x_n+x_{n-1}=(x_{n+1}-x_n)-(x_n-x_{n-1})$$

$$=\left[\frac{\Delta x}{\Delta n}(n)-\frac{\Delta x}{\Delta n}(n-1)\right]\frac{1}{\Delta n}$$

$$\to \frac{\partial^2 x}{\partial n^2}. \tag{31.14}$$

Equation (31.12) then takes the form

$$\xi \frac{\partial x(t,n)}{\partial t}=\frac{3T}{a^2}\frac{\partial^2 x(t,n)}{\partial n^2}+f^r(t,n) \tag{31.15}$$

of a well-known, second-order, linear differential equation: the diffusion equation. The relations (31.6) and (31.7) for the random force are readily generalized in the continuous limit. Specifically, instead of Eq. (31.7), we have

$$\langle f_\alpha^r(t,n)f_\beta^r(t',n')\rangle=2\xi T\delta(n-n')\delta_{\alpha\beta}\delta(t-t'). \tag{31.16}$$

Like any partial differential equation, Eq. (31.15) must be supplemented with boundary conditions. In the given case, these are the conditions at $n=0$ and $n=N$ (i.e., at the chain ends). For the terminal monomer links (the first and the N-th), the expression for f^{ch} becomes modified in comparison with Eq. (31.4), because they experience a force exerted by only one neighboring link in the chain. As a result, the equations analogous to Eq. (31.12) take the following form for these links

$$\xi \frac{\partial x_1}{\partial t}=\frac{3T}{a^2}(x_2-x_1)+f_1^r(t), \quad \xi \frac{\partial x_N}{\partial t}=\frac{3T}{a^2}(x_{N-1}-x_N)+f_N^r(t).$$
$$\tag{31.17}$$

The form of Eq. (31.17) can be made identical with the general equation (31.12) by introducing fictitious links, numbered 0 and $N+1$, and letting

$$x_0-x_1\equiv 0, \quad x_{N+1}-x_N\equiv 0. \tag{31.18}$$

When written in the continuous limit, these conditions are equivalent to the following

$$\left.\frac{\partial x}{\partial n}\right|_{n=0}=0, \quad \left.\frac{\partial x}{\partial n}\right|_{n=N}=0 \tag{31.19}$$

[cf. Eq. (31.13)]. The relations (31.19) play the role of boundary conditions for Eq. (31.15).

*31.4. *Having performed the Fourier transformation of the basic equation of the Rouse model, one can represent the motion of a polymer chain as a superposition of independent Rouse modes.*

We seek the solution of Eq. (31.15) with the boundary conditions in the form

$$x(t,n) = y_0(t) + 2 \sum_{p=1}^{\infty} y_p(t) \cos \frac{\pi p n}{N}, \qquad (31.20)$$

that is, we perform the Fourier transformation relative to the variable n. In the Fourier expansion of the function $x(t,n)$, defined in the region $0 \leqslant n \leqslant N$, we omitted terms proportional to $\sin(\pi p n / N)$ to satisfy the boundary conditions (31.19). The coordinates $y_p(t)$ are expressed via $x(t,n)$ using the inverse Fourier transformation

$$y_p(t) = \frac{1}{N} \int_0^N dn \cos \frac{\pi p n}{N} \cdot x(t,n), \quad p=0,1,2,\ldots \qquad (31.21)$$

Differentiating Eq. (31.21) with respect to t and taking into account that function $x(t,n)$ in the integrand satisfies the Rouse equation (31.15), one can immediately see that the equation for $y_p(t)$ is written in the form

$$\xi \frac{\partial y_p(t)}{\partial t} = -\frac{\xi}{\tau_p} y_p(t) + f_p(t) \quad (p \neq 0),$$

$$\xi \frac{\partial y_0(t)}{\partial t} = f_0(t), \qquad (31.22)$$

where

$$\tau_p = \frac{N^2 a^2 \xi}{3\pi^2 T p^2} \quad (p \neq 0) \qquad (31.23)$$

$$f_p(t) = \frac{1}{N} \int_0^N dn \cos(\pi p n / N) f'(t,\, n). \qquad (31.24)$$

The random forces defined by Eq. (31.24) have the moments

$$\langle f_p(t) \rangle = 0; \quad \langle f_{p\alpha}(t) f_{q\beta}(s) \rangle = \frac{\xi T}{N} (1 + \delta_{p0}) \delta_{pq} \delta_{\alpha\beta} \delta(t-s) \qquad (31.25)$$

This can easily be verified using Eqs. (31.7) and (31.24).

Hence, we see that because forces $f_p(t)$ are independent of one another, the Rouse equation (31.15) decomposes into a set of independent equations (31.22) for the coordinates $y_p(t)$ with $p=0, 1, 2, \ldots$ These coordinates are called the *relaxation* or *Rouse modes*. Thus, motion of a polymer chain in the Rouse model can be represented as a superposition of independent Rouse modes.

31.5. In the Rouse model, the maximum relaxation time of the polymer coil and the diffusion coefficient of the coil as a whole vary with the growth of the number N of chain links as N^{-2} and N^{-1}, respectively; slow intermolecular relaxation and diffusive motion of the coil as a whole conform to the first and fundamental Rouse modes.

Equation (31.22) has a very simple form, and it can be simplified further after a substitution of variables

$$y_p(t) = z_p(t)\exp\left(-\frac{t}{\tau_p}\right), \quad \tau_p \equiv \frac{\xi_p N a^2}{3\pi^2 T p^2}, \quad p \neq 0. \tag{31.26}$$

The solution can be written in the form

$$y_p(t) = \frac{1}{\xi_p} \int_{-\infty}^{t} dt' \exp\left(-\frac{t-t'}{\tau_p}\right) f_p(t'). \tag{31.27}$$

The lower limit of integration is chosen here to be infinite, because the polymer chain in the process of Brownian motion "loses memory" of the infinitely removed past. Therefore, without restricting generality, one can make any assumptions about the corresponding initial conditions. The solution for the mode with $p=0$ is derived from Eq. (31.27) in which $\tau_p = \infty$ is formally set.

Now consider the physical consequences of the solution obtained. Suppose that we need to know how the end-to-end vector relaxes in the Rouse model

$$R(t) = x(t,N) - x(t,0). \tag{31.28}$$

According to Eq. (31.20),

$$R(t) = -4 \sum_{p=1,3,5,\dots} y_p(t). \tag{31.29}$$

Therefore, to determine the correlation function $\langle R(t)R(0)\rangle$, we must calculate $\langle y_p(t)y_p(0)\rangle$. Taking into account Eqs. (31.27) and (31.25), we obtain

$$\langle y_p(t)y_p(0)\rangle = \frac{1}{\xi_p^2} \int_{-\infty}^{t} dt' \int_{-\infty}^{0} dt'' \exp\left(-\frac{t-t'-t''}{\tau_p}\right) \langle f_p(t')f_p(t'')\rangle$$

$$= \frac{3}{\xi_p^2} \int_{-\infty}^{t} dt' \int_{-\infty}^{0} dt'' \exp\left(-\frac{t-t'-t''}{\tau_p}\right) 2\xi_p T\delta(t'-t'')$$

$$= \frac{6T}{\xi_p} \int_{-\infty}^{0} dt'' \exp\left(-\frac{t-2t''}{\tau_p}\right)$$

$$= \frac{3T\tau_p}{\xi_p} \exp\left(-\frac{t}{\tau_p}\right). \tag{31.30}$$

Consequently, allowing for Eqs. (31.29) and (31.26), we obtain

$$\langle R(t)R(0)\rangle = 16 \sum_{p=1,3,5,\ldots}^{\infty} \langle y_p(t)y_p(0)\rangle$$

$$= 48T \sum_{p=1,3,5,\ldots}^{\infty} \left(\frac{\tau_p}{\xi_p}\right)\exp\left(-\frac{t}{\tau_p}\right)$$

$$= \frac{8Na^2}{\pi^2} \sum_{p=1,3,5,\ldots}^{\infty} \frac{1}{p^2}\exp\left(-\frac{tp^2}{\tau_1}\right), \qquad (31.31)$$

where τ_1 denotes the relaxation time complying with the Rouse mode with $p=1$:

$$\tau_1 = N^2a^2\xi/(3\pi^2T). \qquad (31.32)$$

The relation (31.31) describes the dynamics of the end-to-end vector in the Rouse model of the polymer chain. It represents a superposition of relaxation modes with $p=1,3,5,\ldots,\infty$. Certainly, the system with a finite number N of links cannot possess an infinite number of relaxation modes. This is why the upper limit of summation in Eq. (31.31) $p=\infty$ is an artifact of the formal transition to the continuous variable n. For real chains, the summation of type (31.20) or (31.31) should be cut off at large but finite values of $p \sim N$. From the standpoint of slow relaxation processes in the coil, however, such a truncation is quite insignificant. It can be seen from Eq. (31.31) that with the growth of the number p of the mode, the corresponding relaxation time diminishes drastically:

$$\tau_p = \tau_1/p^2 = N^2a^2\xi/(3\pi^2Tp^2), \qquad (31.33)$$

that is, modes with large values of p attenuate rapidly. The modes with small p attenuate most slowly, especially $p=1$ [see Eq. (31.32)]. This mode defines the behavior of the correlation function (31.31) for large values of t. Accordingly, the time τ_1 (31.32) is called the *maximum relaxation time of the polymer coil*; it grows with the number N of chain links as N^2: $\tau_1 \sim N^2$.

It is seen from Eq. (31.33) that the time τ_p with $p>1$ can be regarded as the maximum relaxation time for a chain section of N/p links. The corresponding relaxation mode conforms to motions on length scales $\sim (N/p)^{1/2}a$. Note also that in the general case, the correlation functions of type (31.31) contain the contributions of Rouse modes with both odd and even values of p.

To investigate the diffusive motion of the coil as a whole, one should note that $y_0(t)$ represents a radius vector of the center of mass of the coil. Indeed, according to Eq. (31.21),

$$y_0(t) = (1/N) \int_0^N dn\ x(t,n). \qquad (31.34)$$

Therefore, the rms displacement of the center of mass of the coil during the time t is

$$\langle (y_0(t) - y_0(0))^2 \rangle = \frac{1}{\xi_0^2} \left\langle \left(\int_0^t dt' f_0(t') \right)^2 \right\rangle$$

$$= \frac{1}{\xi_0^2} \int_0^t dt' \int_0^t dt'' \langle f_0(t') f_0(t'') \rangle$$

$$= \frac{6T}{N\xi} t, \tag{31.35}$$

where we used Eqs. (31.27), (31.25), and (31.23). Thus, the diffusion coefficient of the coil as a whole equals [cf. Eq. (31.10)]

$$D_{\text{coil}} = T/(N\xi). \tag{31.36}$$

Comparing Eq. (31.36) and Eq. (31.12), we conclude that in the Rouse model, the value of D_{coil} is smaller than the coefficient of diffusion of an individual monomer link by a factor of N. Accordingly, the Einstein relation (31.11) yields that the friction coefficient for the coil as a whole equals $N\xi$. Clearly, this circumstance is associated with the additivity of frictional forces when the polymer chain moves through an immobile solvent.

*31.6. *The rms displacement of a Rouse chain link grows as $t^{1/4}$ over time intervals less than the maximum relaxation time τ_1, and only for $t > \tau_1$ does it become proportional to $t^{1/2}$ as in the ordinary diffusion of a Brownian particle.*

Let us find how the rms displacement $\langle [x(t,n) - x(0,n)^2] \rangle^{1/2}$ of a certain (n-th) link varies with time t for the Rouse model. Expressing $x(t,n)$ via $y_p(t)$ according to Eq. (31.20) and taking into account that $\langle y_p(t) y_q(t') \rangle = 0$ [see Eqs. (31.25) and (31.27)], we have at $p \neq q$

$$\langle (x(t,n) - x(0,n))^2 \rangle = \langle (y_0(t) - y_0(0))^2 \rangle + 4 \sum_{p=1}^{\infty} \cos^2 \frac{\pi p n}{N} \cdot \langle (y_p(t) - y_p(0))^2 \rangle. \tag{31.37}$$

In this relation, the solution (31.27) is to be inserted and the averaging carried out using Eq. (31.25); this is in complete analogy with the way it was done previously [cf. Eq. (31.20)]. Finally, we obtain

$$\langle (x(t,n) - x(0,n))^2 \rangle = \frac{6T}{N\xi} t + \frac{4Na^2}{\pi^2} \sum_{p=1}^{\infty} \frac{1}{p^2} \cos^2 \frac{\pi p n}{N} \cdot \left[1 - \exp\left(-\frac{tp^2}{\tau_1} \right) \right]. \tag{31.38}$$

Let us now examine the last expression. If $t \gg \tau_1$, then the exponential term within brackets can be ignored, and one can see that the first term in Eq. (31.38) greatly exceeds the second. This means that for sufficiently long time intervals, the rms displacement of the chain link is defined by the diffusion of the coil as a whole, which is physically quite apparent.

Now, let the time interval t be much less than the maximum relaxation time of the polymer coil: $t \ll \tau_1$. Then, the dominating contribution to the sum (31.38) is provided by the terms with large values of p and the sum itself transformed into an integral. Next, consider the rms displacement of a link that is located not too close to the chain ends; the factor $\cos^2(\pi pn/N)$ in the integrand then would represent a rapidly oscillating function, which can be replaced by its mean value $1/2$. Therefore, in the limit $t \ll \tau_1$ and with allowance made for Eq. (31.32), we have

$$\langle (x(t,n) - x(0,n))^2 \rangle \cong \frac{4Na^2}{\pi^2} \int_0^\infty \frac{dp}{2p^2} \left[1 - \exp\left(-\frac{tp^2}{\tau_1} \right) \right] = \left(\frac{12}{\pi} \frac{Ta^2}{\xi} t \right)^{1/2}$$
(31.39)

Note first that in this relation, the dependence on N has disappeared (as expected) for $t \ll \tau_1$. Second, the rms displacement of the link grows with t not by the ordinary law $\sim t^{1/2}$ as for a free Brownian particle, but rather by the law $\sim t^{1/4}$. This is because the monomer link is joined into a common chain together with other links, which effectively slows down the diffusive motion of the given link.

The result (31.39) can be shown to be accurate up to a numeric factor by means of the following simple estimates. If $t \ll \tau_1$, then the time t is the maximum relaxation time for a small chain section of g monomer links, including the given link: $g \sim (Tt/a^2\xi)^{1/2}$ according to Eq. (31.32). The rms displacement of the center of mass of this section is given by Eq. (31.35); it is of order $Tt/(g\xi) \sim (Ta^2t/\xi)^{1/2}$. For time intervals of the order of the maximum relaxation time, however, the rms displacements of the center of mass of the coil and of any link must coincide in order of magnitude and be of the order of the rms size of the coil. This follows from the scaling assumption of uniqueness of the intrinsic size of the coil. Admitting this assumption for the chain section of g links, we come to the estimate $\langle [\Delta x(t,n)]^2 \rangle \sim (Ta^2t/\xi)^{1/2}$ coinciding with Eq. (31.39).

*31.7. The motion of the ideal phantom macromolecule in immobile solvent obeys the laws of the Rouse theory irrespective of the polymer chain model.

Until now, this section has been conducted in the framework of the standard Gaussian bead model. The question arises: how sensitive the obtained results are to the choice of the specific model of the polymer chain? We now show that the basic equation (31.15) remains invariable for the long-term modes in the general case, provided that the volume interactions and topologic constraints are neglected and the solvent is assumed immobile.

Under the assumptions made, the Langevin equation for the n-th link of the macromolecule [of type (31.12)] and for an arbitrary model of a polymer chain takes the form

$$\frac{\partial x_n}{\partial t} = \sum_m A_{nm} x_m + g_n(t),$$
(31.40)

where $g_n(t)$ is the term responsible for random forces and A_{nm} the matrix of coefficients characterizing interactions between the chain links. In accordance with the assumption of the absence of volume interactions and topologic constraints, the matrix elements differ from zero only for close indices n and m. Let us introduce the variable $s \equiv m - n$ and replace in Eq. (31.40) the summation over m for that over s. Then, we move to the continuous limit [cf. Eq. (31.14)] and expand the quantity $x_m = x_{n+s} = x(n+s)$ into a series, assuming the parameter s to be small (which is possible only because links located close to one another in the chain interact). Then, we have

$$\sum_m A_{nm} x_m = \int_{-\infty}^{+\infty} A(s) x(n+s) ds$$

$$= \int_{-\infty}^{+\infty} A(s)[x(n) + s\partial x/\partial n + (s^2/2)\partial^2 x/\partial n^2 + ...] ds$$

$$= a_0 x(n) + a_1 \partial x/\partial n + a_2 \partial^2 x/\partial n^2 + ...;$$

$$a_0 \equiv \int_{-\infty}^{+\infty} A(s) ds, \quad a_1 \equiv \int_{-\infty}^{+\infty} s A(s) ds, \quad a_2 \equiv \frac{1}{2} \int_{-\infty}^{+\infty} s^2 A(s) ds,$$

$$\tag{31.41}$$

where $A(s) \equiv A_{n,n+s}$. The coefficient a_0 equals zero, because Eq. (31.40) should be invariant with respect to the choice of the origin, that is, relative to the substitution $x(n) \rightarrow x(n) + x_0$. Then, $A(s)$ is an even function of s, because both directions along the polymer chain are equivalent. Therefore, the quantity a_1 also equals zero. Finally, in the general case, we obtain the following equation:

$$\frac{\partial x(n,t)}{\partial t} = a_2 \frac{\partial^2 x(n,t)}{\partial n^2} + g(n,t), \tag{31.42}$$

coinciding completely with the basic Rouse equation (31.15) for the standard Gaussian bead model.

The results of the present subsection allow us to speak about a certain universality of dynamic properties of the polymer coil for sufficiently slow processes (which occur on length scales sufficiently greater than that of one monomer link; cf. subsections 12.1 and 13.4, where similar statements about universal equilibrium properties were made). Consequently, use of the standard Gaussian model of the polymer chain in studies of the dynamics of polymer systems does not restrict generality (if we deal with long, flexible chains).

31.8. *In computer simulations of polymer coil dynamics, the Verdier-Stockmayer lattice model is used; in terms of large-scale properties, it is equivalent to the Rouse model.*

In computer simulation of polymer systems, the lattice models (*see* subsection 12.5) are the most convenient. One such model was proposed by P. H. Verdier and W. Stockmayer in 1962 to solve the problems of polymer coil dynamics

(Fig. 6.1). The polymer chain is represented as a random walk of N steps without returns along the cubic lattice with edge a. It is supposed that any chain section that is in the conformation of "a corner" (Fig. 6.1b) can jump with probability w per unit time to the mirror-symmetric position relative to the axis connecting the ends of that section. Motion of the polymer chain is accomplished via these "corner flips."

The computer simulation performed for the described model of coil dynamics confirmed the basic results (31.32), (31.36), and (31.39) of the Rouse model. In terms of the Verdier-Stockmayer model, these results are written as

$$\tau_1 \sim N^2/w, \quad D_{coil} \sim wa^2/N, \quad \langle (x(t,n)-x(0,n))^2 \rangle \sim a^2 (wt)^{1/2}. \quad (31.43)$$

It can be proved rigorously that for the description of long-term, large-scale dynamic properties, the Verdier-Stockmayer and Rouse models reduce to one another. Conceptually, that proof is analogous to that presented in the previous subsection.

*31.9. In a long polymer chain, the long-term dynamic properties are determined by friction of the links against the solvent; the effects of internal friction are negligible in this case.

The dynamic properties in the Verdier-Stockmayer model are determined by the probability w of the "jump" and are independent of the viscosity of the solvent η_s (and of the presence of solvent at all). At the same time, all characteristics of the Rouse model for the chain of beads depend on the coefficient of the friction of the bead against the solvent, ξ, which for spheric beads of radius r_0 is related to η_s by the Stokes formula

$$\xi = 6\pi\eta_s r_0. \quad (31.44)$$

In connection with this discrepancy, there arises the general question: what factor is prevailing in the long-term dynamic properties of real polymer coils, the forces of friction of the links against the solvent or the internal evolution of the

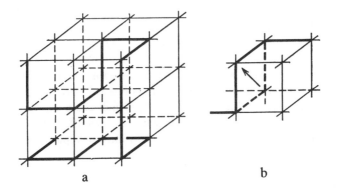

a   b

FIGURE 6.1. Verdier-Stockmayer lattice model of polymer chain dynamics.

polymer chain irrespective of the presence of the solvent (e.g., transitions between rotational-isomeric states)? The latter factor was named the internal friction effect, because during the motion of the polymer chain, friction against solvent is supplemented with the effective friction caused by internal interactions in the macromolecule (e.g., because of barriers between rotational-isomeric states).

This problem was solved by W. Kuhn and G. Kuhn in 1946. When dynamics are studied on the length scales exceeding a monomer link, friction against the solvent is always more substantial than the internal friction. This proof is presented in Ref. 8; we do not discuss it here.

We only stress here that this result argues additionally in favor of using the standard Gaussian model of the macromolecule for the description of dynamic properties of polymer systems, because the idea of friction of a link against solvent can be introduced in this model from the very beginning. This is why we use this model in this chapter.

32. THE ZIMM MODEL: A PHANTOM CHAIN WITH HYDRODYNAMIC INTERACTION

32.1. The Rouse model yields results that differ from experimental observations; one reason is the neglect of the hydrodynamic interaction appearing because of solvent entrainment during motion of a polymer chain.

According to Eqs. (31.32) and (31.36), the maximum relaxation time τ_1 and the diffusion coefficient D_{coil} for the polymer coil in the Rouse model depend on the number N of links in the chain as $\tau_1 \sim N^2$, $D_{coil} \sim N^{-1}$. These relations are not confirmed experimentally. In the θ-solvent, they are

$$\tau_1 \sim N^{3/2}, \quad D_{coil} \sim N^{-1/2}. \tag{32.1}$$

Such discrepancies are not surprising: the Rouse model is based on some crucial assumptions (e.g., ideal and phantom chains, immobile solvent). For the θ-solvent, the assumption of the absence of volume interactions (ideal chain) is satisfied. As for topological constraints (which, when disregarded, imply that the chains are phantom), they define (according to the current views) only very fine effects in dilute solutions, and they have no relation to the results of Eq. (32.1).

Consequently, the basic reason for the discussed discrepancy between theory and experiment is the assumption of an immobile solvent. Clearly, real solvent becomes partially involved in the motion of polymer chain and the velocity field $v(x)$ of the solvent becomes perturbed, which in turn affects the friction forces experienced by the links. Therefore, Eq. (31.5) must be replaced by the equation

$$f_n^{fr} = -\xi(\partial x_n/\partial t - v(x_n)), \tag{32.2}$$

Here, $v(x_n)$ is the perturbation of the solvent velocity field at the n-th link location because of the motion of other links of the polymer chain. Such indirect link interaction (via solvent) is referred to as a *hydrodynamic interaction*.

Taking into account hydrodynamic interaction (i.e., of entrainment of solvent by the links) substantially modifies all of the relations of the previous section. Indeed, let us suppose, for example, that the entrainment is so strong that all of the solvent inside the polymer coil is carried away by the macromolecule during its motion (the so-called absolutely non-draining coil). In this case, the coefficient of diffusion of the coil coincides with the Stokes coefficient of diffusion of a solid sphere whose size is of the order of the coil size R in a solution of viscosity η_s, that is, [cf. Eqs. (31.12) and (31.44)]:

$$D_{coil} = T/\xi_{coil} \sim T/(6\pi\eta_s R). \tag{32.3}$$

For a chain in θ-solvent, $R \sim aN^{1/2}$, so the relation (32.3) yields $D_{coil} \sim N^{-1/2}$ in accordance with Eq. (32.1). This suggests that the model of an absolutely non-draining coil is closer to reality than the Rouse model.[a] To prove this, let us turn to the qualitative theory of the hydrodynamic interaction.

*32.2. The motion of the solvent, entrained by the polymer links, is potential, weakly inertial, and corresponds to small Reynolds numbers; this allows the hydrodynamic Navier-Stokes equation to be simplified and linearized.

We digress for a short time from the polymer physics to present the solution to the following problem from classical hydrodynamics. Imagine an incompressible viscous liquid, a reasonable approximation to a solvent. Suppose that this liquid experiences a certain force $\varphi(x)$ per unit volume in the vicinity of each point x; in the polymer case, $\varphi(x)$ is the friction exerted on the solvent by the polymer links. We must determine the perturbation of the solvent velocity field $v(x)$ at the point x caused by the forces $\varphi(x')$ acting at all other points x'.

The condition of incompressibility of a liquid means that its flow is potential,[39] that is,

$$\text{div } v(x) = 0. \tag{32.4}$$

Apart from the assumption about an incompressible liquid, we make two more simplifications that are fully justified for the case of polymer applications discussed here. First, we neglect the inertial effects for the solvent (i.e., the term $\partial v/\partial t$ in the Navier-Stokes equation), which is possible because the motions considered here are slow and gradual. Second, we omit the nonlinear term $\sim (v\nabla)v$ of the Navier-Stokes equation, assuming the perturbation of the velocity v weak, which corresponds to the motion with a small Reynolds number[39] and is perfectly justified for the polymer problems in question. Under the assumptions made, the Navier-Stokes equation reduces to

$$\eta_s \Delta v - \nabla p + \varphi(x) = 0. \tag{32.5}$$

---

[a]The Rouse model is also called the model of a free draining coil, because the solvent in this model remains immobile and is not entrained with the polymer chain motion.

32.3. *The solvent entrainment effect and the fact that the coil is non-draining are best understood in terms of the simplified model.*

Methodically, it is expedient to begin analyzing the hydrodynamic problem with the following simple model. Suppose that the lower half-space $z < 0$ contains the "friction centers" with coefficient ξ, distributed with a certain constant density c, and moving simultaneously with velocity u in the direction of the x axis (so that the plane $z = 0$ remains immobile). In the upper half-space, there is a free liquid that is immobile at $z \to \infty$. In this case, the velocity v and force $\varphi(x) = \xi(u - v)$ possess only the x component, and they depend only on z. The relation (32.4) is satisfied automatically, and Eq. (32.5) reduces to

$$\eta_s \frac{\partial^2 v}{\partial z^2} + \xi c[u - v(z)] = 0.$$

This equation is easily solved to yield

$$v(z) = u[1 - \exp(z/l)], \quad z < 0, \tag{32.6}$$

where the characteristic "skin depth" equals

$$l = (\eta_s / \xi c)^{1/2}. \tag{32.7}$$

Here, we took into account the boundary conditions $v(z)|_{z=0} = 0$ and $v(z)|_{z \to -\infty} = u$. The obtained result (32.6) means that a stream flowing around a group of obstacles only penetrates it to a finite depth of order l. If the size of the group is much greater than l, then the predominant fraction of the liquid is entrained by the obstacles and essentially does not move relative to them.

For an ordinary Gaussian coil of size $R \sim aN^{1/2}$, we have $c \sim N/R^3$ $\sim a^{-3}N^{-1/2}$, that is, $l \sim (\eta_s a^3/\xi)^{1/2}N^{1/4}$. Because $l \ll R$ for $N \gg 1$, non-draining and involvement of solvent in the motion of the coil is implied.

Note also that the friction coefficient ξ of a link is naturally written as $\xi = 6\pi\eta_s r_D$, where r_D is the radius of the sphere experiencing the same friction. Correspondingly, $\xi/\eta_s = 6\pi r_D$. Certainly, $r_D \sim a$, but the precise value of r_D depends on more specific properties of the link (e.g., on its asymmetry and shape in general).

*32.4. *A remote action of the particle moving in a liquid on the motion of the liquid itself is described by the Oseen tensor.*

Let us now discuss the solution of the hydrodynamic equations (32.4) and (32.5) for arbitrary geometry. This solution can be found most easily via the Fourier transformation:

$$v_k = \int v(x)\exp(ikx)d^3x, \quad p_k = \int p(x)\exp(ikx)d^3x,$$

$$\varphi_k = \int \varphi(x)\exp(ikx)d^3x. \tag{32.8}$$

After the Fourier transformation, Eqs. (32.5) and (32.4) take the form

$$-\eta_s k^2 v_k - ikp_k + \varphi_k = 0,$$

$$kv_k = 0, \qquad (32.9)$$

respectively. The latter equality signifies that the velocity field of an incompressible liquid is transverse. This fact can be used to eliminate from Eq. (32.9) the pressure that is of no interest in our analysis; the scalar multiplication of both sides by k yields

$$p_k = -ik\varphi_k/k^2, \quad k^2 \equiv |k|^2.$$

This result can be inserted into Eq. (32.9) to obtain

$$v_k = \frac{1}{\eta_s k^2} \left(\varphi_k - \frac{k(k\varphi_k)}{k^2} \right). \qquad (32.10)$$

Through the inverse Fourier transformation, we obtain

$$v(x) = \int d^3x' \hat{H}(x - x') \varphi(x') \qquad (32.11)$$

in vector notation or

$$v_\alpha(x) = \int d^3x' H_{\alpha\beta}(x - x') \varphi_\beta(x') \qquad (32.12)$$

in tensor notation, where $\alpha,\beta = x,y,z$. The summation is taken over repeated indices, and the tensor $H_{\alpha\beta}(r)$ is

$$H_{\alpha\beta}(r) = \frac{1}{(2\pi)^3} \int d^3k \frac{1}{\eta_s k^2} \left[\delta_{\alpha\beta} - \frac{k_\alpha k_\beta}{k^2} \right] \exp(-ikr). \qquad (32.13)$$

The tensor $H_{\alpha\beta}(r)$ is called the *Oseen tensor*. According to Eq. (32.12), it defines the α component of the velocity field in the liquid at a point separated by a distance r from the point source of the external force acting in the direction β.

The expression (32.13) for the Oseen tensor can be simplified substantially by integrating over k. Note that $\hat{H}$ is the second-order tensor, because it depends on one vector r only and, therefore, must take the form

$$H_{\alpha\beta}(r) = A\delta_{\alpha\beta} + B \frac{r_\alpha r_\beta}{r^2}, \qquad (32.14)$$

where A and B are so far unknown scalar quantities. They can easily be found from the relations following from Eq. (32.14):

$$H_{\alpha\alpha} = 3A + B, \quad H_{\alpha\beta} r_\alpha r_\beta = (A + B)r^2, \qquad (32.15)$$

where, as before, the summation is taken over repeated indices. Taking into account Eqs. (32.13), Eq. (32.15) yields

$$3A + B = \frac{1}{(2\pi)^3} \int d^3k \, \frac{2}{\eta_s k^2} \exp(-ikr),$$

$$A + B = \frac{1}{(2\pi)^3} \int d^3k \, \frac{1 - \dfrac{(kr)^2}{k^2 r^2}}{\eta_s k^2} \exp(-ikr). \tag{32.16}$$

The integrals in Eq. (32.16) are easily evaluated to give $A = 1/(8\pi\eta_s r)$, $B = 1/(8\pi\eta_s r)$. The expression for the Oseen tensor therefore takes the following simple final form

$$H_{\alpha\beta}(r) = \frac{1}{8\pi\eta_s |r|} \left[\delta_{\alpha\beta} + \frac{r_\alpha r_\beta}{r^2} \right]. \tag{32.17}$$

*32.5. The equation of motion of an ideal phantom polymer chain for which the hydrodynamic interaction is taken into account (the Zimm equation) is nonlinear; its approximate solution can be obtained after a preliminary averaging of the Oseen tensor.

Let us return to the polymer coil. The force with which it acts on the solvent during its motion is the friction force (32.1), that is,

$$\varphi(x) = - \sum_n f_n^{(fr)} \delta(x - x_n). \tag{32.18}$$

On the other hand, the equation of motion of the polymer links retain their previous form (31.2):

$$f_n^{(fr)} + f_n^{(ch)} + f_n^{(r)} = 0. \tag{32.19}$$

In fact, the last two terms can be omitted. This means that the motion of the solvent is adjusted so that the friction force would become zero (i.e., the solvent would be entrained). To prove this formally, we reason in the following way. From our equations, one can easily eliminate the quantities $f^{(fr)}$, $\varphi(x)$, and $v(x)$. First, $f^{(fr)}$, found from Eq. (32.19), is inserted into Eq. (32.18). The expression thus obtained for $\varphi(x)$ is then inserted in Eq. (32.11) to obtain $v(x)$. Next, we substitute $v(x)$ into Eq. (32.1), and the result obtained for $f^{(fr)}$ is eventually inserted into Eq. (32.19) again. Despite the cumbersome description, this procedure is in fact very simple. It transforms the first term in Eq. (32.19) so that this equation takes the form

$$f_n^{(ch)} + f_n^{(r)} = \xi \left\{ \frac{\partial x_n}{\partial t} - \sum_{m \neq n} \hat{H}(x_n - x_m) [f_m^{(ch)} + f_m^{(r)}] \right\}. \tag{32.20}$$

It should be noted that the quantity $v(x)$ in Eq. (32.2) is a perturbation of the velocity field of the solvent at the n-th link location, caused by the motion of all the links except the n-th one. That is why the summation in Eq. (32.20) is taken over all values of m, except for $m=n$. The summation can be extended formally to all values of the index m, to include the last terms in Eq. (32.20). Indeed, having determined

$$H_{\alpha\beta}^{(nn)} = (1/\xi)\delta_{\alpha\beta}, \quad H_{\alpha\beta}^{(nm)} = H_{\alpha\beta}(x_n - x_m) \quad (n \neq m)$$

we can obtain

$$\frac{\partial x_n}{\partial t} = \sum_{m=1}^{N} \hat{H}^{(nm)}[f_m^{(ch)}(x_m) + f^{(r)}(x_m)]$$

$$= \sum_{m=1}^{N} \hat{H}^{(nm)}\left[\frac{3T}{a^2}(x_{m+1} - 2x_m - x_{m-1}) + f_m^{(r)}(t)\right]. \quad (32.21)$$

In the continuous limit, as shown in subsection 31.3, $f_m^{(ch)} = \partial^2 x(m, t)/\partial m^2$, and consequently,

$$\frac{\partial x(n,t)}{\partial t} = \int_0^N dm \hat{H}_{nm}\left[\frac{3T}{a^2}\frac{\partial^2 x(m, t)}{\partial m^2} + f^{(r)}(m, t)\right]. \quad (32.22)$$

This result shows that the terms with $m=n$ are in fact eliminated, as we expected. The same conclusion can also be seen in further discussions. Equation (32.22) was first derived by B. Zimm in 1956. Because of the dependence of the Oseen tensor $\hat{H}$ on r [Eq. (32.17)], that is, on $x_n - x_m$, the Zimm equation is nonlinear in $x(n, t)$ and permits no accurate analytic solution.

The most effective way to approximate the solution of the Zimm equation lies in the replacement of the $\hat{H}$ tensor, which depends on the instantaneous link coordinates $x(n, t)$, by the tensor $\langle \hat{H} \rangle$ averaged over the distribution function $P(x_1, x_2, \ldots, x_N, t)$. Moreover, when we examine polymer chains near thermodynamic equilibrium (and in this book, we confine ourselves just to this case) the averaging is performed over the equilibrium distribution function. Such an approximation is called the *preaveraging approximation*, and in a certain sense, it is analogous to the mean field approximation. (More detailed description of the conditions of its applicability can be found in Refs. 5 and 11.)

While averaging the Oseen tensor (32.17), it should first be noted that the distribution of the vector $r = x_n - x_m$ is isotropic, and its length and orientation are statistically independent of each other. This yields

$$\langle H_{\alpha\beta}^{(nm)}(r) \rangle = \frac{1}{8\pi\eta_s}\left\langle \frac{1}{|r|}\left[\delta_{\alpha\beta} + \frac{r_\alpha r_\beta}{r^2}\right]\right\rangle$$

$$= \frac{1}{8\pi\eta_s}\left\langle \frac{1}{|r|}\right\rangle\left\langle\left[\delta_{\alpha\beta} + \frac{r_\alpha r_\beta}{r^2}\right]\right\rangle$$

$$= \frac{1}{6\pi\eta_s}\left\langle \frac{1}{|r|}\right\rangle \delta_{\alpha\beta}. \quad (32.23)$$

The quantity $\langle 1/|r_{nm}|\rangle$ is just a function of $|n-m|$ and for a Gaussian coil is given by Eq. (5.4): $\langle 1/|r_{nm}|\rangle = (1/a)[6/|n-m|\pi]^{1/2}$. As long as the singularity at $n \to m$ is integrable, it is clear that the contribution of the point $n=m$ is vanishingly small. This means that the two last terms in Eq. (32.20) can be neglected, as mentioned earlier.

After substituting the result (32.23) into Eq. (32.22), we obtain the Zimm equation in the preaveraging approximation:

$$6\pi\eta_s \frac{\partial x(n,\ t)}{\partial t} = \int_0^N dm \left\langle \frac{1}{|r_{n-m}|} \right\rangle \left[\frac{3T}{a^2} \frac{\partial^2 x(m,\ t)}{\partial m^2} + f^{(r)}(m,\ t) \right]. \tag{32.24}$$

In contrast to Eq. (32.24), this equation is linear in $x(m,\ t)$.

Comparing Eq. (32.24) with the Rouse equation (32.15), we see that the allowance for hydrodynamic interactions leads to the appearance of an effective long-range interaction between the links n and m of the chain. It is proportional to $|n-m|^{-5/2}$ by integrating twice by parts the term containing $\partial^2 x(m,\ t)/\partial m^2$ in Eq. (32.24). Even though Eq. (32.24) can be written in the form (31.40), it nevertheless does not belong to the class of Rouse equations, because the coefficient a_2 [see Eq. (31.41)] does not converge. It is the long-range character of interaction along the chain that results in a non-draining coil in the Zimm model.

*32.6. The interaction of Rouse modes in a chain with hydrodynamic interaction is fairly weak; neglecting this enables one to find the hydrodynamic radius of the coil for the Zimm model (i.e., the diffusion coefficient and the maximum relaxation time of the chain).

Rewriting the Zimm equation (32.24) in terms of the Rouse modes (31.21) and (31.24), we obtain

$$\frac{\partial y_p(t)}{\partial t} = \sum_{q=0}^{\infty} h_{pq} \left[-\sqrt{6\pi} \frac{Tq^2}{N^{3/2}a^3\eta_s} y_q(t) + \frac{(2-\delta_{q0})}{\sqrt{6\pi^3}} \frac{N^{1/2}}{a\eta_s} f_q(t) \right], \tag{32.25}$$

where

$$h_{pq} = \frac{1}{N^{3/2}} \sqrt{\frac{\pi}{6}} \int_0^N dn \int_0^N dm \cos\left(\frac{\pi pn}{N}\right) \cos\left(\frac{\pi qm}{N}\right) \left\langle \frac{a}{|r_{n-m}|} \right\rangle$$

$$= \int_0^1 du \int_0^1 dv \cos(\pi up)\cos(\pi vq) \frac{1}{\sqrt{|u-v|}}. \tag{32.26}$$

In contrast to the Rouse case, the modes y_q with different q are not decoupled and thus interact, because the matrix h_{pq} is not diagonal. The analysis shows, however, that its deviation from diagonal form is not very significant, and even though there is no small parameter in the problem, one can write with adequate numeric accuracy

$$h_{pq} \approx h_{qq}\delta_{pq}. \tag{32.27}$$

To prove this, recall that for the Gaussian coil $\langle a/|r_{nm}|\rangle = [6/|n-m|\pi]^{1/2}$ according to Eq. (5.4). For $p=0$, numeric integration yields $|h_{01}/h_{00}| \approx 0.04$, showing that already the first nondiagonal element is approximately 25 times smaller than the diagonal one. As q grows, the decrease becomes less dramatic: $h_{02}/h_{01} \approx 0.4$; $h_{03}/h_{02} \approx 0.6$; $h_{04}/h_{03} \approx 0.75$. For the opposite limiting case $p, q \gg 1$, the asymptotic behavior can be easily found analytically[b]:

$$h_{pq} \cong \frac{\delta_{pq}}{\sqrt{2q}}. \tag{32.28}$$

Using Eq. (32.27), called the *Kirkwood-Riseman approximation*, we obtain at $p=0$

$$\frac{\partial y_0}{\partial t} = \left[\frac{h_{00}}{\sqrt{6\pi^3}} \frac{1}{N^{1/2}a\eta_s} \right] N f_0(t). \tag{32.29}$$

Because y_0 is the coordinate of the center of mass of the coil (31.34) and Nf_0 the total force acting on the coil, the coefficient in Eq. (32.29) equals D_{coil}/T:

$$D_{\text{coil}} = T \left[\frac{h_{00}}{\sqrt{6\pi^3}} \frac{1}{N^{1/2}a\eta_s} \right] = \frac{T}{6\pi\eta_s} \frac{1}{N^2} \int_0^N dn \int_0^N dm \left\langle \frac{1}{|r_{n-m}|} \right\rangle \tag{32.30}$$

If one defines the hydrodynamic radius R_D of the coil as the radius of a solid sphere with diffusion coefficient D_{coil}, that is, $D_{\text{coil}} = T/(6\pi\eta_s R_D)$, then

$$R_D = \frac{T}{6\pi\eta_s D_{\text{coil}}} = N^2 \left[\sum_{n=0}^{N} \sum_{m=0}^{N} \left\langle \frac{1}{|r_{n-m}|} \right\rangle \right]^{-1} \tag{32.31}$$

We have already cited this definition [*see* Eq. (5.3)], and the value of R_D for the Gaussian coil has been calculated [*see* Eq. (5.6)]: $R_D = aN^{1/2}(3\pi/128)^{1/2}$. Hence, $D_{\text{coil}} \sim aN^{-1/2}$ in accordance with the non-draining coil model and the experimental result (32.1).

For $p \neq 0$, we have according to Eqs. (32.25) and (32.28)

---

[b] To do this, the substitution $v - u = s$ should be made in Eq. (32.26). Then

$$h_{pq} = \int_0^1 du \left[\cos(\pi up)\cos(\pi uq) \int_{-u}^{1-u} ds \cos(\pi sq) \frac{1}{\sqrt{|s|}} \right.$$

$$\left. - \cos(\pi up)\sin(\pi uq) \int_{-u}^{1-u} ds \sin(\pi sq) \frac{1}{\sqrt{|s|}} \right].$$

At $q \gg 1$, the integrals with respect to s can be extended to the limits $-\infty$ and $+\infty$ (because these integrals converge). Then the second integral is zero, and the first converts to a standard integral. Eventually, one obtains Eq. (32.28)

$$\frac{\partial y_p}{\partial t} = -\frac{1}{\tau_p} y_p(t) + \frac{1}{\sqrt{3\pi^3 p}} \frac{N^{1/2}}{a\eta_s} f_p(t), \tag{32.32}$$

$$\tau_p = \sqrt{\frac{3}{\pi}} \frac{N^{3/2} a^3 \eta_s}{T p^{3/2}}. \tag{32.33}$$

It can be seen that the maximum relaxation time of the coil is $\tau_1 \sim N^{3/2}$ in the Zimm model in accordance with Eq. (32.1). The relaxation time of the mode p equals $\tau_p = \tau_1 p^{-3/2}$. To confirm the concept of a non-draining coil, it should be pointed out that $\tau_1 \sim \eta_s R_D^3/T$ (i.e., τ_1 corresponds to the characteristic time of rotational relaxation of a solid sphere of radius R_D).

33. DYNAMIC PROPERTIES OF REAL POLYMER COILS

33.1. *The hydrodynamic interaction leads to an absolute non-draining of a real polymer coil with excluded volume.*

Now we begin to consider real polymer coils, allowing for both the hydrodynamic interactions and excluded volume effects. Strictly speaking, in a real situation, the topologic constraints on the motion of polymer chains (i.e., non-phantom nature of the chains) should also be taken into account, but as pointed out earlier, these effects are insignificant in dilute solutions and, in any case, exert no influence on the basic conclusions of this section.

If the excluded volume effects are taken into account, then the approach in subsection 32.3 becomes invalid, because the equation of motion of the n-th link of the chain (32.9) now must also contain the forces of volume interactions between links. If these forces are included explicitly, then the equation becomes very complicated.

Even in the absence of excluded volume forces, however, we used for the solution of Eq. (32.19) in subsection 32.3 the preliminary averaging approximation. Thus, with the same level of accuracy, it can be assumed that as long as the average link concentration in the coil volume is low (*see* subsection 5.2), the equation of motion of the polymer coil with excluded volume would have the same structure (32.26) as the Zimm equation. The Oseen tensor $\hat{H}_{nm}$, however, is averaged not by means of the Gaussian distribution function (32.25) but by using the distribution function characteristic for a coil with excluded volume (*see* subsection 19.7). Such an approximate approach to the coil dynamics is called the *linearization approximation*.

Here, do not perform any averaging of the Oseen tensor for the coil with excluded volume or solve the obtained equation of motion in the linearization approximation. The detailed information can be found elsewhere.[5,11] Instead, we try to answer only the following question: does the conclusion about absolute non-draining of the coil still hold true in the presence of excluded volume?

The qualitative answer is clear from the estimate (32.7) of the depth l to which the flow penetrates into the coil. For a swollen coil, $R \sim aN^{3/5}$ and

$c \sim N/R^3 \sim a^{-3}N^{-4/5}$, that is, $l \sim (\eta_s a^3/\xi)^{1/2}N^{2/5}$. Because $l \ll R$ this situation corresponds to non-draining.

The problem can be analyzed more accurately using the formula (32.31). By verifying its derivation once again, one can make sure that it holds for a swollen coil as well. For a coil with excluded volume,

$$\left\langle \frac{a}{|r_{n-m}|} \right\rangle = |n-m|^{-v}, \tag{33.1}$$

where $v=3/5$ is the critical exponent of the excluded volume problem (*see* subsection 16.1). Hence,

$$R_D \sim aN^v \sim R \tag{33.2}$$

or

$$D_{coil} \sim (T/6\pi\eta_s a)N^{-v}, \tag{33.3}$$

where R is the spatial size of the swollen coil. Thus, the hydrodynamic radius of a real coil (as for the ideal one) is of the order of the coil size. The maximum relaxation time of the swollen coil[11] is of order

$$\tau_1 \sim \eta_s(aN^v)^3/T \sim \eta_s R^3/T, \tag{33.4}$$

that is, of the order of the rotational diffusion time of a solid sphere of radius R. Therefore, the hydrodynamic interaction is strong enough to ensure a non-draining situation not only for a Gaussian coil but also for a (still more loose) swollen one.

33.2. The dynamic properties of a polymer coil, as well as its equilibrium properties, are determined by the characteristic size, so the relevant problems can be tackled using the scaling concept.

So far, we have used repeatedly the scaling concept and the method of scaling estimates in solving the problems of equilibrium statistics. We have noted that this concept is based on the claim for uniqueness of the characteristic size of the polymer coil $R \sim aN^v$ and on the scale-invariant (self-similar) structure of the coil on smaller scales. This chapter shows that this statement remains valid for dynamic problems as well.

Indeed, the structure of dynamic modes of the polymer coil is such that the single selected mode conforms to the maximum relaxation time (i.e., to motions on the scale of a whole coil). For such motions, the value of R is the only characteristic scale: because of the qualitative validity of the non-draining coil model, the characteristic scale (associated with entrainment of solvent by the polymer chain) coincides with the spatial scale of the coil.

Uniqueness of the characteristic size of the polymer coil automatically leads to uniqueness of the characteristic time, that is, the maximum relaxation time of diffusion of the coil as a whole τ_1. For example, the characteristic time for the maximum relaxation time of the coil as a whole (i.e., the time taken for diffusion over a distance of the order of the coil size R^2/D_{coil}) coincides with τ_1. Actually, according to Eqs. (33.3) and (33.4),

$$R^2/D_{coil} \sim R^3 \eta_s/T \sim \tau_1. \tag{33.5}$$

The fact that in dynamic problems there appears no additional characteristic spatial size allows scaling concepts to be applied in the conventional sense for the solution of these problems. Later, this method will be used repeatedly.

*33.3. *In experiments on inelastic light scattering by a dilute polymer solution, one observes both diffusive motion of whole coils (for small-angle scattering) and intermolecular chain dynamics (for large-angle scattering); the dynamic structure factor has the Lorentz form at small angles and an essentially non-Lorentz form at large angles.*

Until now, we have been considering dynamic properties of polymer coils without reference to experimental techniques. Of course, the diffusion coefficient D_{coil} of the coil as a whole can be measured if, having produced in a dilute solution of coils a certain concentration gradient, one observes its evolution [the corresponding experiments confirm Eq. (33.3)]. The most abundant information about the quantity D_{coil} as well as about τ_1 and the internal motions in the polymer coil, however, can be obtained from experiments on inelastic light scattering.

The scattering of an optical, monochromatic (frequency ω_0 and wavelength λ) beam, usually emitted by a laser source, by a polymer solution was studied in these experiments. Suppose we observe the scattering pattern at the angle ϑ to the initial beam direction. Then, the modulus of the characteristic wave vector of scattering equals $|k| = (4\pi/\lambda)\sin(\vartheta/2)$ (cf. subsection 5.5). Because of motions in the scattering system, the scattered light is no longer monochromatic. Let $S(k,\omega)$ denote its intensity at the frequency $\omega_0 + \omega$ (for scattering at a given angle); then, it can easily be shown[40] that the directly measured quantity $S(k,\omega)$ is proportional to

$$S(k,\omega) \sim \int_{-\infty}^{+\infty} dt \, \exp(i\omega t) G(k,t), \tag{33.6}$$

where $G(k,t)$ is the so-called dynamic structural factor of the polymer solution, defined for $t > 0$ by the relation

$$G(k,t) = \frac{1}{N} \sum_{n,m} \langle \exp[ik(x_n(t) - x_m(0))] \rangle, \tag{33.7}$$

the summation being taken over all links of the polymer system, that is, for one chain: $1 \leqslant n \leqslant N$, $1 \leqslant m \leqslant N$ [cf. the definition of the static structure factor (5.11)]. For $t < 0$, we define $G(k,t) = G(k,-t)$.

Clearly, the value of $G(k,t)$ substantially depends on the structure of the dynamic modes of the polymer coil. When the measurements are performed at the given angle ϑ (i.e., for the given value $k \equiv |k|$), the dynamic structural factor (33.7) will correspond to light scattering by motions with wavelengths $\sim k^{-1}$. Thus, varying the observation angle ϑ, one can study motions in a polymer coil taking place on various length scales.

For example, for $kR \gg 1$ (i.e., for large angle measurements), the internal motions in the polymer coil (i.e., the motions taking place on scales less than the coil size R) are studied. At $kR \sim 1$, we deal with the motions proceeding on scales of order R, which correspond to the first relaxation modes. Finally, for $kR \ll 1$, the motion of the coil as a whole is investigated; for such long wavelengths (small observation angles), the polymer coils behave as point scatterers.

The dynamic structural factor (33.7) can be calculated precisely for the Rouse model following the methods presented in Sec. 31. Here are the asymptotic forms of the exact result.[11] For $k^2 N a^2 \gg 1$,

$$G(k,t) = \frac{12}{k^2 a^2} \int_0^\infty du \, \exp\left\{ -u - \left(\frac{|t|}{\tau_k} \right)^{1/2} r\left(\left(\frac{|t|}{\tau_k} \right)^{-1/2} u \right) \right\}, \qquad (33.8)$$

where

$$\tau_k \equiv \frac{12\xi}{T a^2 k^4}, \quad r(x) = \frac{2}{\pi} \int_0^\infty dy \, \frac{\cos(xy)}{y^2} [1 - \exp(-y^2)]; \qquad (33.9)$$

and for $k^2 N a^2 \ll 1$,

$$G(k,t) = N \exp(-k^2 T t / N\xi). \qquad (33.10)$$

Once the hydrodynamic and volume interactions are taken into account, however, $G(k,t)$ cannot be calculated analytically, and the preaveraging approximation and some other approximations must be applied. Here, we try to establish only the fundamental properties of the dynamic structural factor for real coils with excluded volume.

Suppose that initially $kR \sim kaN^\nu \ll 1$. As already mentioned, for such long wavelengths, polymer coils can be considered as structureless point scatterers, and x_n and x_m in the exponent of Eq. (33.7) can be replaced by the coordinate y_0 of the center of mass [see Eq. (31.34)]. Then, we obtain (for $t > 0$)

$$G(k,t) \cong N \langle \exp[ik(y_0(t) - y_0(0))] \rangle. \qquad (33.11)$$

The dependence $y_0(t)$ describes the diffusion process with diffusion coefficient D_{coil}. The distribution of the quantity $y_0(t) - y_0(0)$ is consequently Gaussian, with variance $(2D_{\text{coil}}t)^{1/2}$, that is, for $t > 0$

$$G(k,t) = N \int d^3r \, \exp(ikr) \cdot (4\pi D_{\text{coil}}t)^{-3/2} \exp(-r^2/4D_{\text{coil}}t)$$

$$= N \exp(-D_{\text{coil}} k^2 t). \qquad (33.12)$$

Thus, in the long-wavelength limit, the dynamic structural factor diminishes exponentially with time. Equation (33.12) is of a general nature. For the Rouse model, it coincides with Eq. (33.10), because $D_{coil} = T/N\xi$ for this model [*see* Eq. (31.36)]. For real coils, the value of D_{coil} in Eq. (33.12) is defined by the relation (33.3).

According to Eq. (33.6), the Fourier transformation of the dynamic structure factor $S(k,\omega)$ is measured directly in experiments on inelastic light scattering. Inserting Eq. (33.12) into Eq. (33.6), we obtain [because $G(k,t)=G(k,|t|)$]

$$S(k,\omega) \sim \frac{D_{coil}k^2}{(D_{coil}k^2)^2 + \omega^2} \text{ for } kR \ll 1. \tag{33.13}$$

The dependence of $S(\omega)$ for a given k has a characteristic Lorentz shape (Fig. 6.2). The characteristic width of this curve is

$$\Delta\omega \sim D_{coil}k^2 \sim Tk^2/(\eta_s R). \tag{33.14}$$

Thus, by measuring the value of $\Delta\omega$ for $kR \ll 1$, one can determine the coefficient D_{coil} of diffusion of the polymer coil. Note also that the characteristic relaxation time of a dynamic mode with wave vector k equals $\tau_k \sim 1/(D_{coil}k^2)$ at $kR \ll 1$ [*see* Eq. (33.12)]. The dependence $\tau_k \sim 1/k^2$ is obeyed whenever we deal with a purely diffusive process.

Suppose that we increase the observation angle (by increasing k) and move into the region $kR \gg 1$ of internal motions in the coil. Let us now examine how the dynamic structural factor $G(k,t)$ and intensity of scattered light $S(k,\omega)$ behave in this case. First, we find the characteristic relaxation time τ_k for motions with $kR \gg 1$. To do this, we use scaling considerations (*see* subsection 33.2). We have seen that $\tau_k \sim 1/D_{coil}k^2$ if $kR \ll 1$. At the same time, given the unique characteristic coil size R, the following relation should be valid for τ_k throughout the region of variation of k [cf. Eqs. (19.5), (19.9), and so on]:

$$\tau_k = (1/D_{coil}k^2)\varphi(kR). \tag{33.15}$$

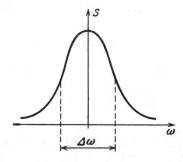

FIGURE 6.2. Dependence $S(\omega)$ for $kR < 1$ (Lorentz curve).

For $kR \gg 1$, we have $\tau_k \sim (1/D_{coil}k^2)(kR)^x \sim (\eta_s R/Tk^2)(kR)^x$. The exponent x must be chosen in this case from the condition that for $kR \gg 1$, the time τ_k should be independent of N as far as we deal with internal motions of the polymer coil. As a result, we obtain $x = -1$. Hence,

$$\tau_k \sim \eta_s/(Tk^3) \quad \text{for} \quad kR \gg 1. \tag{33.16}$$

The relaxation time τ_k is seen to decrease with the growth of k as k^{-3} but not as k^{-2}, because internal motions in the coil do not represent simple diffusion any longer (as distinguished from motions in a polymer solution for $kR \ll 1$). Consequently, the shape of the spectral curve $S(\omega)$ of inelastic scattering will not precisely be Lorentzian for $kR \gg 1$. Still, a certain curve with maximum at $\omega = 0$ (of the type shown in Fig. 6.2) will be observed, whose width

$$\Delta\omega \sim 1/\tau_k \sim Tk^3/\eta_s \quad \text{for} \quad kR \gg 1. \tag{33.17}$$

Measurements performed in the region $kR \gg 1$ confirmed accurately the dependence (33.17).

As for the dynamic structural factor $G(k, t)$, it will not be described in this case by a simple exponential function, even for $t \gg \tau_k$ [cf. Eq. (33.8)], which yields for the Rouse model for $t > 0$,

$$G(k,t) = G(k,0)\exp\left[-\frac{2}{\pi^{1/2}}\left(\frac{t}{\tau_k}\right)^{1/2}\right] \quad \text{for} \quad t \gg \tau_k. \tag{33.18}$$

In this limit, $G(k,t)$ can be calculated analytically for a chain with excluded volume using the method of the renormalization group.

APPENDIX TO SECS. 32 AND 33. VISCOSITY OF A DILUTE POLYMER SOLUTION AND THE VISCOMETRY METHOD

A common method of obtaining experimental information on polymer solutions is based on a measurement of their macroscopic viscosity. The concepts regarding the dynamic properties of the coils presented earlier make it possible to examine the macroscopic viscosity η of a dilute polymer solution in good or θ-solvent, or (more exactly) the contribution to the viscosity caused by the presence in the solution of a small number of coils. In experiments, this contribution is usually analyzed via the so-called intrinsic viscosity $[\eta]$, defined by the relation

$$[\eta] = \lim_{\rho \to 0} \frac{1}{\rho} \frac{\eta - \eta_s}{\eta_s} \tag{33.19}$$

and obviously obtained by the extrapolation of viscometry data to the zero concentration. Here, ρ is the density of polymer in the solution, that is,

$\rho=cM/(NN_A)$, where c is the concentration (the number of links in a unit volume of the solution), M the molecular mass of the chain, N the number of links in the chain, and N_A the Avogadro constant.

To find the viscosity η of a dilute solution of coils or its intrinsic viscosity $[\eta]$, recall that in hydrodynamic terms, each coil (Gaussian or swollen) behaves as a solid sphere of radius R_D (32.30). In classical hydrodynamics, the viscosity of a suspension, (i.e., the liquid in which small spheric particles are suspended) was studied by Einstein in 1906 (see Ref. 38, Sec. 22). If the volume fraction ϕ of these particles is small, then the viscosity of the suspension equals[c]

$$\eta=\eta_s\left(1+\frac{5}{2}\phi\right).\tag{33.20}$$

Substituting a sphere of radius R_D for each coil, we can readily obtain $\phi=(4/3)\pi R_D{}^3(c/N)$, or

$$[\eta]=\frac{10\pi}{3}\frac{R_D^3 N_A}{M}.\tag{33.21}$$

Thus, by measuring the intrinsic viscosity, we can in principle determine the hydrodynamic radius of the coil. In practice, however, the accuracy of such measurements is not very good; for example, it is insufficient to find the critical exponent v of the excluded volume [for good solvent, Eq. (33.21) yields $[\eta]$ $\sim N^{3v-1}$]. In many cases, however, the measured value of $[\eta]$ and Eq. (33.21) can be used to find the molecular mass or chain length (when, for example, we consider a θ-solvent).

Measurement results of the intrinsic viscosity of a dilute solution can also be used to estimate the shape of the particles (e.g., of complicated nonspheric globules or relatively short chains). In this case, instead of Eq. (33.20), we obtain

$$\eta=\eta_s(1+A\phi),$$

where the coefficient A depends on the shape of the particles and grows in a definite way[39] from the initial value 5/2 either on flattening the spheres into disks or stretching them into rods.

Sometimes, shear viscosity is measured in a transient regime. The Cartesian velocity components of a moving liquid can then be written as

$$v_x(r,\ t)=\varkappa(t)r_y,\quad v_y=v_z=0.\tag{33.22}$$

If the shear velocity, that is, the quantity $\varkappa(t)$, is low, then the shearing stress σ_{xy} must vary linearly with $\kappa(t)$:

---

[c]Note that this expression is the first term of the expansion in a power series of ϕ, while the expansion itself is analogous to the virial expansion (12.3). Specifically, the term $\sim\phi$ is from hydrodynamic pair interactions of suspended particles.

$$\sigma_{xy}(t) = \int_{-\infty}^{t} dt' G(t-t')\varkappa(t'). \tag{33.23}$$

The quantity $G(t)$ is called the *shear relaxation modulus*. Allowance for the time lag made in Eq. (33.23) is essential, because as we have seen, polymer coils are characterized by very long relaxation times (e.g., $\sim N^2$ in the Rouse model, or $\sim N^{3/2}$ in the Zimm model). For dilute polymer solutions, when the contribution of the polymer to the total viscosity provides only a small correction, Eq. (33.23) can be rewritten in the form

$$\sigma_{xy}(t) = \eta_s \varkappa(t) + \int_{-\infty}^{t} dt' G^{(p)}(t-t')\varkappa(t'), \tag{33.24}$$

where $G^{(p)}(t)$ is the dynamic characteristic of an individual polymer chain.

Equation (33.24) is valid for an arbitrary small value of $\varkappa(t)$. In particular, the stationary case considered earlier corresponds to $\varkappa(t) = \varkappa = \text{const}$. Consequently,

$$\int_{0}^{\infty} G^{(p)}(t)dt = \sigma_{xy}/\varkappa(t') - \eta_s = \eta - \eta_s.$$

Of course, investigation of viscosity in transient conditions provides much richer information about the internal dynamics of the coils. This can also be accomplished by using an oscillating flow of the type

$$\varkappa(t) = \varkappa_0 \cos(\omega t).$$

The behavior of a dilute polymer solution in such a flow has been described elsewhere.[11]

34. DYNAMICS OF CONCENTRATION FLUCTUATIONS IN SOLUTIONS OF OVERLAPPED POLYMER COILS

34.1. *The dynamics of concentration fluctuations in solutions of overlapped polymer coils can be examined, assuming the chains to be phantom, even though in studies of diffusion and viscosity of such solutions the uncrossability of chains cannot be ignored.*

Having considered the basic concepts associated with the dynamic properties of both an individual polymer chain and dilute solutions of chains, we move to studies of the dynamics of semidilute and concentrated polymer solutions (and polymer melts as well) in which individual polymer coils are strongly overlapped with one another. As mentioned in subsection 23.1, the concentration of the polymer solution in this range exceeds the overlap concentration c^* of polymer coils.

As a rule, on exceeding the concentration c^*, the dynamic properties of the polymer solution change substantially: the viscosity grows abruptly, the coefficients of diffusion of macromolecules decrease, and the effects of memory of previous flow become clearly pronounced. It is easy to realize that these effects are connected with the uncrossability of chains (see subsection 11.1; Fig. 1.16), because in the system of entangled coils, the forbidding of mutual crossings dramatically reduces the set of possible motions of macromolecules. Consequently, the uncrossability of chains is significant when the diffusion and viscosity of semidilute and concentrated polymer solutions and melts are investigated (see Secs. 35 and 36).

At the same time, there are some dynamic properties that can be described without taking into account the uncrossability of polymer chains. Concentration fluctuations in a solution of entangled coils belong to this category. The corresponding effects result from the simultaneous motion of many polymer chains; as a rule, the processes involved proceed on not-too-large time and length scales. This is why the uncrossability of chains, which shows most clearly when the motion of one chain is considered over sufficiently long time intervals, is insignificant for the fluctuation dynamics in polymer solutions.

The fluctuation dynamics in polymer solutions can be studied experimentally by the inelastic light scattering technique. Indeed, this method allows a direct measurement of the Fourier transform (33.6) of the sum of the type (33.7), in which the summation for the system of many chains is extended to all monomer links of the solution. Let us now introduce the microscopic concentration of links at the point x:

$$c_\Gamma(x) = \sum_{a,n} \delta(x - x_{an}), \qquad (34.1)$$

where the sum is extended to all links of all the chains (a is the number of the chain and n the number of the link). After the Fourier transformation with respect to the coordinate x, we obtain

$$c_\Gamma(k) = \frac{1}{V} \int d^3x \exp(ikx) c_\Gamma(x) = \frac{1}{V} \sum_{a,n} \exp(ikx_{an}), \qquad (34.2)$$

where V is the volume of the system. Comparing Eqs. (34.2) and (33.7), we conclude that for the dynamic structural factor of the solution, one may write

$$G(k,t) \sim \langle c_\Gamma(k, t) c_\Gamma(-k, 0) \rangle, \qquad (34.3)$$

that is, the function $G(k, t)$ [as well as $S(k, \omega)$], determined by inelastic light scattering, describes the dynamics of concentration fluctuations with wave vector k or wavelength $\sim 1/|k|$ [cf. Eqs. (23.6) and (23.7)].

In this section, we consider fluctuation dynamics and inelastic light scattering in solutions of entangled polymer coils. To simplify the notation, we also consider in both this and subsequent sections the case of flexible polymer chains in good solvents far from the θ-temperature ($\tau \sim 1$). In this case, the corresponding standard Gaussian chain of beads is characterized by only two parameters: 1) the number N of links in the chain, and 2) the microscopic size a of a link. The generalization to a more general case is performed with no difficulty.

34.2. *In a semidilute polymer solution, the hydrodynamic interactions are screened; the screening radius coincides in order of magnitude with the correlation radius of the solution.*

To study the dynamics of concentration fluctuations in a polymer solution, we use the method of scaling estimations. This method is applicable because of the uniqueness of the characteristic size of the system. For dilute solutions, this is the size of the polymer coil, which is simultaneously the characteristic size associated with hydrodynamic interactions, that is, with entrainment of solvent by the links of a polymer chain (*see* subsection 33.2).

The characteristic size for equilibrium properties of a semidilute polymer solution is its correlation length ξ (*see* subsection 25.6). Thus, to apply the method of scaling estimations in its usual form, it is necessary that the characteristic size associated with hydrodynamic interactions also be of order ξ (i.e., that a substantial screening of the hydrodynamic interaction would occur in the semidilute solution).

Such a screening does exist, and its cause is easy to understand. According to subsection 32.2, hydrodynamic interaction between the links is determined by the Oseen tensor, which diminishes with separation r between these links as $1/(\eta_s r)$ [*see* Eq. (32.17)]. It is this slow decrease that ensures the long-range character of hydrodynamic interactions and complete entrainment of the solvent located inside a polymer coil.

In the semidilute solution, Eq. (32.17) is valid for the hydrodynamic interaction of links on small ($<\xi$) scales, because the presence of the links of other chains is insignificant on these length scales. On scales exceeding ξ, however, the Oseen tensor should coincide with the expression that is obtained in macroscopic hydrodynamics for a solution with viscosity η. In other words, it should be proportional to $1/(\eta r)$. Because the viscosity η of a semidilute polymer solution is usually much higher than the solvent viscosity η_s, we conclude that on scales greater than ξ, the hydrodynamic interaction between links will diminish drastically. This phenomenon is referred to as the *screening of hydrodynamic interactions*; it implies that such interactions between links separated by a distance exceeding ξ can always be neglected.

Two important consequences follow. First, dynamic behavior of a macromolecule in semidilute solutions obeys on long length scales the Rouse model (if topologic constraints can be ignored), not the Zimm model. Second, the characteristic size associated with entrainment of solvent during the motion of the polymer chain coincides in order of magnitude with the correlation length ξ. The uniqueness of the characteristic size allows one to use the method of scaling estimations to investigate the fluctuation dynamics of the semidilute polymer solution.

34.3. *In the dynamic behavior of a semidilute polymer solution, one can distinguish (and indicate on the "dynamic" diagram of states) three main regimes of fluctuation dynamics, which correspond to diffusion of individual coils, internal motions of the polymer chain, and cooperative motion of many entangled chains.*

It was shown in subsection 33.3 that the motions can be studied on different length scales by changing the scattered wave vector k in experiments on inelastic light scattering by a semidilute polymer solution; depending on the ratio of the values $1/k$ and R, the different expressions for the scattering function $S(k, \omega)$ were obtained. In the concentrated solution, the additional variable c (the solution concentration) supplements the variable $k \equiv |k|$. The various dynamic regimes of the polymer solution corresponding to different functions $S(k, \omega)$, that is, to different fluctuation dynamics, are conveniently shown on the so-called dynamic diagram of states plotted for variables kR and c/c^* in Fig. 6.3 (cf. the diagram of states of equilibrium properties of a polymer solution; Fig. 4.4).

As shown in subsection 33.3, a dilute solution may exist for $c \ll c^*$ in two regimes of dynamic behavior depending on whether the inequality $kR \ll 1$ (regime I in Fig. 6.3) or $kR \gg 1$ (regime II) is satisfied. In regime I, the dynamic structure factor (33.12) is determined by diffusion of the coils as a whole; consequently, long-wavelength concentration fluctuations exist in this regime, which disperse via diffusion of individual coils [see Eq. (34.3)]. In regime II, we have short-wavelength fluctuations, whose dynamics are determined by internal motions in macromolecules.

Suppose that we gradually increase the concentration of a polymer solution in regime II, keeping the wave vector k fixed (i.e., keeping the observation angle constant). The question is: up to what concentration c will Eq. (33.16) hold true for the characteristic time of the fluctuation mode with wave vector k? Obviously, the concentration fluctuations will spread out according to the mechanism (33.16) of internal motions in the polymer coil, until the chain can be treated on scales $\sim 1/k$ as isolated (i.e., non-interacting with the other chains). From Figure 4.3 and the uniqueness of the characteristic size ξ in a semidilute solution, one can conclude that this will continue until $k\xi \sim 1$.

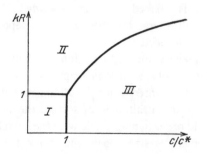

FIGURE 6.3. Dynamic diagram of states of polymer solution.

Indeed, if $k\xi > 1$, that is, $1/k < \xi$, the chain section of size $\sim 1/k$ is smaller than the blob size ξ; therefore, this section does not "perceive" the other chains (*see* subsection 25.6). If $k\xi < 1$, then the fluctuation wavelength exceeds the blob size, and such fluctuations obviously would disperse via cooperative motion of many polymer chains (regime III in Fig. 6.3). Thus, the boundary of regime II from the side of high concentrations (Fig. 6.3) is found from the condition $k\xi \sim 1$ or, taking into account Eq. (25.13), as

$$kR \sim (c/c^*)^{3/4}. \tag{34.4}$$

This relation defines the boundary between regimes II and III only for $kR > 1$. For $kR < 1$, the wavelength of the fluctuation exceeds the coil size in a dilute solution, and the dispersal of fluctuations via cooperative motion of entangled chains therefore already begins for the overlap concentration of the coils c^* (i.e., the boundary between regimes I and III is $c \sim c^*$).

34.4. The dispersal of long-wavelength concentration fluctuations in a solution of overlapped coils is a simple diffusive process; the corresponding effective coefficient of cooperative diffusion grows with the concentration of the solution.

Let us examine the fluctuation dynamics in regime III, where on the one hand, $c > c^*$ (i.e., the coils strongly penetrate each other), and on the other hand, $k\xi > 1$ (i.e., the concentration fluctuations are sufficiently large, and their dispersal proceeds through the motion of many chains). It is better to use the scaling method for this purpose.

Suppose we gradually increase the concentration c of polymer in solution in regime I. Because c^* is the only characteristic concentration, the characteristic time τ_k of the fluctuation mode with wave vector k is defined by the scaling formula

$$\tau_k = \frac{1}{D_{\text{coil}}k^2} \varphi(c/c^*), \tag{34.5}$$

where $\varphi(x \ll 1) \cong 1$ [*see* Eq. (33.12)]. For $c \gg c^*$, that is, in regime III, we have $\tau_k \sim (\eta_s R/Tk^2)(c/c^*)^y$. The exponent y is chosen here from the condition that τ_k is independent of N for $c \gg c^*$. (The time of fluctuation dispersal via cooperative motion of the system of strongly overlapped chains is obviously independent of the length of the chains.) Hence, we obtain [taking into account Eqs. (16.1) and (25.13)] $y = -3/4$ and, finally,

$$\tau_k \sim (\eta_s a/Tk^2)(ca^3)^{-3/4} \sim (\eta_s \xi/Tk^2). \tag{34.6}$$

The characteristic relaxation time of fluctuations, measured by the method of inelastic light scattering, is seen to diminish with the growth of concentration. This is a manifestation of the fact that the polymer solution becomes more "elastic" with concentration growth and responds faster to external perturbations.

Equation (34.6) can be rewritten as $\tau_k \sim 1/(D_{\text{coop}}k^2)$, where

$$D_{\text{coop}} \sim T/(\eta_s \xi), \tag{34.7}$$

that is, the value of D_{coop} is of the order of the diffusion coefficient of an isolated blob of size ξ. This relation can be compared with the expression $\tau_k \sim 1/(D_{\text{coil}} k^2)$, which was obtained in subsection 33.3 for regime I, where the dynamics of concentration fluctuations depended on the ordinary diffusion of coils as a whole and the dynamic structure factor $G(k, t)$ was expressed as a simple exponential (33.12). It is seen that in regime III, as in regime I, $\tau_k \sim 1/k^2$; as noted in subsection 33.3, this is an indication that in both cases, the dispersal of concentration fluctuations is a routine diffusion process.

In terms of the Fourier transform of concentration $c(k)$ (34.2), the last statement means that the function $c(k)$ obeys the equation

$$\partial c(k)/\partial t = -c(k)/\tau_k + r_k(t), \tag{34.8}$$

where $r_k(t)$ is the appropriately normalized stochastic term [cf. Eq. (31.12)]. It can easily be shown that Eq. (34.8) holds true in regime I; specifically, Eq. (33.12) for the dynamic structure factor follows from Eq. (34.8) [with allowance made for Eq. (34.3)]. For regime III, this equation can also be obtained as the result of an independent calculation.[11]

It follows from Eq. (34.8) that in regime III, the dynamic structural factor (34.3) diminishes with t according to the simple exponential law

$$G(k, t) \sim \exp(-t/\tau_k) \sim \exp(-D_{\text{coop}} k^2 t), \tag{34.9}$$

where the last equality is written taking into account Eq. (34.7) [cf. Eq. (33.12)]. By analogy with ordinary diffusion, the coefficient D_{coop} is called the *coefficient of cooperative diffusion of the solution of overlapped polymer coils*. It can be seen from Eq. (34.6) that D_{coop} grows with concentration.

In regime III, the scattering function $S(k, \omega)$ (33.6) obeys Eq. (33.13) after the substitution $D_{\text{coil}} \to D_{\text{coop}}$. This means that the spectrum of inelastic scattering in regime III is a Lorentz curve with width $\Delta\omega \sim 1/\tau_k$.

Finally, it should be noted that the coefficient of cooperative diffusion D_{coop} has no relation to the coefficient of self-diffusion D_{self} of a macromolecule as a whole among other chains. This is seen from the fact that the value of D_{coop} grows with concentration c, while D_{self} decreases because of the effect of topologic constraints (Fig. 6.4). Only at $c \sim c^*$ are these two coefficients matched and become of the order of the diffusion coefficient D_{coil} of an individual polymer coil.

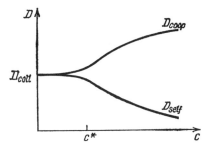

FIGURE 6.4. Coefficients of cooperative diffusion, D_c, and self-diffusion, D_s, as a function of concentration of a polymer solution, c.

35. REPTATION MODEL

35.1. *Topological constraints arising from the uncrossability of chains in a system of strongly entangled macromolecules lead to the formation of an effective tube along each macromolecule so large-scale motions of the macromolecule resemble diffusive creep inside the channel of the tube; such motion is called reptation.*

As noted in Sec. 11, the prohibition against chain crossing in a system of ring chains restricts the number of its possible conformations to one topological type, which is formed during the preparation of the system. Such "perpetual" topological restrictions naturally are absent in a system of linear chains. Quite obviously, however, many conformations may appear or disappear only via complicated motions, which require much time. Accordingly, the prohibition against chain crossing appears over relatively short time intervals, almost like real topological constraints; traditionally, they are briefly referred to as *topological constraints*.

While considering in the previous section the dynamics of concentration fluctuations, we did not take into account topological constraints, because the considered phenomena were not associated with motion of macromolecules relative to one another over long distances. As noted in subsection 34.1, however, the topological constraints are quite significant in most dynamic phenomena proceeding in systems of entangled polymer coils. In particular, they define such important characteristics of a polymer solution or melt as the viscosity η, coefficient of self-diffusion of a single polymer chain D_{self}, spectrum of relaxation times, and so on.

Current molecular concepts about the influence of topological constraints on the motion of an individual macromolecule in a concentrated system of other polymer chains are based on the following model. Let us consider one test polymer chain (Fig. 6.5), supposing for a moment that the other chains of the melt or solution are frozen (i.e., immobile). How then can the given test macromolecule move in the "frozen" surrounding? The fundamental point is that because the chains cannot cross one another, the test chain is contained within

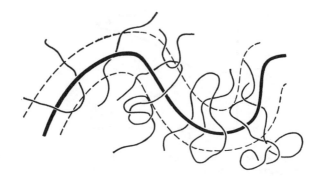

FIGURE 6.5. A polymer chain in a concentrated system of other chains.

some kind of a tube, formed by the frozen surrounding, so that motions in the direction perpendicular to the tube axis are obstructed and the only feasible motion is a diffusive creep along the tube (Fig. 6.5). It can be seen quite well in the case of the two-dimensional model illustrated in Figure 6.6; the frozen surrounding is provided on the plane by the fixed obstacles, which cannot be crossed by the chain during its motion.

Now, let the surrounding chains "melt." A competitive mechanism of the test chain motion then appears, because the surrounding chains can recede from the test one, that is, some topological constraints, specifically, those forming the tube (Figs. 6.5 and 6.6) can relax. This mechanism, however, called *tube renewal*, is insignificant in most cases, because it leads to relaxation times much greater than the time of creeping of the chain along the tube.[8] This is why the mechanism described earlier, based on the concept that the chain moves in a tube, is the basic mechanism of motion for macromolecules in a concentrated system of other chains.

Motions accomplished through creeping along a tube are called *reptations*. The corresponding model of the dynamics of polymer solutions and melts is

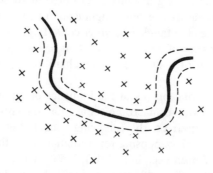

FIGURE 6.6. A polymer chain in the "frozen" surrounding (a two-dimensional case).

frequently referred to as the *reptation model*. This model was proposed by P. G. de Gennes in 1971 and developed in publications of M. Doi and S. F. Edwards.

35.2. *The effective thickness of the tube in the reptation model substantially exceeds the monomer size; the axial line of the tube (the primitive path) itself becomes coiled in a Gaussian state.*

Let us discuss in more detail the tube concept that was introduced in the previous subsection. For clarity, we consider a solution, or a melt, of standard Gaussian chains consisting of N monomer links of size a and a friction coefficient for a link of μ. As we confine the discussion to flexible polymer chains, the bead chain has the form shown in Fig. 2.4a ($v \sim a^3$).

According to Figure 6.5, the tube is determined by the condition that the given chain cannot move "through" the other chains (i.e., by contacts with the other chains). Because the number of such contacts grows with the solution concentration c, the effective thickness of the tube becomes smaller as this occurs.

Consider the highest concentration, namely, a polymer melt with $ca^3 \sim 1$. The tube thickness will be the smallest in this case. Each link in the melt comes into contact with several links of neighboring chains; therefore, assuming each such contact to be equivalent to a part of an impenetrable wall, the tube thickness in the melt would be $\sim a$. On the other hand, visualizing each polymer chain in the melt to be confined within an effective tube of diameter $\sim a$, one intuitively realizes that the restrictions imposed on possible motions of the chain are exaggerated in this case.

Thus, it should be kept in mind that in the reptation model, the effective tube is only a model conception. The tube "walls" are by no means formed by direct contacts with other chains. This circumstance is additionally illustrated in Figure 6.7. The contact between the two macromolecules in Figure 6.7a brings about only weak topologic constraints and therefore does not contribute to the formation of an effective tube, whereas in the situation illustrated in Figure 6.7b, the additional topologic constraints on possible motions of the chains are quite appreciable.

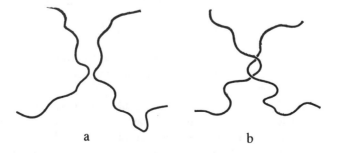

a   b

FIGURE 6.7. Contacting macromolecules forming (a) and not forming (b) "an entanglement."

Contacts of the type shown in Figure 6.7b are frequently called "entanglements." (The quotation marks are used to distinguish this qualitative term from the strictly defined topologic entanglement of ring macromolecules discussed in subsection 11.1). Using this terminology, one can say that the effective tube is formed not by all contacts with other chains, but rather by only a small fraction of them, in fact, only those that correspond to "entanglements." To allow for this circumstance quantitatively, the additional parameter N_e is introduced, which equals the average number of links in the chain between two consecutive "entanglements" of the given macromolecule with other chains.

A consistent calculation of the parameter N_e is extremely complicated. It requires the solution of a more general problem of topologic characteristics of a system of strongly entangled polymer chains (*see* Sec. 11); this is why in the modern dynamic theory of polymer liquids, the parameter N_e appears as a phenomenologic one. It characterizes the ability of a polymer chain to become entangled with the other chains. Clearly, this parameter should depend, for example, on chain stiffness, the presence of short side branches, and so on. Because the dynamic characteristics of the melt depend on N_e, the value of this parameter can be found experimentally. Typical values of N fall within the interval of 50 to 500. Regardless, $N_e \gg 1$ for any chain.

Let us now assess the structure of the effective tube in a polymer melt. As long as there exists an average number N_e of links between two successive "entanglements" of a given chain, the characteristic scale $d \sim a N_e^{1/2}$ also exists, related to the spatial distance between these "entanglements." (Recall that in a polymer melt, chains obey Gaussian statistics; *see* subsection 24.2). It can easily be inferred that this characteristic scale corresponds to the effective tube diameter, because the tube is formed just by the "entanglements." Taking into account that $N_e \gg 1$, we conclude that $d \sim a N_e^{1/2} \gg a$, that is, the tube diameter in the reptation model should be considered, even for a polymer melt, as substantially exceeding the size a of the monomer link.

When studying a solution of concentration c, it is advisable to use the conclusion made in subsection 25.7: the semidilute solution of polymer chains in good solvent can be treated as a melt of blobs of size ξ, and each blob comprises g links of the chain [*see* Eqs. (25.12) and (25.13)]. If one uses this notion, then all of the conclusions drawn earlier remain valid, and the tube diameter is

$$d \sim \xi N_e^{1/2} \sim a N_e^{1/2}(ca^3)^{-3/4}. \tag{35.1}$$

When applying Eq. (35.1), it should be kept in mind that the parameter N_e can also depend on c (i.e., the ability of blob chains to become "entangled" may depend on the blob size).

The conformation of a polymer chain in the tube is comprehensively depicted in Figure 6.8. On length scales $r < d$, the chain is insensitive to entanglements and performs a random walk around the axial line of the tube. Restrictions on the possible motions of the polymer chain appear only on scales $r > d$. The axial line of the tube is called the *primitive path of the macromolecule* (cf. subsection

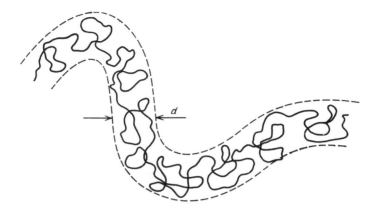

FIGURE 6.8. Conformation of a macromolecule in a tube.

11.6). This is the shortest line connecting the ends of the polymer chain, which topologically relates to its surrounding as the macromolecule itself does. In other words, the primitive path and trajectory of the macromolecule can be transformed from one to another for fixed ends by a continuous deformation without crossing other chains.

It is seen from Figure 6.8 that the primitive path constitutes a roughening of the chain trajectory to the scale $r \sim d$. However, as the statistics of the polymer chain in the melt is Gaussian on all scales and in the solution on scales $r > \xi$ (*see* subsection 25.8), in particular, for $r \sim d$ [$d > \xi$; *see* Eq. (35.1)], the statistics of the primitive path is also Gaussian (i.e., on large scales, the primitive path is coiled into a Gaussian state). The length of the effective segment associated with the primitive path equals d. One can see this, for example, by equating the expressions for the rms end-to-end distance of the initial polymer chain with that of the primitive path. The total contour length of the primitive path equals

$$L \sim \begin{cases} Nd/N_e \sim Na/N_e^{1/2} & \text{for the melt,} \\ \dfrac{N}{N_e g} d \sim \dfrac{Na}{N_e^{1/2}} (ca^3)^{1/2} & \text{for the semidilute solution,} \end{cases} \quad (35.2)$$

because the polymer chain in the melt can be depicted as a sequence of N/N_e subcoils, each of which comprises N_e links and covers a section of length $\sim d$ along the primitive path (Fig. 6.8). In a solution, the only difference is that the coil of size d comprises $N_e g$ links [*see* Eqs. (25.12) and (35.1)].

Consequently, we may refine the interpretation of Figures 6.5 and 6.6. It should be assumed that the topologic constraints on the motions of the given polymer chain, illustrated in these figures, correspond to the primitive paths of neighboring chains, not to their real trajectories, because the contacts of the primitive paths form the "entanglements" and tube walls.

35.3. *The friction coefficient for reptation along a tube is proportional to the number N of links in the macromolecule; the maximum relaxation time corresponds to the time taken by the chain to creep out of the initial tube and in the reptation model is proportional to N^3.*

We now discuss in a more detailed fashion the reptation mechanism of the motion of a macromolecule described in subsection 35.1. First, we calculate the friction coefficient for the macromolecule creeping along the tube. Subsection 34.2 showed that the hydrodynamic interaction of links is screened in a concentrated system of links, with the screening radius being of the order of the correlation radius ξ. Therefore, in the melt where $\xi \sim a$, the hydrodynamic interaction is totally screened, and the friction forces of each link are summed so that the resulting coefficient μ_t of friction of the chain creeping along the tube is higher by a factor of N than the coefficient μ of friction for an individual link: $\mu_t = N\mu$. The same reasoning can be applied to a semidilute solution, the only difference being that the summation is taken over the friction forces acting on each blob. Each of these forces should be calculated proceeding from the assumption about complete entrainment of the solvent inside a blob (*see* subsection 34.2). As a result, we obtain

$$\mu_t \sim \begin{cases} N\mu \sim N\eta_s a & \text{for the melt,} \\ \dfrac{N}{g}\eta_s\xi \sim N\eta_s a(ca^3)^{1/2} & \text{for the semidilute solution.} \end{cases} \tag{35.3}$$

In both cases, $\mu_t \sim N$. Hence, the diffusion coefficient D_t for reptation along the tube can be calculated according to the Einstein relation (31.11):

$$D_t = T/\mu_t.$$

In the process of reptational diffusion along the tube axis (i.e., along the primitive path), the chain leaves sections of the initial tube (in which it resided at the initial moment $t=0$) and creates new sections. This process is shown in Figure 6.9. The new sections of the tube are created by the motion of the ends of the polymer chains. Because the motion of the ends leaving the tube is random and uncorrelated with the initial tube conformation, it is natural to expect that the memory of the initial conformation would be erased completely when the chain entirely abandons the initial tube. The average time τ^* taken by the chain to creep out of the initial tube is easy to evaluate on the basis of Eqs. (35.2) and (35.3):

$$\tau^* \sim \frac{L^2}{D_t} \sim \frac{L^2\mu_t}{T} \sim \begin{cases} \dfrac{a^2\mu}{N_e T}N^3 & \text{for the melt,} \\ \dfrac{a^3\eta_s}{N_e T}N^3(ca^3)^{3/2} & \text{for the semidilute solution.} \end{cases}$$

$$\tag{35.4}$$

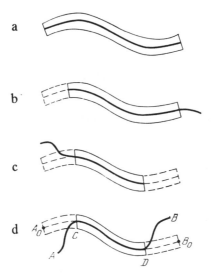

FIGURE 6.9. Four consecutive stages of a chain creeping out of the initial tube. As the chain moves, the terminal sections of the tube disintegrate (b, c, and d; *broken lines*) at the moments when the chain ends are located near these sections (b and c).

Because the memory of the initial conformation of each chain and, consequently, of the whole polymer system is erased after the time interval τ^*, this quantity can be identified with the maximum relaxation time of the polymer melt or solution. In the next subsection, we prove the validity of such an identification by specific calculations of the correlation function.

35.4. The correlation function for the end-to-end vector of a chain in the reptation model is defined by the fraction of links that remain in the initial tube after the time interval t; this function can be represented as a sum of exponentially diminishing terms.

Consider the correlation function $\langle R(t)R(0)\rangle$ for the vector R connecting the ends of a polymer chain in a semidilute solution or melt. Note that the vector R also connects the ends of the primitive path. The correlator is easier to calculate not in terms of the coordinates of the polymer chain links, but rather of those of the primitive path. Let $x(s, t)$ denote the coordinates of a point removed a distance s along the primitive path from one of its ends at moment t. Then, $R(t)=x(L,t)-x(0,t)$.

Examine Figure 6.9 once again. At the moment t shown in Figure 6.9d, part of the primitive path CD coincides with part of the initial primitive path (at $t=0$), while the parts AC and DB are new sections of the tube. One may write

$$R(0)=A_0C+CD+DB_0, \quad R(t)=AC+CD+DB. \quad (35.5)$$

It should be noted again that the new sections of the tube are generated through random motions of the ends of the polymer chain, so the vectors AC and DB by no means correlate with $R(0)$. The same can also be said about the correlations of the vectors A_0C and DB_0 with CD, because the primitive path obeys Gaussian statistics. Consequently,

$$\langle R(t)R(0)\rangle = \langle CD^2\rangle = d\langle\beta(t)\rangle, \tag{35.6}$$

where $\beta(t)$ is the contour length of the primitive path section CD remaining from the initial primitive path at the moment t.

To calculate $\langle\beta(t)\rangle$, we consider the point removed by contour length s from the end of the initial primitive path (hereafter called "the point s"). This point ceases belonging to the primitive path of the given chain when one of the chain ends reaches this point. Letting $\chi(s, t)$ denote the probability that the point still belongs to the initial tube at the moment t, one can write

$$\langle\beta(t)\rangle = \int_0^L ds\chi(s, t). \tag{35.7}$$

Let us now introduce the function $\vartheta(\sigma, t; s)$ to define the probability that at the moment t, the primitive path shifts by reptation over a distance σ along the tube, provided that the point s still belongs to the initial tube. During reptation, the primitive path performs diffusive motion along the tube with diffusion coefficient $D_t = T/\mu_t$ [see Eq. (35.3)]; therefore, for $\vartheta(\sigma, t; s)$, the following equation is valid:

$$\partial\vartheta/\partial t = D_t\partial^2\vartheta/\partial\sigma^2 \tag{35.8}$$

with the initial condition

$$\vartheta(\sigma, 0; s) = \delta(\sigma). \tag{35.9}$$

The boundary conditions for $\vartheta(\sigma, t; s)$ are specified by the requirement that the point s should still belong to the initial tube. This occurs if the value of σ varies within the limits $s - L < \sigma < s$. At $\sigma = s$ or $\sigma = s - L$, one of the chain ends reaches the point s, and the initial tube disintegrates near this point. Hence, it is clear that the solution of Eq. (35.8) satisfies the set condition provided that the boundary conditions

$$\vartheta(s, t; s) = 0, \quad \vartheta(s - L, t; s) = 0. \tag{35.10}$$

are imposed.

The solution of Eq. (35.8) with the given initial and boundary conditions takes the form[41]

$$\vartheta(\sigma, t; s) = \sum_{p=1}^{\infty} \frac{2}{L}\sin\frac{\pi ps}{L}\sin\frac{\pi p(s-\sigma)}{L}\exp\left(-\frac{p^2t}{\tau^*}\right), \tag{35.11}$$

$$\tau^* \equiv L^2/(D_t\pi^2) \tag{35.12}$$

[cf. Eqs, (35.12) and (35.4)]. The function $\chi(s, t)$ introduced earlier is explicitly connected with $\vartheta(\sigma, t; s)$ as

$$\chi(s, t) = \int_{s-L}^{s} d\sigma \vartheta(\sigma, t; s), \tag{35.13}$$

because the point s will belong to the initial tube irrespective of the magnitude of the displacement σ of the primitive path in the interval $s - L < \sigma < s$. Therefore, taking into account Eqs. (35.6), (35.7), (35.11), and (35.13), we obtain

$$\langle R(t)R(0) \rangle = Ld \sum_{p=1,3,5,\dots} (8/\pi^2 p^2) \exp(-p^2 t/\tau^*). \tag{35.14}$$

Note that the product Ld in Eq. (35.14) is nothing but the mean square of the end-to-end distance $\langle R^2 \rangle$ (*see* subsection 35.2). Consequently,

$$\langle R(t)R(0) \rangle = \langle R^2 \rangle \sum_{p=1,3,5,\dots} (8/\pi^2 p^2) \exp(-p^2 t/\tau^*). \tag{35.15}$$

The correlator sought is represented as a sum of terms exponentially diminishing with time. The most "long-lived" term has a characteristic relaxation time τ^*, in complete agreement with the evaluation (35.4). Thus, we conclude that the maximum relaxation time in the system of strongly entangled polymer chains corresponds to the average time for the chain to creep entirely out of the initial tube.

The equality (35.15) can be compared with the similar relation (31.31) for the Rouse model. Both expressions have an identical structure. The only distinction consists in the substitution of τ^* for the Rouse maximum relaxation time τ_1. It is seen from Eq. (31.22) for τ_1 and Eq. (35.14) for τ^* that $\tau_1 \sim N^2$ and $\tau^* \sim N^3$. Consequently, $\tau^* \gg \tau_1$ for long chains. Thus, the presence of topologic constraints leads to a substantial slowing of the relaxation processes.

The unusual properties of polymer liquids (in particular, viscoelasticity) mentioned at the beginning of this chapter, result from the fact that the maximum relaxation time of polymer melts and concentrated solutions are proportional to the cube of the length of the macromolecule (i.e., very large). If the characteristic time of an external action is less than τ^*, then the relaxation has no time to set in and the polymer substance behaves as an elastic body. Viscous flow appears only for very slow actions with a characteristic time exceeding τ^*. The viscoelasticity of polymer liquids is considered in more detail in Sec. 36.

*35.5. The coefficient of self-diffusion of a macromolecule in the reptation model diminishes with the growth of the number of links as N^{-2}.

We now determine the coefficient D_{self} of translational self-diffusion of a macromolecule as a whole in the polymer melt or semidilute solution. First, we make the simplest estimate of the value and then confirm the estimate by rigorous calculation.

To evaluate D_{self}, note that during the time τ^* taken by the chain to creep entirely out of the initial tube, it is natural to expect that the center of mass of the chain shifts over a distance of the order of the size R of the macromolecule. On the other hand, displacements of the chain during different time intervals of duration τ^* are statistically independent. Therefore, for long time intervals, diffusive motion of the center of mass of the chain sets in. According to Eqs. (25.16) and (35.4), we obtain for the diffusion coefficient

$$D_{\text{self}} \sim \frac{R^2}{\tau^*} \sim \begin{cases} \dfrac{N_e T}{N^2 \mu} & \text{for the melt,} \\[2ex] \dfrac{N_e T}{N^2 a \eta_s} (ca^3)^{-7/4} & \text{for the semidilute solution.} \end{cases} \tag{35.16}$$

The value of D_{self} is seen to diminish with the number N of links in the chain as N^{-2}. This dependence agrees well with relevant experimental data.

A more consistent quantitative calculations of the coefficient D_{self} requires the time dependence of the mean square of the displacement of a point on the primitive path, that is, the correlator $\langle (x(s, t) - x(s, 0))^2 \rangle$. First, let us write the so-called fundamental equation of the dynamics of the primitive path. Suppose that during the time interval between the moments t and $t + \Delta t$, the points of the primitive path move along the tube by the contour length $\Delta\sigma(t)$. Then,

$$x(s, t + \Delta t) = x(s + \Delta\sigma(t), t). \tag{35.17}$$

Equation (35.17) indicates that in shifting along the tube over the length $\Delta\sigma(t)$, the point s reaches the location at which the point of the primitive path with coordinate $s + \Delta\sigma(t)$ was residing at the moment t. Because motion along the tube is described by the diffusion coefficient D_t, the displacement $\Delta\sigma(t)$ is assumed to be a random quantity with the moments

$$\langle \Delta\sigma(t) \rangle = 0, \quad \langle (\Delta\sigma(t))^2 \rangle = 2 D_t \Delta t. \tag{35.18}$$

Equation (35.17) together with Eq. (35.18) is the fundamental equation of dynamics for the primitive path. Note that it is invalid near the ends of the primitive path when the argument $s + \Delta\sigma(t)$ runs beyond the interval limits 0 and L, because chain ends creeping out of the tube generate new sections of the primitive path that are unrelated to the initial tube. This fact must be allowed for while formulating the boundary conditions for Eq. (35.17).

Let us return to the correlator $\langle (x(s, t) - x(s, 0))^2 \rangle$. It is convenient to perform calculations for the function

$$\Phi(s, s', t) = \langle (x(s,t) - x(s', 0))^2 \rangle, \tag{35.19}$$

assuming s' to be a fixed parameter and putting $s' = s$ in the final expression. With allowance made for the fundamental equation (35.17), we have

$$\Phi(s, s', t+\Delta t) = \langle [x(s+\Delta\sigma(t),t)-x(s', 0)]^2 \rangle = \langle \Phi(s+\Delta\sigma(t), s', t) \rangle, \tag{35.20}$$

where the averaging in the latter expression must be carried out over the random variable $\sigma(t)$ with regard to Eq. (35.18). Expanding into a Taylor series, we obtain

$$\langle \Phi(s+\Delta\sigma(t),s',t) \rangle = \left\langle \left(1+\Delta\sigma\frac{\partial}{\partial s}+\frac{(\Delta\sigma)^2}{2}\frac{\partial^2}{\partial s^2}\right)\Phi(s, s', t) \right\rangle$$

$$=\left(1+\langle\Delta\sigma\rangle\frac{\partial}{\partial s}+\frac{\langle(\Delta\sigma)^2\rangle}{2}\frac{\partial^2}{\partial s^2}\right)\Phi(s, s', t)$$

$$=\left(1+D_t\Delta t\frac{\partial^2}{\partial s^2}\right)\Phi(s, s', t). \tag{35.21}$$

It follows from Eqs. (35.20) and (35.21) that the function $\Phi(s, s', t)$ obeys the diffusion equation

$$\frac{\partial}{\partial t}\Phi(s, s', t)=D_t\frac{\partial^2}{\partial s^2}\Phi(s, s', t). \tag{35.22}$$

The initial condition for Eq. (35.22) is determined proceeding from the fact that at $t=0$, the quantity $\Phi(s, s', 0)$ is the mean square of the spatial distance between the points s and s' along the primitive path. The statistics of the primitive path are Gaussian on large length scales, and the length of the effective segment of the primitive path equal d. Therefore, we have

$$\Phi(s, s', t)|_{t=0}=|s-s'|d \quad \text{at} \quad |s-s'|\gg d. \tag{35.23}$$

Generally speaking, for $|s-s'|\lesssim d$, the condition (35.23) changes, but in the case $s=s'$, (which is of special interest to us), this equality becomes valid again.

To derive the boundary conditions, note that the derivative $\partial\Phi/\partial s$ can be written at $s=L$ as

$$\frac{\partial}{\partial s}\Phi(s, s', t)|_{s=L}=2\langle u(L, t)(x(L, t)-x(s', 0))\rangle, \tag{35.24}$$

where $u(s, t)=\partial x(s, t)/\partial s$ is a unit vector tangential to the primitive path at the point s. The right-hand side of Eq. (35.24) can be written as the following sum:

$$2\langle u(L, t)(x(L, t)-x(s'', t))\rangle+2\langle u(L, t)(x(s'', t)-x(s', 0))\rangle, \tag{35.25}$$

where s'' is the arbitrary quantity varying in the interval $0<s''<L$. Such a representation is more convenient to use. While averaging the second term in Eq. (35.25), one can make use of the fact that the new chain sections generated by the chain ends do not correlate at all with the conformation of the initial prim-

itive path. Accordingly, the direction of the primitive path at the terminal point $u(L, t)$ does not correlate with $x(s'', t) - x(s', 0)$, so the second term in Eq. (35.25) equals zero. As for the first term, it can be written as

$$2\langle u(L, t)(x(L, t) - x(s'', t))\rangle = \frac{\partial}{\partial s} \langle (x(s,t) - x(s'', t))^2 \rangle |_{s=L}$$

$$= \frac{\partial}{\partial s} d(s - s'') |_{s=L} = d \qquad (35.26)$$

[cf. Eq. (35.23)]. Hence, we obtain the boundary condition at $s = L$ (and the analogous condition at $s = 0$):

$$\frac{\partial}{\partial s} \Phi(s, s', t) |_{s=L} = d, \quad \frac{\partial}{\partial s} \Phi(s, s', t) |_{s=0} = -d. \qquad (35.27)$$

Equation (35.22) is a partial differential equation of the diffusion type. Solving it by a standard technique[41] with allowance made for the initial condition (35.23) and the boundary conditions (35.27), we obtain

$$\Phi(s, s', t) = |s - s'| d + 2 \frac{d}{L} D_t t$$

$$+ \sum_{p=1}^{\infty} \frac{4Ld}{\pi^2 p^2} \cos \frac{\pi p s}{L} \cos \frac{\pi p s'}{L} \cdot \left[1 - \exp\left(-\frac{t p^2}{\tau^*} \right) \right]. \qquad (35.28)$$

Specifically, assuming $s = s'$, we obtain

$$\langle (x(s, t) - x(s, 0))^2 \rangle = \Phi(s, s, t)$$

$$= 2 \frac{d}{L} D_t t + \sum_{p=1}^{\infty} \frac{4Ld}{\pi^2 p^2} \cos^2 \frac{\pi p s}{L} \cdot \left[1 - \exp\left(-\frac{t p^2}{\tau^*} \right) \right]. \qquad (35.29)$$

Hence, we immediately find the diffusion coefficient D_{self}. For sufficiently large values of t ($t \gg \tau^*$), the first term in Eq. (35.29) appreciably exceeds the second one, so the mean square of the displacement begins growing proportionally with time. This corresponds to ordinary diffusion with the coefficient

$$D_{\text{self}} = \lim_{t \to \infty} \frac{\Phi(s, s, t)}{6t} = \frac{dD_t}{3L} = \frac{d}{3L} \frac{L^2}{\pi^2 \tau^*} \sim \frac{R^2}{\tau^*} \qquad (35.30)$$

in full correspondence with the estimate (35.16) [in writing one of the equalities in Eq. (35.30), we used the formula (35.12)].

*35.6. *In the motion of a chain link among entangled macromolecules, four qualitatively different regimes can be distinguished, corresponding to different scales and described by different dependences of the rms displacement* $\langle(\Delta x)^2\rangle^{1/2}$ *of the link on time: 1) Rouse motion of the chain section between entanglements,* $\langle(\Delta x)^2\rangle^{1/2} \sim t^{1/4}$; *2) one-dimensional analogue of Rouse motion of chain links along the tube,* $\langle(\Delta x)^2\rangle \sim t^{1/8}$; *3) reptation of the chain along the tube,* $\langle(\Delta x)^2\rangle^{1/2} \sim t^{1/4}$; *and 4) self-diffusion on scales greater than the chain size,* $\langle(\Delta x)^2\rangle^{1/2} \sim t^{1/2}$.

The result (35.29) allows one to analyze not only the diffusion behavior of a reptating macromolecule for large values of t ($t \gg \tau^*$) but also the mean square of the displacement of a point of the primitive path for smaller values of t. Indeed, for $t \ll \tau^*$, the crucial contribution to the sum (35.29) is provided by the terms with large values of p. In this case, the rapidly oscillating factor $\cos^2(\pi ps/L)$ can be replaced by its mean value $1/2$ and the sum itself transformed into an integral [cf. the derivation of Eq. (31.39)]. Then, we obtain

$$\Phi(s, s, t) = \int_0^\infty dp \, \frac{4Ld}{\pi^2 p^2} \cdot \frac{1}{2}\left[1 - \exp\left(-\frac{tp^2}{\tau^*}\right)\right] = 2d\left(\frac{D_t t}{\pi}\right)^{1/2}. \quad (35.31)$$

Thus, the mean square of the displacement of a point of the primitive path grows proportionally to $t^{1/2}$ for $t \ll \tau^*$.

It would be wrong to conclude from this, however, that for $t \ll \tau^*$, the mean square of the displacement of the n-th link of the polymer chain $\langle(x(t, n) - x(0, n))^2\rangle$ [cf. Eq. (31.39)] also obeys the law (35.31). The primitive path appears as a result of a roughening of the real trajectory of the polymer chain up to the scale $\sim d$ (*see* subsection 35.2). Consequently, laws governing the motion of the chain and the primitive path may differ substantially on small time and length scales.

Specifically, topologic constraints must not affect the motions in the melt on scales less than $d \sim a N_e^{1/2}$ (*see* subsection 35.2). Because the hydrodynamic interactions are totally screened (*see* subsection 34.2), the mean square of the displacement of the n-th link should be the same as for the Rouse model [*see* Eq. (31.39)].[d] The result (31.39) remains valid until the mean square of the displacement defined by this formula becomes equal to d^2. This happens at $t \sim \tau_A$, where $d^2 \sim (Ta^2\tau_A/\mu)^{1/2}$ [*see* Eq. (31.39)]. Thus, we obtain

$$\tau_A \sim d^4\mu/(Ta^2) \sim N_e^2\mu a^2/T. \quad (35.32)$$

Comparing Eqs. (35.32) and (31.32), we conclude that τ_A is the Rouse maximum relaxation time for a chain of N_e links.

Hence, for $t < \tau_A$, the mean square of the link displacement is determined by Eq. (31.39) (i.e., grows with t as $t^{1/2}$). Note that despite the identical depen-

---

[d]For a semidilute solution, one must substitute the correlation length ξ, for a and $\eta_s\xi$ for μ in Eq. (31.39). Because this case can be fully examined by analogy, we restrict this subsection to the analysis of a polymer melt only.

dence on t in relation (31.39), these formulas differ substantially [if only that in Eq. (35.31), in contrast to Eq. (31.39), there is a dependence on N].

What happens for longer times $t > \tau_A$? The topologic constraints (i.e., the presence of the effective tube) cannot be neglected any longer. It is necessary, however, to consider that the tube inhibits motion of the links only in the direction perpendicular to the primitive path. Therefore, for $t > \tau_A$, the Rouse modes continue during the motion along the primitive path, determining the mean square displacement $\langle (s(t, n) - s(0, n))^2 \rangle$ of a link along the tube, where $[s(t, n)]$ is the coordinate of the n-th link counted along the primitive path. Having written the one-dimensional analogue of Eq. (31.38), we obtain

$$\langle (s(t, n) - s(0, n))^2 \rangle = \frac{2T}{N\mu} t + \frac{4Na^2}{3\pi^2} \sum_{p=1}^{\infty} \frac{1}{p^2} \cos^2 \frac{\pi pn}{N} \cdot \left[1 - \exp\left(-\frac{tp^2}{\tau_1} \right) \right].$$

(35.33)

Suppose first that $\tau_A < t < \tau_1$. Then, terms with large values of p prevail in the sum (35.33). Similar to the derivation of Eqs. (31.39) and (35.31), we obtain

$$\langle (s(t, n) - s(0, n))^2 \rangle \sim (Ta^2t/\mu)^{1/2}.$$

(35.34)

The spatial displacement of a link is found proceeding from the fact that the statistics of the primitive path is Gaussian on scales exceeding d. Therefore [cf. Eq. (35.23)],

$$\langle (x(t, n) - x(0, n))^2 \rangle \sim d\langle |s(t, n) - s(0, n)| \rangle$$

$$\sim d\langle (s(t, n) - s(0, n))^2 \rangle^{1/2}$$

$$\sim d(Ta^2t/\mu)^{1/4}.$$

(35.35)

Thus, the mean square of the link displacement grows proportional to $t^{1/4}$ at $\tau_A < t < \tau_1$. Such unusual behavior results from a superposition of two effects. First, for $t < \tau_1$, the motion along the primitive path still does not evolve into the ordinary diffusion regime, so $\langle (\Delta s)^2 \rangle \sim t^{1/2}$ [cf. Eq. (31.39)]. Second, the primitive path itself is entangled into a spatial Gaussian coil, which leads to an additional decrease in the exponent of the dependence $\langle (\Delta x)^2 \rangle (t)$.

Now let $\tau_1 < t < \tau^*$. Then, the first term responsible for the diffusion of the chain as a whole along the tube becomes dominant in the sum (35.33). Therefore,

$$\langle (x(t, n) - x(0, n))^2 \rangle \sim d\langle (s(t, n) - s(0, n))^2 \rangle^{1/2} \sim d(Tt/N\mu)^{1/2}.$$

(35.36)

It is seen that in this case, $\langle (\Delta x)^2 \rangle \sim t^{1/2}$. As $D_t \sim T/(N\mu)$, Eqs. (35.31) and (35.36) are consistent with each other. This should be expected, because for $t > \tau_1$, when the diffusion of the chain as a whole along the tube prevails, the displacements of a link and a point of the primitive path are equal.

Finally, for $t > \tau^*$, the self-diffusion of the chain as a whole, determined by the first term in Eq. (35.29), predominates. Consequently,

$$\langle (x(t, \, n) - x(0, \, n))^2 \rangle \sim (d/L) \, D_t t \sim D_{\text{self}} t. \qquad (35.37)$$

Formula (35.33) is inapplicable in this case, because for $t > \tau^*$, a random creeping of the chain ends out of the tube should be taken into consideration, as was done in subsection 35.5.

Hence, there are four different regimes of the dependence $\langle (\Delta x)^2 \rangle (t)$: for short times ($t < \tau_A$), this quantity is proportional to $t^{1/2}$, $\langle (\Delta x)^2 \rangle \sim t^{1/4}$ for $\tau_A < t < \tau_1$, $\langle (\Delta x)^2 \rangle \sim t^{1/2}$ for $\tau_1 < t < \tau^*$, and $\langle (\Delta x)^2 \rangle \sim t$ for $t > \tau^*$. All of these regimes are shown schematically in Figure 6.10. For a melt of phantom chains, Rouse dynamics would be valid; According to subsection 31.6, there would be only the two regimes: $\langle (\Delta x)^2 \rangle \sim t^{1/2}$ for $t < \tau_1$, and $\langle (\Delta x)^2 \rangle \sim t$ for $t > \tau_1$.

As seen from Figure 6.10, the effects of the topologic constraints can already be observed experimentally in the time interval $\tau_A < t < \tau_1$ (i.e., over intervals substantially shorter than the maximum relaxation time τ^*). The reptation model receives experimental confirmation when the exponent in the dependence $\langle (\Delta x)^2 \rangle (t)$ equals 1/4 in this interval. Thus, the correctness of the reptation model can basically be proved by computer simulation of the motion of macro-molecules in a polymer melt or concentrated solution over time intervals much less than τ^*. Such computer experiments have been performed yielding good agreement with the results given earlier, at least for a macromolecule moving in a frozen surrounding of other chains (*see* subsection 35.1).

35.7. In contrast to the prediction $\tau^* \sim N^3$ of the reptation model, the dependence $\tau^* \sim N^{3.4}$ is experimentally observed for the maximum relaxation time; the cause of this discrepancy may be associated with fluctuations of the contour length of the primitive path.

We noted earlier that the reptation model yields the self-diffusion coefficient $D_{\text{self}} \sim N^{-2}$ for a macromolecule, which agrees well with experiment. This is not

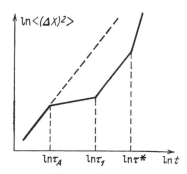

FIGURE 6.10. Dependence of $\langle (\Delta x)^2 \rangle$ on t for a polymer chain in the reptation model on a double logarithmic scale.

the case, however, for the corresponding dependence $\tau^* \sim N^3$ for the maximum relaxation time (*see* subsection 35.3). Usually, the exponent of the experimentally observed dependence is slightly higher: $\tau \sim N^{3.4}$.

There are several explanations for this discrepancy. The generally accepted one is associated with fluctuations of the contour length L of the primitive path. Indeed, until now (specifically, while deriving the expression for τ^*) we assumed that the value of L is constant and determined by Eq. (35.2). In a real situation, however, the length of the primitive path should fluctuate, because the Rouse modes continue in the direction along the primitive path (*see* subsection 35.6). Figure 6.11 illustrates two examples, one with and one without such fluctuations. The presence of fluctuations reduces the lifetime of the initial tube (i.e., the maximum relaxation time τ^*). The rigorous calculation carried out by M. Doi in 1981 showed that

$$\tau_F^* \sim \tau^* \left(1 - \frac{\text{const}}{(L/d)^{1/2}} \right) \sim \frac{a^2 \eta}{N_e T} N^3 \left(1 - \frac{\text{const}}{(N/N_e)^{1/2}} \right) \qquad (35.38)$$

where τ_F^* is the maximum relaxation time obtained with allowance made for the fluctuations of the value of L. The last equality is written for the polymer melt with Eqs. (35.2) and (35.4) taken into account.

It can be seen from Eq. (35.38) that the asymptotic dependence $\tau_F^* \sim N^3$ persists as $N \to \infty$. However, for finite values of the ratio N/N_e, there are some deviations [e.g., the second term in Eq. (35.38)]. Plotting the dependence $\tau_F^*(N)$, defined by Eq. (35.38), for $N/N_e \sim 100$ in double logarithmic coordinates, the apparent exponent may prove to be close to the experimentally observed value of 3.4 for certain values of the factor const in Eq. (35.38).

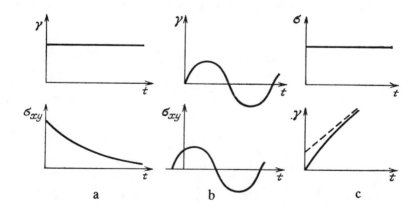

FIGURE 6.11. Simple examples of shear strain. (a), Stepwise strain. (b), Harmonic strain. (c), Flow because of stepwise strain.

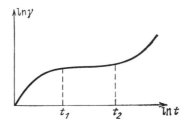

FIGURE 6.12. Typical dependence $\gamma(t)$ for the beginning of the flow of a polymer melt under constant shear stress.

35.8. Motion mechanisms of polymer chains of the "tube renewal" type are insignificant for melts and concentrated solutions of linear macromolecules.

Another probable cause of experimental deviations in the dependence $\tau^*(N)$ from theoretical predictions based on the reptation model is the "tube renewal" process mentioned in subsection 35.1. Recall that this process occurs because the macromolecules surrounding the given chain (i.e., the chain under study) can move away from it. In this process, the topological constraints that have been imposed by these macromolecules on the motion of the chain disappear, and the tube is partially destroyed.

Let us evaluate the effects of tube renewal in the process of motion of polymer chains in a polymer melt. We will partially complicate the problem by considering the motion of one long chain of N_0 links in a melt of chemically identical but shorter chains composed of N links. Then, for $N = 1$, the problem is reduced to that of one chain in a low-molecular-weight solvent. In this case, the rebuilding of the tube (i.e., the motion of chain segments in various directions, not just along the primitive path) succeeds. As N grows, the topologic constraints become more noticeable and, starting from certain value of N^*, reptation (and not tube renewal) becomes the foremost mechanism of motion of the test chain in the melt.

To evaluate N^*, one must determine the maximum relaxation time appearing because of renewal of the tube. Figure 6.12 shows that when one of the surrounding chains moves away and stops imposing the topologic constraints on the given macromolecule, there appears an opportunity for tube segment of size $\sim d$ to shift in the direction perpendicular to the primitive path over a distance of order d. How often do such events occur? When $N > N_e$, the foremost mechanism of motion for macromolecules in the melt of chains with N links is reptation.[e] Therefore, the time taken by a chain to leave the surroundings of the given chain (and by the accompanying topologic constrain to relax) equals

---

[e]Strictly speaking, our reasoning proves that it is this mechanism and not rebuilding of the tube that prevails in a monodisperse melt for $N > N_e$. Thus far, it may be regarded as an assumption to be confirmed later.

$\tau^*(N)$, that is, the maximum relaxation time (35.12) for the melt of N-link chains.

If one considers such events throughout the primitive path of a macromolecule of N_0 links, it becomes clear that rebuilding of the conformation of the tube because of its renewal may generally be described as the motion of a polymer chain of N_0/N_e links with link size d in the Verdier-Stockmayer lattice model (*see* subsection 31.8; Fig. 6.1). The characteristic time of the "corner jump" in this model is taken to equal $\tau^*(N)$. Taking into account Eq. (31.43), expressing the maximum relaxation time τ_{ren} for tube renewal, we obtain

$$\tau_{ren} \sim \left(\frac{N_0}{N_e}\right)^2 \tau^*(N) \sim \left(\frac{N_0}{N_e}\right)^2 N^3 \left(\frac{a^2\eta}{N_e T}\right). \tag{35.39}$$

The time τ_{ren} is to be compared with the maximum relaxation time $\tau^*(N_0)$ of reptation motion of the chain of N_0 links. According to Eq. (35.4), $\tau^*(N_0) \sim N_0^3 (a^2\eta/N_e T)$. For $N_0 < N^3/N_e^2$, $\tau^*(N_0) < \tau_{ren}$; consequently, during a time interval equal to that taken by the chain of N_0 links to creep out of the initial tube, essentially no renewal of the tube occurs, which means that the reptation mechanism of motion predominates. When $N_0 > N^3/N_e^2$, $\tau^*(N_0) > \tau_{ren}$, and the dynamics of the polymer chain is accomplished via tube renewal of the tube. Hence, it can be concluded that the boundary value of N^* corresponding to the crossover between the regimes mentioned here is of order $N^* \sim N^3/N_e^2$.

It can thus be inferred that for the motion of the chain of N links in a melt of similar chains ($N_0 = N$), $\tau^*(N) \ll \tau_{ren}(N)$, $N \ll N^*$, when $N \gg N_e$ (i.e., reptation is the basic type of motion of macromolecules and tube renewal is immaterial). This justifies the main assumptions of the reptation model (*see* subsection 35.1).

It should be noted, however, that the motions involved in renewal of the tube can lead to some corrections to the expressions derived using the orthodox reptation model (cf. the corrections from fluctuations of the contour length of the tube, examined in subsection 35.7). This may appear in various experimentally observed deviations from theoretic predictions for finite values of N, in particular, in a variation of the apparent exponent in the dependence $\tau^*(N)$ (cf. subsection 35.7).

From the previous analysis of the motion of one long macromolecule in a melt of shorter ones, it might be surmised that for $N_0 > N^* = N^3/N_e^2$, the prevailing mechanism of motion of a macromolecule involves tube renewal; however, this is not the case. The estimates cited did not consider that the conclusion about the total screening of the hydrodynamic interactions in the melt (*see* subsection 34.2) is generally imprecise for $N_0 \gg N$. In fact, when the length N of surrounding chains is fairly short, they will be entrained (according to Sec. 32) by the motion of the coil of N_0 links, that is, the motion of that coil can be described by the Zimm model as the motion of a hydrodynamically impenetrable sphere of radius $R(N_0) \sim a N_0^{1/2}$. According to Eq. (32.33), the corresponding maximum relaxation time is $\tau'(N) \sim N_0^{3/2} a^3 \eta(N)/T$, where $\eta(N)$ is the viscosity of "the solvent," which in this case is a melt of short polymer chains of N links. We will show in subsection 36.4 that $\eta(N) \sim (\eta_s/aN_e^2)N^3$ for $N \gg N_e$; therefore,

$$\tau' \sim \frac{\eta_s a^2}{N_e^2 T} N^3 N_0^{3/2}. \tag{35.40}$$

Comparing Eqs. (35.39) and (35.40) with the expression $\tau^*(N_0) \sim (\eta_s a^2/N_e T)N_0^3$, we conclude that renewal of the tube does not prevail for any ratio of N_0 to N. For $N_0 < N^2/N_e^{2/3}$, a chain of N_0 links moves by reptation, while for $N_0 > N^2/N_e^{2/3}$, the basic mechanism of motion for a coil as a whole is associated with the entrainment of all short N-link macromolecules residing inside the coil.

35.9. The maximum relaxation time for melts of branched macromolecules grows exponentially with the length of the branches.

To conclude this section, we consider a system to which the reptation model cannot be applied in its conventional form. Suppose we have a melt of branched star macromolecules (*see* subsection 10.2) with the number of branches $f = 3$. If the branches are long enough, then each becomes strongly entangled with the branches of other macromolecules. In this example, the linear tube does not appear, so conventional reptation becomes impossible.

The only feasible mechanism for the displacement of branched macromolecules that are strongly entangled with other similar macromolecules is the motion along the tube formed by the two branches by a length equal to the distance d between entanglements, which requires that the third branch fold so that its end becomes a neighbor to the branching point. Then, because of the motion of that end, the initiation of a new tube for the third branch becomes possible, which allows the tube of the first two branches to shift (on average) over a distance $\sim d$.

Hence, any displacement of a star macromolecule requires realization of the extremely improbable folded conformation for one of the branches. Entropy losses from the transformation of the macromolecule from the most probable conformation of a Gaussian coil to the folded conformation are then proportional to the number N of links in the branch, because the number of folds is proportional to N. The activation barrier that must be overcome to make the process possible consequently has the height $\Delta U/T \sim \text{const} N$, where const is independent of N. Making use of the obvious analogy with the theory of activation processes, one can deduce that the maximum relaxation time for the melt of star macromolecules takes the form

$$\tau^* = \tilde{\tau}(N)\exp(\text{const } N), \tag{35.41}$$

where the factor $\tilde{\tau}(N)$ is a power function of N and const a positive constant independent of N.

A similar relation for τ^* can be obtained for the melts of branched macromolecules of any other type (not necessarily stars), provided that the length N of the branches is sufficiently long $(N \gg N_e)$. Hence, the maximum relaxation time for melts of branched macromolecules grows exponentially with N. As the relation $D_s \sim R^2/\tau^*$ (where R is the spatial size of the macromolecule) can always be written for the self-diffusion coefficient D_s, one may conclude that the value of D_s diminishes exponentially with the growth of N. This behavior generally agrees with experimental observations.

APPENDIX TO SEC. 35. DNA GEL ELECTROPHORESIS

One of the most important applications of the reptation model is associated with the penetration of charged (*see* subsection 37.4) macromolecules of DNA through a polymer network (gel) under the influence of an electric field. This phenomenon (referred to as *gel electrophoresis*) underlies a fruitful experimental method. As the motion velocity (or more exactly, the mobility) depends strongly on the length and other properties of the macromolecule, a polydispersed mixture of DNA chains separates after a certain period of electrophoretic motion into essentially monodispersed fractions localized at different regions of the gel. This is exceptionally valuable for solving certain problems of molecular biology.

At present, the theory of gel electrophoresis is not complete, and we give here only the simplest evaluations. Let us assume that the persistent length of the double helix of DNA is much greater than the size of the network mesh through which the electrophoresis proceeds. Although this limiting case often does not comply with real conditions, its analysis is useful to clarify the general situation. At the same time, it is very simple, because the chain contour coincides with the primitive path: because of high stiffness, the DNA macromolecule does not form any loops extending from the tube to neighboring cells of the network.

The mechanism of reptation in this limiting case is especially simple, being reduced to the motion of the DNA macromolecule as a stiff entity along the tube.[f] Because the tube thickness is approximately the size of the network mesh, the subcoils discussed in subsection 35.2 are degenerate; they are shorter than the persistent length. Such a mechanism of motion implies that the effective external force exerted on a DNA chain by an electric field is obtained by summing (over all chain sections) those force components that are directed along the tube:

$$F = \int_0^L u_s E(Q/L)ds = (Q/L)ER,$$

where Q and L are the charge and the length of the DNA chain, respectively,

---

[f]Here, a relevant graphic analogy would be a long train moving along a railroad twisting through a mountain area.

$(Q/L)ds$ the charge of a section of length ds, u_s the unit vector tangent to the chain at the point s, $R = \int_0^L u_s ds$ [see Eq. (3.4)] the end-to-end vector, and E the external field. The coefficient of friction for motion along the tube equals $\mu_t = L\eta$ according to Eq. (35.3), where η is a value having the dimensionality of viscosity and also making allowance for the friction of the DNA macromolecule against the subchains of the network. The reptation motion velocity equals $v_t = F/\mu_t$.

Displacement of the chain along the tube by the length element ds can be pictured as a transposition of this element from one chain end to the other (i.e., by the vector R). The center of mass of the chain shifts accordingly by a distance Rds/L. Hence, one easily obtains the expression for the velocity of electrophoretic motion of the center of mass of the DNA chain. After averaging, we find

$$ v = \frac{Q}{L\eta} \frac{\langle R_\parallel^2 \rangle}{L^2} E, \tag{35.42} $$

where $R_\parallel$ is the component of R directed along the field E.

Let us now examine the basic relation (35.42). If the external force is very weak and essentially does not affect the Gaussian form of the DNA coils, then $R_\parallel^2 = Ll/3$ [see Eq. (3.3)]. In this case, the velocity $v \sim 1/L$ (i.e., strongly depends on the chain length L).

In stronger fields, the DNA chains stretch appreciably along the direction of the field, because the leading end of the DNA chain, forming new sections of tube, moves along the field with a higher probability than transverse to it, not to mention against it. In this case, as $L \to \infty$, it must be that $R_\parallel \sim L$. Therefore, for reasons of dimensionality, $R_\parallel \sim EQl/T$, and we finally obtain

$$ v = \frac{Q}{3L\eta} \left[\frac{l}{L} + \text{const} \cdot \left(\frac{EQl^2}{LT} \right)^2 \right] E. $$

The velocity is seen to become independent of L as the field grows (because $Q \sim L$). In other words, the proportional dependence between v and $1/L$ saturates as $1/L \to 0$. It is easy to realize that this effect occurs in a more realistic situation when the size of the network mesh is not small with respect to the persistent length of the DNA chain (but of course still smaller than the size of the DNA coil).

This circumstance drastically curtails the potential for gel-electrophoretic separation of DNA fractions. To overcome this difficulty, the following elegant approach is taken: the external field is periodically switched off (or turned by 90°). The time interval during which the field is on (just as the time interval during which the field is off) must apparently be compatible with the time of the total change of the initial tube, that is, both time intervals must be of order $\tau^* \sim L^3$ [see Eq. (35.4)]. In this case, the proper phoretic motion proceeds continuously under the most favorable Gaussian statistics of the tubes.

36. VISCOELASTICITY OF POLYMER MELTS

36.1. *In the linear case, the viscoelastic properties of a polymer melt for shear flow are determined by the relaxation modulus of elasticity.*

As mentioned at the very beginning of this chapter, one of the most fundamental and important properties of polymer liquids (melts and concentrated solutions) consists in their viscoelasticity, that is, in that such liquids behave differently under fast and slow external forces; in the former case as elastic bodies and in the latter as viscous liquids. This section considers the simplest molecular theory of viscoelasticity for shear flow [*see* Eq. (33.22)] in polymer melts. In so doing, we assume that the shear velocity $\varkappa(t)$ is not high so that the relation between the shear stress σ_{xy} and $\varkappa$ is linear and can be written in the form (33.23).

For convenience, let us rewrite Eq. (33.23) as a strain–stress relation. The dimensionless strain for shear flow, counted from the state at the moment $t=0$, can be expressed as

$$\gamma(t) = \int_0^t dt' \varkappa(t').$$

(36.1)

Then, integrating by parts, one can rewrite Eq. (33.23) in the form

$$\sigma_{xy}(t) = \int_{-\infty}^t dt' G(t-t')(\partial\gamma(t')/\partial t') = \int_{-\infty}^t dt' \frac{\partial G(t-t')}{\partial t'}(\gamma(t)-\gamma(t')).$$

(36.2)

Let us consider the three most typical examples (Fig. 6.11). First, suppose that at the moment $t=0$, we subject a melt sample to a shear strain that remains constant as time goes on (Fig. 6.11a). Then,

$$\gamma(t) = \begin{cases} 0 & \text{when } t<0, \\ \gamma_0 & \text{when } t>0. \end{cases}$$

(36.3)

Inserting Eq. (36.3) into Eq. (36.2), we obtain

$$\sigma_{xy}(t) = \gamma_0 G(t).$$

(36.4)

Equation (36.4) helps to clarify the physical meaning of the relaxation modulus $G(t)$ of elasticity specified by Eq. (33.23): the function $G(t)$ determines the stress relaxation for a stepwise shear strain (Fig. 6.11a). This allows the modulus $G(t)$ to be found easily experimentally.

The other example often studied in experiments is shown in Figure 6.11b. The sample is subjected to a harmonic external action

$$\gamma(t) = \gamma_0 \cos \omega t = \gamma_0 \operatorname{Re}(e^{i\omega t})$$

(36.5)

Inserting Eq. (36.5) into Eq. (36.2), we obtain

$$\sigma_{xy}(t) = \gamma_0 \, \mathrm{Re}(G^*(\omega)e^{i\omega t}), \tag{36.6}$$

where $G^*(\omega)$, a so-called complex modulus of elasticity, is defined by the equality

$$G^*(\omega) \equiv i\omega \int_0^\infty dt G(t) e^{-i\omega t} = G'(\omega) + iG''(\omega). \tag{36.7}$$

The real part $G'(\omega)$ of the expression (36.7) is called the *storage modulus*, whereas its imaginary part $G''(\omega)$ is called the *loss modulus*. From Eqs. (36.5) to (36.7), it is immediately seen that by measuring the ratio of the amplitudes of stress and strain and the phase shift between them under a harmonic external action, one can find the corresponding Fourier components of the relaxation modulus $G(t)$ of elasticity.

Finally, suppose the melt is subjected to a constant shear stress σ_0 at the moment $t=0$ to initiate its flow (Fig. 6.11c). According to Eq. (36.2), the dependence $\gamma(t)$ then is found from the integral equation

$$\sigma_0 = \int_0^t dt' G(t-t')(d\gamma(t')/dt'), \tag{36.8}$$

whose solution can be written in the form

$$\gamma(t) = \sigma_0 \int_{-i\delta-\infty}^{-i\delta+\infty} \frac{d\omega}{2\pi} \frac{e^{i\omega t}}{\omega G^*(\omega)}, \tag{36.9}$$

where the integration is taken within the complex plane and δ is an arbitrary positive constant. The behavior of the function $\gamma(t)$ [see Eq. (36.9)] for large values of t is determined by small values of ω. Taking into account the definition (36.7), for $\omega t \ll 1$ we obtain

$$G^*(\omega) = i\omega \int_0^\infty dt G(t)(1 - i\omega t + ...) = i\omega\eta + \omega^2 J\eta^2 + ..., \tag{36.10}$$

where

$$\eta \equiv \int_0^\infty dt G(t), \quad J\eta^2 \equiv \int_0^\infty dt G(t)t. \tag{36.11}$$

Substituting Eq. (36.10) into Eq. (36.9) and taking only the first two terms, we obtain

$$\gamma(t) = \sigma_0(t/\eta + J). \tag{36.12}$$

Hence, we obtain $\sigma_0 = \eta(d\gamma(t)/dt) = \eta\varkappa(t)$, that is, the constant η can be identified with the steady-state shear viscosity of the polymer melt. The quantity J is called a steady-state compliance. If the second term in Eq. (36.12) predominates under some conditions, then we have $\sigma_0 = (1/J)\gamma$, that is, a linear strain–stress dependence, that is typical for a conventional clastic body with elasticity modulus $E=1/J$ (Hooke's law).

The basic definitions and relations given here can be applied not only to polymer melts but to other substances as well. It has already been mentioned that the specifics of a polymer liquid consists in its pronounced viscoelasticity. This notion is illustrated in Figure 6.12, where the typical dependence $\gamma(t)$ is shown for the beginning of the flow of a polymer melt under the influence (starting at the moment $t=0$) of a constant shear stress σ_0 (cf. Fig. 6.11c). It is seen that $\gamma(t) \cong$ const in the interval $t_1 < t < t_2$ (i.e., the melt behaves as an elastic body). The melt starts flowing only for $t > t_2$. It therefore is natural to assume that under a harmonic external influence with $\omega t_2 > 1$, the response of the melt is similar to that of an elastic solid [i.e., the storage modulus exceeds substantially the loss modulus, cf. Eq. (36.10)], while for $\omega t_2 < 1$, the response corresponds to that of a viscous liquid (the loss modulus prevails). The described behavior is called *viscoelasticity*.

From the discussion of this subsection, it follows that viscoelastic properties of the polymer melt for different types of deformation are specified entirely by the relaxation modulus $G(t)$ of elasticity. Let us now calculate $G(t)$ on the basis of microscopic molecular considerations.

36.2. The relaxation modulus of elasticity for melts of relatively short chains can be calculated within the scope of the Rouse model; for long chain melts, the reptation concept is applied.

We showed in Sec. 35 that reptation is the basic type of motion of polymer chains in the melt only for $N > N_e$. If $N < N_e$, then the topologic constraints are inessential (i.e., one can disregard effects connected with the non-phantom nature of the chain). On the other hand, the hydrodynamic interactions are screened as before (i.e., the arguments of subsection 34.2 remain valid). The same can be said about volume interactions (*see* subsection 24.2), so the motion dynamics of chains in the melt for $N < N_e$ can be described within the scope of the Rouse model. Because the parameter N_e can reach a few hundred (*see* subsection 35.2), the condition $N < N_e$ is quite compatible with the assumption about long polymer chains ($N \gg 1$).

On the other hand, when $N > N_e$, topologic constraints strongly affect the properties of the melt. An abrupt change of properties is indeed observed in experiments at $N \sim N_e$. Figure 6.13 shows a typical dependence of the steady-state shear viscosity η of polymer melts on log N. For $N < N_e$, we obtain $\eta \sim N$, whereas for $N > N_e$, this dependence becomes substantially stronger: $\eta \sim N^{3.4}$. Thus, it is clear that the analysis of the viscoelastic properties of polymer melts for $N > N_e$ calls for an account of topologic constraints and the application of the reptation concept.

*36.3. Stresses in the melt of Rouse macromolecules are determined by orientations of the chains under the influence of an external force.

First, we calculate the relaxation modulus $G(t)$ of elasticity for a melt of macromolecules with $N < N_e$, when each chain of the melt can be considered to

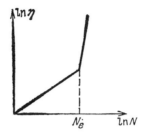

FIGURE 6.13. Typical dependence $\eta(N)$ for a polymer melt on a double logarithmic scale.

be within the scope of the Rouse model. Subsection 36.1 showed that $G(t)$ can be determined by examining the case when the melt is subjected to a stepwise shear strain (Fig. 6.11a). Then, the function $G(t)$ will equal σ_{xy}/γ_0 [see Eq. (36.4)], that is, be determined by the stress relaxation.

To calculate $G(t)$, it thus is necessary to know a molecular expression for the stress tensor in the non-equilibrium deformed state of the polymer melt. Because in the Rouse model no other effective forces between the links (except the interactions between neighbors along the chain) are taken into account, it is clear that only these interactions cause stresses in the melt of Rouse macromolecules. This circumstance is illustrated in Figure 6.14.

It should be recalled that the stress tensor $\sigma_{\alpha\beta}$ is a α component of the force per unit area perpendicular to the axis β (α, $\beta = x, y, z$). One can see from Figure 6.14 that in the case of a melt of N-link Rouse chains with c links per unit volume, one can write

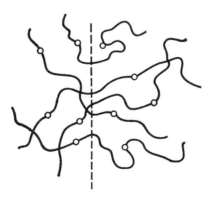

FIGURE 6.14. Force (acting on the plane shown by the *broken line*) is summed from the bond tensions (of the threads between the beads). In a deformed melt, this force is not necessarily directed along the direction perpendicular to the plane.

$$\sigma_{\alpha\beta} = \frac{c}{N} \sum_{n=1}^{N-1} \langle f_{n\alpha}(x_{n+1,\beta} - x_{n\beta}) \rangle, \tag{36.13}$$

where $f_{n\alpha}$ is the force with which the $(n+1)$-th link acts on the n-th one. The formula (36.13) is obtained most easily by replacing every bond between two neighboring links with a spring stretched by the force f_n and calculating the number of such springs crossing the unit area oriented perpendicular to the axis β. Taking into account that according to Eq. (31.3)

$$f_{n\alpha} = \frac{3T}{a^2} (x_{n+1,\alpha} - x_{n\alpha}), \tag{36.14}$$

and passing to the continuum limit (31.13), we obtain

$$\sigma_{\alpha\beta} = \frac{c}{N} \frac{3T}{a^2} \int_0^N \left\langle \frac{\partial x_{n\alpha}}{\partial n} \frac{\partial x_{n\beta}}{\partial n} \right\rangle dn. \tag{36.15}$$

As the vector $\partial x / \partial n = x_{n+1} - x_n$ assigns the orientation of the Rouse chain at a given point, Eq. (36.15) can be interpreted as an indication that the stress tensor is determined by the average orientation of the chains that inevitably appear in the non-equilibrium deformed state.

*36.4. The steady-state shear viscosity and steady-state compliance of a melt of Rouse macromolecules are proportional to the number N of links in the chain.

The exact solution for the Rouse model was obtained in Sec. 31. Using this solution, we can also calculate the integral (36.15). Let us introduce the Rouse coordinates $y_p(t)$ [see Eq. (31.20)]. Then, the expression (36.15) can be rewritten in the form

$$\sigma_{\alpha\beta} = \frac{c}{N} \frac{3T}{a^2} \sum_{p=1}^{\infty} \sum_{q=1}^{\infty} 4 \frac{\pi p}{N} \frac{\pi q}{N} \langle y_{p\alpha}(t) y_{q\beta}(t) \rangle \int_0^N dn \sin \frac{\pi p n}{N} \sin \frac{\pi q n}{N}$$

$$= \frac{3cT}{Na^2} \sum_{p=1}^{\infty} \frac{2\pi^2 p^2}{N} \langle y_{p\alpha}(t) y_{p\beta}(t) \rangle. \tag{36.16}$$

Before calculating $\langle y_{p\alpha}(t) y_{p\beta}(t) \rangle$, it should be noted that the problem analyzed here differs somewhat from that discussed in Sec. 31, because we now examine the dynamics of macromolecules not in immobile liquid but in shear flow (33.22). This is why the friction force acting on the n-th link is given not by Eq. (31.50) but rather by the following expression:

$$f_n^{\text{fr}} = -\mu(\partial x_n / \partial t - v(r, t)) \tag{36.17}$$

where the function $v(r, t)$ is defined by the equalities (33.19). As a result, the equation for the x component of the vector y_p is written in the form [cf. Eq. (31.22)]

$$\mu_p \frac{\partial y_{px}(t)}{\partial t} = -\frac{6\pi^2 T p^2}{Na^2} y_{px}(t) + f_{px}(t) + \mu_p \varkappa(t) y_{py}(t). \qquad (36.18)$$

As to the equation for the y component, it has the usual form [cf. Eq. (31.22)]

$$\mu_p \frac{\partial y_{py}(t)}{\partial t} = -\frac{6\pi^2 T p^2}{Na^2} y_{py}(t) + f_{py}(t). \qquad (36.19)$$

Now, we multiply Eq. (36.18) by $y_{py}(t)$ and Eq. (36.19) by $y_{px}(t)$, sum the equalities obtained, and then average the result to obtain

$$\mu_p \frac{\partial}{\partial t} \langle y_{px} y_{py} \rangle = -\frac{12\pi^2 T p^2}{Na^2} \langle y_{px} y_{py} \rangle + \mu_p \varkappa(t) \langle y_{py}^2 \rangle, \qquad (36.20)$$

where we used $\langle y_{p\alpha} f_{p\beta} \rangle = 0$ for $\alpha \neq \beta$ [see Eq. (31.25)]. For small values of $\varkappa(t)$ the quantity $\langle y_{py}^2 \rangle$ in the last term of equality (36.20) can be replaced in the first approximation by the quantity $\langle y_{py}^2 \rangle_0$, where the subindex $\langle\langle 0 \rangle\rangle$ denotes averaging in an immobile liquid (i.e., at $\varkappa = 0$). From Eq. (31.27), we have $\langle y_{py}^2 \rangle_0 = Na^2/(6\pi^2 p^2)$ $(p \neq 0)$. Consequently, Eq. (36.20), with account taken for Eq. (31.23), acquires the final form

$$\frac{\partial}{\partial t} \langle y_{px} y_{py} \rangle = -\frac{6\pi^2 T p^2}{N^2 \mu a^2} \langle y_{px} y_{py} \rangle + \varkappa(t) \frac{Na^2}{6\pi^2 p^2}, \quad p \neq 0. \qquad (36.21)$$

Equation (36.21) can be solved by the same method used in subsection 31.5, where the solution of Eq. (31.22) was derived. As a result, we obtain

$$\langle y_{px} y_{py} \rangle = \frac{Na^2}{6\pi^2 p^2} \int_{-\infty}^{t} dt' \varkappa(t') \exp\left[-\frac{2(t-t')}{\tau_p}\right], \quad p \neq 0, \qquad (36.22)$$

where the quantities τ_p are defined by Eq. (31.26). Substituting Eq. (36.22) into Eq. (36.16), we find the components of the stress tensor σ_{xy} for shear flow (33.19) of a melt of Rouse chains

$$\sigma_{xy} = \frac{cT}{N} \sum_{p=1}^{\infty} \int_{-\infty}^{t} dt' \varkappa(t') \exp\left[-\frac{2(t-t')}{\tau_p}\right]. \qquad (36.23)$$

Taking into account the definition (33.20) of the relaxation modulus of elasticity, we conclude that

$$G(t) = \frac{cT}{N} \sum_{p=1}^{\infty} \exp\left(-\frac{2t}{\tau_p}\right) = \frac{cT}{N} \sum_{p=1}^{\infty} \exp\left(-\frac{2tp^2}{\tau_1}\right), \qquad (36.24)$$

where τ_1 is the maximum relaxation time of the Rouse coil, defined by Eq. (31.32).

The formula (36.24) defines the function $G(t)$ for the melt of macromolecules in the Rouse model. Hence, using the results of subsection 36.1, we can obtain the viscoelastic characteristics of such a melt. Taking into account Eq. (36.11), for the steady-state shear viscosity η and steady-state compliance J we have

$$\eta = \int_0^\infty dt\, G(t) = \frac{cT}{N} \frac{\tau_1}{2} \sum_{p=1}^\infty p^{-2} = \frac{cT}{N} \frac{\pi^2 \tau_1}{12} = \frac{c\mu}{36} Na^2, \qquad (36.25)$$

$$J = \frac{1}{\eta^2} \int_0^\infty dt\, G(t)t = \frac{N}{cT} \left(\sum_{p=1}^\infty p^{-4} \right) \left(\sum_{p=1}^\infty p^{-2} \right)^{-2} = \frac{2N}{5cT}. \qquad (36.26)$$

Both quantities are seen to be proportional to N. This circumstance is confirmed experimentally for melts of macromolecules with $N < N_e$.

According to Eq. (36.4), the function $G(t)$ [see Eq. (36.24)] describes the stress relaxation for a stepwise shear strain of a melt of Rouse chains; this function is shown in Figure 6.15. For $t \ll \tau_1$, the sum (36.24) can be transformed into an integral. Consequently, we obtain

$$G(t) \cong \frac{cT}{N} \int_0^\infty \exp\left(-\frac{2tp^2}{\tau_1} \right) dp = \frac{cT}{N} \frac{1}{2^{3/2}} \left(\frac{\tau_1}{t} \right)^{1/2}, \quad t \ll \tau_1. \qquad (36.27)$$

If $t \gg \tau_1$, however, then the term with $p=1$ predominates in the sum (36.24), and

$$G(t) \cong \frac{cT}{N} \exp\left(-\frac{2t}{\tau_1} \right), \quad t \gg \tau_1. \qquad (36.28)$$

The function $G(t)$ is seen to diminish with t in both limits, and there is no time interval in which $G(t) \approx \text{const}$. It therefore can be presumed that there is no frequency interval of external action in which the melt of Rouse chains behaves as an elastic body, that is, such a melt does not possess any pronounced viscoelastic properties in the sense described at the end of subsection 36.1 (Fig. 6.15).

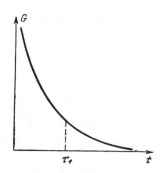

FIGURE 6.15. Dependence $G(t)$ for a melt of Rouse chains.

*36.5. *Melts of long polymer chains possess pronounced viscoelastic properties caused by reptation-type thermal motion of the macromolecules; in terms of the reptation model, the viscosity of such melts is proportional to N^3 and the steady-state compliance independent of chain length.*

Now let us calculate the relaxation modulus $G(t)$ of elasticity for melts of long polymer chains $(N > N_e)$ when the motion of macromolecules is basically accomplished through reptation. Suppose the melt is subjected to a stepwise shear strain as shown in Figure 6.11a; according to Eq. (36.4), the function $G(t)$ then will describe the stress relaxation.

In subsection 35.6, we obtained the basic characteristic times associated with diverse motions in a melt of entangled polymer coils. The shortest, τ_A, is the Rouse relaxation time for a chain of N_e links. For $t < \tau_A$, the topologic constraints do not affect the chain motion. It therefore is natural to assume that the function $G(t)$ would coincide for $t < \tau_A$ with that for a melt of Rouse chains. Because $\tau_A \ll \tau_1$, we can use Eq. (36.27) and write in the general case

$$G(t) = (cT/N)(\tau_1/t)^{1/2}2^{-3/2}, \quad t \ll \tau_A. \tag{36.29}$$

For $t > \tau_A$ the topological constraints are essential, and the calculation of $G(t)$ becomes more complicated. As soon as we deal with linear viscoelasicity, however, $G(t)$ can be derived from the following considerations. At short times $(t \sim \tau_A)$, the chains of the melt are located within the initial tubes exposed to strain (Fig. 6.16). Then, in the process of reptating motion of the macromolecules, new, "relaxed" tube sections emerge. It is essential that the residual orientation because of the initial strain imposed on the melt only remains in the segments of the initial tube, because the motion of the chain ends creeping out of the tube is quite random. The stress in the melt is caused only by that residual orientation (cf. subsection 36.3); therefore, this stress can readily be presumed

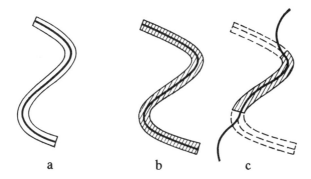

FIGURE 6.16. (a), Initial equilibrium tube before deformation. (b), Deformed tube immediately after deformation. (c), After time interval t since deformation. The deformed section remaining from the initial tube is shaded; tube sections formed afresh are in equilibrium (they lose memory of the initial deformation).

to be proportional to the fraction $\psi(t)$ of links remaining in the initial tubes at moment t, that is, that the relaxation modulus of elasticity takes the following form for $t > \tau_A$:

$$G(t) = G_N^{(0)}\psi(t), \tag{36.30}$$

where G_N^0 is a certain constant that will be defined later; the function $\psi(t)$ was calculated in subsection 35.4 to be equal to $\langle\beta(t)\rangle/L$. According to Eqs. (35.6) and (35.14), we obtain

$$\psi(t) = \sum_{p=1,3,5,\ldots} (8/\pi^2 p^2)\exp(-p^2 t/\tau^*). \tag{36.31}$$

The constant G_N^0 is derived from the condition that the relation (36.30) must gradually pass into the formula (36.29) at $t \sim \tau_A$. For $t \sim \tau_A \ll \tau^*$, $\psi(t) = 1$, and consequently,

$$G_N^{(0)} \sim (cT/N)(\tau_1/\tau_A)^{1/2} \sim cT/N_e, \tag{36.32}$$

where the latter equality is written with allowance made for Eqs. (31.32) and (35.32). Eqs (36.29) to (36.32) totally define the function $G(t)$ for all values of t. A typical plot of $G(t)$ is shown in Figure 6.17.

The relaxation modulus of elasticity is virtually constant within a broad interval of values of t ($\tau_A < t < \tau^*$; $\tau_A \ll \tau^*$, the plateau region). As $\tau^* \sim N^3$ and τ_A is independent of N, the plateau region broadens as the chain length grows, spanning for $N \gg N_e$ several orders of magnitude of t. Throughout this region, $G(t) \approx$ const, that is, a sample of the melt behaves as an elastic body [according to Eq. (36.2), the stress is proportional to the strain provided that $G(t) =$ const]. The height of the plateau [i.e., the quantity G_N^0 is independent of N, in accordance with Eq. (36.32)]. At the same time, $G_N^0 \sim N_e^{-1}$; this points to the fact that the elasticity modulus depends on topologic constraints or entanglements, whose number in a unit volume is proportional to N_e^{-1} (see subsection 35.2).

For $t > \tau^*$, $G(t)$ begins diminishing, and the flow becomes inhibited (i.e., a viscous response appears in the melt). Thus, one can conclude that polymer melts with $N \gg N_e$ possess pronounced viscoelastic properties (see subsection

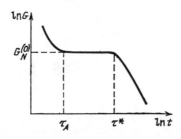

FIGURE 6.17. Dependence $G(t)$ for a melt of polymer chains in a reptation model.

36.1). This is confirmed by Fourier analysis of the function $G(t)$. Calculating the complex modulus of elasticity (36.7), we obtain the storage modulus $G'(\omega)$ and loss modulus $G''(\omega)$ for $\omega\tau_A \gg 1$:

$$G'(\omega) = G''(\omega) = G_N^{(0)}(\pi\omega\tau_A/2)^{1/2}; \tag{36.33}$$

and for $\omega\tau_A \ll 1$:

$$G'(\omega) = G_N^{(0)} \sum_{p=1,3,5,\ldots} \frac{8}{\pi^2 p^2} \frac{(\omega\tau^*)^2}{p^4 + (\omega\tau^*)^2},$$

$$G''(\omega) = G_N^{(0)} \sum_{p=1,3,5,\ldots} \frac{8}{\pi^2} \frac{\omega\tau^*}{p^4 + (\omega\tau^*)^2}. \tag{36.34}$$

The dependences $G'(\omega)$ and $G''(\omega)$ described by Eqs. (36.33) and (36.34) are illustrated in Figure 6.18. It is seen that $G'(\omega) \gg G''(\omega)$ for $1/\tau^* \ll \omega \ll 1/\tau_A$ (i.e., the polymer melt behaves as an elastic body in this interval of frequencies of external action). If $\omega \ll 1/\tau^*$, however, then $G''(\omega) \gg G'(\omega)$ (i.e., the response associated with viscous flow of the melt prevails).

The steady-state shear viscosity and steady-state compliance of the melt can be calculated using formulas (36.11). Because the contribution to the integrals (36.11) in the region $t \ll \tau_A$ is vanishingly small for $N \gg N_e$, we obtain

$$\eta = G_N^{(0)} \int_0^\infty \psi(t)\,dt = (\pi^2/12) G_N^{(0)} \tau^* \sim ca^2 \mu N^3 N_e^{-2}, \tag{36.35}$$

$$J = (G_N^{(0)})^{-1} \left[\int_0^\infty t\psi(t)\,dt \right] \left[\int_0^\infty \psi(t)\,dt \right]^{-2} = 6/(5G_N^{(0)}) \sim N_e/(cT), \tag{36.36}$$

where we used the estimates (35.4) and (36.32). Thus, in accordance with the theory based on the concept of reptation in polymer melt, the quantity J is independent of the number N of links in the chain with $N \gg N_e$, and the dependence $\eta(N)$ obeys the law $\eta \sim N^3$.

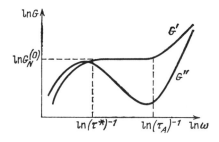

FIGURE 6.18. Dependences $G'(\omega)$ and $G''(\omega)$ for a melt of polymer chains in a reptation model.

36.6. *The dependence* $\eta \sim N^{3.4}$ *is observed experimentally for the viscosity of the melt; deviations from predictions of the reptation model can be associated with fluctuations of the contour length of the tube and the tube renewal process.*

For real polymer melts with $N \gg N_e$, the value of J proves to be independent of N with high accuracy. The formula (36.36) [or (36.32)] can be used for the experimental determination of the parameter N_e. On the basis of such measurements, one can conclude that in melts, the value of N_e falls within the interval 50 to 500 (*see* subsection 35.2).

As a rule, experimental data do not agree well with the predicted dependence $\eta \sim N^3$. For most real melts, $\eta \sim N^{3.4}$. This is not surprising, because according to Eq. (36.35), $\eta \sim G_N^{(0)} \tau^*$, with the quantity $G_N^{(0)}$ being independent of N and the experimental dependence $\tau^*(N)$ taking the form $\tau^* \sim N^{3.4}$ (*see* subsection 35.7). Consequently, experimental deviations of the dependence $\eta(N)$ from the predictions of the reptation model are determined by similar deviations for the dependence $\tau^*(N)$. (A possible physical reason of these deviations was discussed in subsection 35.7.)

CHAPTER 7

Biopolymers

To a large extent, the main reason for interest in macromolecular physics probably stems from the fact that the physical properties of polymers underlie many secrets of animate nature. An ambition to advance toward an understanding of the molecular foundations of biology stimulates researchers in the area of polymer physics and learning the basic laws of polymer science is mandatory for majors in biophysics or molecular biology.

37. BASIC PROPERTIES OF BIOLOGICAL POLYMERS

37.1. *Among many polymers of biologic origin, DNA's, RNA's, and proteins are the most remarkable.*

Indeed, macromolecules of various biopolymers (nucleic acids, proteins, polysaccharides, and others) play an important role in all biologic phenomena and processes. Many molecular-biologic phenomena are associated with quite ordinary properties and characteristics of polymers that were considered in previous chapters (chain structure, flexibility, volume interactions, topological constraints, and so on). Accordingly, substances like polysaccharides (cellulose, chitin, starch, and others) are studied in the conventional chemistry of high-molecular-weight compounds and in polymer physics together with comparable synthetic substances.

At the same time, in a number of biologic processes (which are definitely the most fundamental in animate nature), an important role belongs to certain specific features of the structure of biopolymer molecules themselves (DNA's, RNA's, and proteins). The analysis of these distinguishing features belongs to a marginal area between biophysics and the physics of polymers; this chapter briefly outlines these features.

37.2. *Referring to the structure of a polymer, one should discriminate among the primary, secondary, and tertiary structures.*

The main characteristic of biomacromolecules consists in the biologic functions that they perform. One can regard a protein or DNA chain not only as a molecule of some substance, but also as a quaint machine, or automaton, executing certain operations.[6,7,42] From the physical viewpoint, biopolymer molecules must possess a very uncommon hierarchic structure to function properly. First, it is well-known that each chain of the biopolymer must have a definite sequence of links of different types; this sequence is formed during biological synthesis and is called the *primary structure of the biopolymer*. Second, there should exist an opportunity to form some sort of short-range order in the spatial arrangement of chain elements because of the interaction of nearby links in the chain. Usually, the short-range order in biopolymers manifests in the form of helic turns (Fig. 7.1a) or small pins (Fig. 7.1b). These elements of short-range order are called the *secondary structure*. Third, a biopolymer chain as a whole must possess a more or less defined spatial or tertiary structure (e.g., a coil or globular one), and this defines the long-range order (or disorder) in the arrangement of links.

A principal difference in the secondary and the tertiary structures of the biopolymer should be noted. Secondary structures are governed by interactions of links located near one another along the chain. The secondary structural elements, turns of helices and hair-pins, are small with respect to the whole chain; thus, formation of the secondary structure is not connected with volume interactions. Conversely, tertiary structures are determined by volume interactions, that is, the interactions between links located far from one another along the chain. The tertiary structural elements involve the whole chain, and the appearance of a tertiary structure differing from a Gaussian coil is necessarily connected with the non-ideality of a polymer.

Because of volume interactions in a globule, some elements of short-range order can be stabilized that usually are not stable in a coil chain without volume interactions. This situation is typical, for example, for proteins: in the analysis of the link distribution with respect to rotational-isomeric states in globular proteins, an extremely high abundance of a certain left-hand helic bend becomes evident, which is scarce in the coil state. The question of whether we should call this a secondary structure is, generally speaking, immaterial, being a matter of terminology, but we reply to this question in the negative, which means that in accordance with our terminology, the new elements of short-range order do appear during the formation of tertiary structure. The set of types of the secondary structure, however, does not change.

Sometimes it is possible to speak of the *quaternary structure of biopolymers*. This term has no accurate or generally accepted meaning and is most often used for the description of either some type of tertiary structure (when one chain forms several globules interconnected by short bridges) or a structure of the complexes of several chains.

This chapter considers in turn all hierarchic levels of structure of biopolymers; in so doing, the different biopolymers (i.e., DNA's, RNA's, and proteins) are studied in parallel. Such an approach, which is worthwhile from the physical viewpoint, should not obliterate the principal difference in the biologic functions of DNA's, RNA's, and proteins. Recall that DNA plays an instructive role; the DNA of each cell specifies the set and type of proteins being synthesized in that cell. As for the proteins, they carry out the executive functions by performing

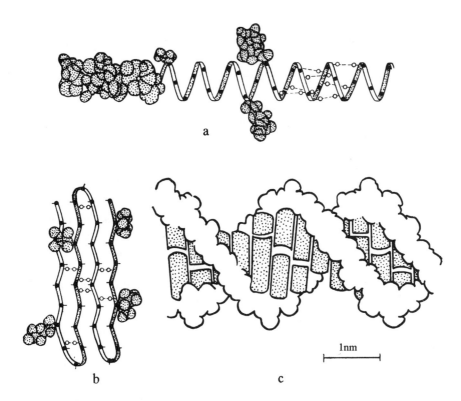

1nm

FIGURE 7.1. A sketch of secondary structures of biopolymers. (a), α Helix of protein. Helic conformation of main polypeptide chain is stabilized with hydrogen bonds directed along the axis of the helix (*right*). With atoms of the main chain circumscribed with van der Waals radii, the α helix looks like a solid rod with a diameter of approximately 0.6 nm (*left*). Side groups of various amino acid residues are linked by covalent bonds to α carbon atoms of the main chain (in black); for example, three of these groups are shown in the middle part of the figure. On average, one turn of the helix comprises 3.6 links, projection of the residue on the axis of the helix equals 0.15 nm, and the helic pitch is 0.54 nm. (b), β Structure of protein. Folded conformation of the main polypeptide chain is stabilized by hydrogen bonds between the neighboring "pins." With atoms of the main chain circumscribed with van der Waals radii, the structure looks like a solid layer with a thickness of approximately 0.5 nm. Side groups of amino acid residues are located on the upper and lower sides of the layer. The distance between the pins in the transverse direction equals 0.46 nm. (c), Double helix of DNA in right-hand *B* form. Two main sugar-phosphate strands are wound on the outside, and the nitrogen bases linked to them by covalent bonds are located inside. The bases have a flat shape, are oriented perpendicular to the axis of the helix, and form complementary hydrogen-bonded pairs. There are 10 base pairs per turn of the helix, the projection of the pair on the axis of the helix equals 0.34 nm, and the pitch of the helix equals 3.4 nm.

specific operations (catalytic, photo- and electrochemical, mechanochemical, and so on). Finally, the functions of RNA's are associated with DNA-controlled protein synthesis. This difference in the functions of the various biopolymers is connected with definite distinction in structures, which are noted and stressed in the appropriate context.

37.3. *DNA's, RNA's, and proteins are heteropolymers, the sequence of link types (i.e., the primary structure of each biopolymer chain) is strictly fixed and is similar to a sequence of letters in the meaningful text describing the function of the given chain, and the "texts" of DNA's and proteins are interrelated by the genetic code.*

Regardless of the variety of executive functions, all proteins are uniform in chemical terms. A molecule of any protein represents a uniform basic chain composed of identical peptide groups; a side radical connected to each link forms an amino acid residue together with the peptide group. In real proteins, there may be 20 different amino acids, and the proteins differ from one another by the set of amino acids and the order of their distribution along the chain (i.e., the primary structure).

Just as uniform in their chemical nature are all DNA's. These are composed of a uniform sugar-phosphate backbone chain, with each saccharide unit carrying the side groups, nucleotides (nitrogen bases) of four types. There are two pyrimidine bases, cytosine (C) and thymine (T), and two purine ones, adenine (A) and guanine (G). RNA chains are similarly built, with uracil (U) substituted for thymine in side groups and different types of saccharide units featured in the backbone chain.

The specifics of the kinetic properties of biopolymers is such that they can be described in the most natural fashion by the cybernetic language adopted in molecular biology, similar to the one used in the theory of digital automats. In particular, the strictly fixed primary structure of the biopolymer is readily compared to the text written in the appropriate molecular code. For example, because the primary structure chemically assigns the distinctiveness of a protein and, specifically, its executive function, the analogy can be continued and the protein text said to describe (or code) the protein function. At the same time, the processes of protein function are only indirectly associated with the text, being directly determined by the secondary and the tertiary structures. Consequently, the secondary and tertiary structures formed by a specific protein chain because of linear and volume interactions between the links should depend strongly on the primary structure.

The function of DNA consists in defining both the types and the number of proteins that are synthesized by a living cell, and the primary structure of a DNA directly encodes the primary structures of proteins by means of the so-called genetic code. Although the biological processes of protein synthesis are three-dimensional and appreciably associated with volume interactions, the genetic code itself is essentially linear: each consecutive triad of nucleotides in the DNA chain defines in a unique way a subsequent amino acid in the protein chain.

Hence, the primary structure of every biopolymer is fixed during its synthesis, while the secondary and the tertiary structures and, consequently, the function depend on link interactions, primarily by their charges and solubility in water. Let us now consider these factors.

37.4. DNA's and RNA's are polyelectrolytes, with each link carrying an elementary negative charge; proteins are polyampholytes (i.e., their links carry opposite charges); and the magnitude of the net charge depends on the conditions of the medium.

Each phosphate group of the basic chain of DNA or RNA dissociates in aqueous solution, and each carries a negative charge. Nucleic acids thus are strongly charged polyelectrolytes (polyanions).

More diverse are the polyelectrolytic properties of proteins. First, the terminal links of the basic chain are ionized in an aqueous solution. One terminal link is positively charged (a so-called C end), and the other carries a negative charge (a N end). Second (and most importantly) many amino acid residues can also be ionized. Some of them are acid (negatively charged) and some basic (positively charged). The values of the ionization energies lie in such a range that the charge magnitude of each residue and of the chain as a whole strongly depends on the medium, or more exactly, on the electrolytes present in it. (More detailed information on amino acids can be found in the fundamental monograph, Ref. 43). For a description of these effects, the following notation, adopted in chemistry, is used.

A change in the state of ionization of a molecule or atomic group is treated as a chemical reaction, the separation or attachment of a hydrogen ion (proton) H^+:

$$A \rightleftarrows H^+ + B. \qquad (37.1)$$

One may imply here that $B \equiv A^-$ (A is a dissociating acid) or $A \equiv B^+$ (B is a base). Either way, an equilibrium reaction constant can always be defined as

$$K = [H^+][B]/[A], \qquad (37.2)$$

where *square brackets* denote the concentration of the given substance. It is customary to use the designation

$$pK = -\lg K. \qquad (37.3)$$

It should be noted that the logarithm in the relation (37.3) is decimal. It can easily be shown that the value of the pK is equal up to a factor to the change in free energy during the reaction (37.1), that is, to the ionization potential. Accordingly, the quantity pK is sometimes referred to as the *ionization potential*.

Aside from reactions of the type (37.1), which change the charge state of amino acids, the H^+ ions also take part in other reactions (e.g., the dissociation of low-molecular-weight acids and bases present in the same solution). Therefore, both the magnitude and the sign of the charge of each link in the protein chain depend on the medium (specifically, on the chemical potential of the H^+ ions). Clearly, the state of ionization is determined by comparing this chemical potential with the ionization potential pK. Because the concentrations $[H^+]$ are normally low enough, the value of pH (which is proportional to the chemical potential for $[H^+] \to 0$) usually is used instead of the chemical potential:

$$pH = -\lg[H^+]. \qquad (37.4)$$

In pure water[a] $pH = 7$, in acid solutions $pH < 7$, and in basic solutions $pH > 7$.

In two possible cases, which are 1) when A is a neutral state of the investigated molecule or group and B a negatively charged state, and 2) when B is neutral and A positively charged, the mean fraction q of charges equals $[B]/([A]+[B])$ and $[A]/([A]+[B])$, respectively. According to Eqs. (37.1) to (37.4), one can write

$$q = [1 + 10^{\mp(pH-pK)}]^{-1}. \qquad (37.5)$$

As expected, the dependence of q for the molecule or group on the pH value of the medium (this dependence is also called the *titration curve*) is strongest when $pH \sim pK$. The acid [the upper sign in Eq. (37.5)] is substantially charged for $pH > pK$ and the base (the lower sign) for $pH < pK$.

The ionization potentials of various amino acid residues fall within the range $pH = 3-13$. Accordingly, the variation of pH in this range is accompanied by a gradual change in the total charge of the protein chain, starting from a positive value at low pH and moving to negative values at high pH. For each protein, there is a so-called isoelectric point $pH \equiv pI$, at which the total charge of the molecule equals zero.

It should be noted that the degree of charge of the protein molecule as a whole by no means must equal the sum of the expressions (37.5) for all amino acid residues of the given chain. Obviously, such an equality would be mandatory only for the ideal chain. For a real macromolecule (especially a globular one), the ionization potential of each group depends on the spatial surroundings of that group, so the degree of charge depends on the values and signs of the neighboring charges. Still, titration curves for real globular proteins are described (at least qualitatively) by summing the expressions (37.5). This is because the ionized groups in globular proteins are always located on the surface of the globule (*see* also Sec. 44).

37.5. *The quality of an aqueous solution for the uncharged links of biopolymers is determined by their polarity: approximately half of the amino acid residues are polar (i.e., hydrophilic and readily soluble in water) and the other half nonpolar (i.e., hydrophobic and poorly soluble in water).*

Specific features of biopolymers are associated with the fact that they are usually immersed in water, a highly polar liquid with a high dielectric permittivity of approximately 80. This is exactly why the links of nucleic acids dissociate, correspondingly, DNA and RNA molecules are readily dissolved in water.

The situation is more complicated in the case of proteins whose links dissociate only partially: 16 out of 20 of the amino acid residues are uncharged (in water with $pH = 7$). Clearly, their solubility in water is determined by the pres-

---

[a]Hereafter, the concentration is assumed to be measured in moles per liter.

ence or absence of a dipole moment. In fact, five residues are polar (hydrophilic), eight are nonpolar (hydrophobic), and the remaining seven occupy an intermediate place.

These circumstances drastically change the character of the volume interactions in proteins. In many cases, the thermodynamic gain because of the appearance of the aggregates of hydrophobic particles with a density sufficient for displacing water (which corresponds to the gaps between the particles ~ 0.3 nm) is a key factor in stabilizing the protein structure. In this range of densities n, the chemical potential $\mu^*(n)$ of the hydrophobic link depends weakly on n, being approximately equal to $\mu^*(n) \approx -(1\text{–}3)$ kcal/mole $\approx -(0.04\text{–}0.12)$ eV, depending on the specific sort of amino acid residue and external conditions.

This effect is frequently referred to as a *hydrophobic interaction*. The significant characteristic of hydrophobic interactions consists of their dependence on temperature. The introduction of nonpolar particles into water substantially restricts the freedom of fluctuations of the network formed by hydrogen bonds between water molecules, or in other words, diminishes the entropy of water. This entropy contribution to the hydrophobic effect is quite appreciable. Consequently, the hydrophobic interaction energy (or more exactly, the quantity $|\mu^*(n)|$) grows with temperature in many cases.

38. PRIMARY STRUCTURES OF BIOPOLYMERS

38.1. *The primary structure of any specific biopolymer can be determined technically; however, the main problem consists in a "linguistic" analysis of this "text."*

Experimental methods for determining the primary structures of biopolymers now have a high degree of efficiency; for proteins and DNA's, the measurements are even partially automated. Existing databases store the primary structures of several thousands of proteins and hundreds of DNA's, and these numbers are rapidly growing. Analysis of these data essentially constitutes a linguistic problem. For example, one can find the fractional content of various amino acids in a specific protein or of nucleotides in a DNA, just as one finds the frequency of appearance of various letters in a particular language. Specifically, the amino acid content varies drastically from one protein to another, and it differs appreciably even among proteins executing similar functions in different species. One can also introduce a cruder characteristic (much like the frequency of vowels or consonants) and clarify, for example, that the fraction of hydrophobic residues in water-soluble proteins typically is close to 0.5 while that in insoluble proteins is higher. Finally, one can study binary (or even more complicated) correlation functions of primary structures, for example, the quantity $\langle c_\alpha(t+\tau)c_\beta(t)\rangle$, where t is the coordinate (the number of a link) along the chain and $c_\alpha(t)$ the fraction of links of type α near the point t; the averaging is taken over t. The analogous investigation of the correlators of a conventional text, being much more informative than the mere determination of the frequency of letters, allows one, for example, to distinguish verses from prose, to determine the metre of a verse, and so on.

In biopolymeric texts, some interesting correlation characteristics can also be found. For example, in DNA's, there exist "palindrome"[b] sections with frequencies much higher than a random noncorrelated sequence might have. Such palindrome sections may include many tens of nucleotides and play an essential role in the physical properties of DNA's (*see* subsection 42.10).

The obvious major inadequacy of a purely linguistic approach to the analysis of biopolymers is that it reflects only the ideal-chain properties and makes no allowance for the effects of volume interactions. The investigation of the various characteristics of the secondary and tertiary structures of biopolymers, however, generally speaking calls for the knowledge of various properties of the texts themselves.

38.2. *Current views on biological evolution include a gradual change in the set of primary structures of biopolymers as one of its most significant elements.*

The idea of evolution holds a central place in biology, and a meaningful description of physical properties of biopolymers is unthinkable without evolutionary concepts. It should immediately be emphasized that the molecular (i.e., basically physical) interpretation of biological evolution and the origin of life is immensely complicated. Many relevant problems have not yet been solved, and many the others have not even been formulated. Still, our further discourse requires that we mention at least a few facts and views in this area.

A comparison of the primary structures of proteins that execute identical functions in different organisms leads to the following interesting conclusions. First, the difference in primary structures is generally greater for organisms that are farther separated in their evolutionary development; regarding the proteins responsible for "newer" (in evolutionary terms) functions, the differences are greater than for proteins executing "old" functions. Second, even within one species, there are individual differences in the primary structure of proteins; some of these differences are essentially imperceptible in functioning while some others are not.

These facts underlie the contemporary conception about the molecular grounds for evolution. Mutations transmitted from one generation to another are the changes[c] of the primary structure of DNA; in accordance with the genetic code of the transformed DNA, the biological synthesis of new proteins occurs. It is assumed that this process may produce proteins that execute their functions in a more efficient way or even proteins that are capable of performing new functions. At this stage, feedback mechanisms should come into effect (which Darwin called *natural selection*): the host organisms of the improved protein that possess the selection privileges transmit the improvement encoded in their DNA to descendants leading thus to its imprinting.

---

[b]The palindrome is a symmetric portion of the text, which reads the same backward or forward.
[c]Definitely, these may not only be the actual changes of individual nucleotides but of all kinds of permutations, block substitutions, and so on.

38.3. *The principal difference in the physical properties of various biopolymers is essential for evolution: the secondary structure of DNA (a double helix) depends relatively weakly on the link sequence, and the secondary and tertiary structures of proteins depend very strongly on the primary structure.*

Of course, the simple map of evolution sketchily drawn here has vast holes. To begin with, the following points are uncertain. First, what is the statistics of mutations (i.e., what are the relevant correlations in time, space, along the chains, and so on)? Second, how does the feedback operate (i.e., how do only improving mutations survive)? The answers to these questions are not known, and the questions themselves lie outside the scope of this book anyway. However, two circumstances are clear and definitive for the physics of biopolymers:

1. The main motive of the secondary structure of DNA, the double helix, is essentially independent of the primary structure, which makes mutations possible. Otherwise, evolution would produce DNA with the lowest energy and not the one encoding the "best" proteins.

2. The distribution of diverse secondary structures along a protein chain and the tertiary structure of the protein is entirely determined by the link sequence, which is necessary to maintain evolutionary feedback.

It should be noted, however, that the statement about the invariance of the DNA secondary structure with respect to a change in the primary one is not precise. Under certain conditions, so-called non-canonic structures may also appear at some (occasional) sections of DNA. On the whole, however, the statement holds true over the major portion of DNA.

38.4. *The number of possible primary structures grows exponentially with the chain length; this likens biopolymers to disordered systems and plays a key role in current ideas about the prebiologic stage of evolution.*

When we talk about linguistic analysis, we meant studies of already existing primary structures. The evolutionary viewpoint makes one raise the question of their origin. The discussion of that issue allows one to formulate the problems of the physics of biopolymers in the most correct form and to define the status of this discipline in the physics of the condensed state.

At present, spontaneous synthesis of fairly complex organic compounds (including amino acids and nucleotides) is well known to be possible in the ancient abiotic environment. Both experimental and other authentic information on the subsequent stages of development is quite scarce, but irrespective of what those stages were, they must have resulted in the formation of polymer chains.[d] Inevitably, the following problem then arises.

The total number of different sequences of N links obviously equals $k^N = \exp(N \ln k)$, where k is the number of different types of links ($k = 20$ for

---

[d]There are various speculations about which polymers appeared first (RNA's, DNA's, or proteins) and how the genetic code evolved.

proteins, and $k=4$ for nucleic acids). The dependence on k is insignificant here, whereas the exponential growth with N is of cardinal importance. For example, the number k^N proves to be incredibly large in typical proteins with $N \sim 100$, exceeding, for example, the number of electrons in the universe. This is why the chemical synthesis of heteropolymers as a rule yields a blend of chains with a random assortment of different primary structures.

This situation is quite analogous to that in disordered systems such as substitution alloys, glasses, and so on. For instance, the number of permutations of impurity atoms in a crystal is extremely great (i.e., always incomparably greater than the number of available samples). Therefore, the individual structure of any finite sample is always unique and random. The system "remembers" the structure that it had at the moment of preparation. Nevertheless, if a disordered crystal is melted and then cooled, its initial structure (i.e., the arrangement of impurities) will not be recovered (i.e., the structure will be forgotten). This kind of memory defines the peculiar physical properties of disordered systems, their energy spectra, kinetic coefficients, and so on. (More detailed information on the physics of disordered systems can be found in Refs. 44 and 45.)

Similarly, the destruction of a heteropolymer chain with a subsequent rejoining of the same links (outside biological synthesis) fails to reproduce their initial sequence. The chain remembers its primary structure only during its lifetime. Such memory is in fact a manifestation of the linear memory, which is quite common for macromolecules but plays a special role for heteropolymers.

Of course, the existence of linear memory is by no means sufficient for understanding prebiological evolution. Some hypotheses consider systems whose chains may not only be synthesized but dissociate; the chains are assumed to possess certain catalytic and autocatalytic properties. In such a system, a "leader" can generally appear, representing one or several interrelated primary structures that by virtue of being synthesized faster than others leaves no "food" (monomers) for the competitors.[e] On the other hand, stability of such a leader is difficult to maintain: the enormous variety of primary structures must contain many potential leaders. The widely adopted hypothesis presumes that the system can somehow remember and fix kinetically a randomly chosen leader. Even though it is unclear how this can happen, we encounter here (as in any other evolutionary phenomenon) the memorizing of a random choice.

Meanwhile, the specifics of biological systems from the physical standpoint generally consists in the kinetic fixing or memorizing of certain structural features. It is this fixing that likens the structures of biologic systems with the construction design of artificial machines and automats and that allows one to speak about the performance of specific functions by them. As I. M. Lifshitz showed for the first time in 1968, this suggests that the macromolecules of biopolymers that should be regarded as the first stage in the hierarchy of biologic structures: because of linear memory, the physical properties of macromolecules,

---

[e] The role of the leader may be played by a set of primary structures of mutually catalyzing polymers of various types (e.g., RNA's and polypeptides) (Ref. 50).

even if not possessing what can be called the biological specifics, are nevertheless closely associated to it.[24] This is exactly why the hypotheses about prebiologic evolution deal with the primary structures of macromolecules as the subject of research.

38.5. *Depending on the mode in which the primary structure is specified, two formulations of the problems in the physics of biopolymers are possible.*

Summarizing this section, one can draw the following conclusions. The physics of biopolymers deals with chains whose primary structure remains fixed during thermal motion. The problem of describing any phenomenon in such a system (helix formation, globule–coil transition, and so on) can be formulated in two ways.

First, one may look for an algorithm to find the characteristics of an investigated phenomenon for each specific and completely known primary structure. Such a formulation is intended for research of existing biopolymers. For example, we may study the melting curve (*see* Sec. 41) of a definite section of DNA with known primary structure or predict secondary and tertiary structures of a globular protein (*see* Sec. 44) having known its primary structure. The picture of the investigated phenomenon frequently does not depend on all of the details of the primary structure; and it is then relevant to resort to linguistic analysis to find the necessary rough characteristics of the text.

Second, one may try to characterize the phenomenon by some probability quantities (variances, means, and so on), assuming the primary structure to be randomly formed according to some statistical distribution. Such a formulation is oriented toward evolutionary problems.

It should be noted that the primary structures of real biopolymers are quite complex (i.e., they cannot be reduced to repetitions or modifications of a certain segment). Generally, the complexity of a sequence depends on the length of the minimal algorithm allowing for total restoration of the given sequence. Hence, the most complex sequence is basically random and uncorrelated, because in this case, the minimal algorithm coincides with the sequence itself. Consequently, even in studies of existing biopolymers with no evolutionary problems in mind, the random primary structure is frequently regarded as a fairly adequate model for real complex primary structures.

39. SECONDARY STRUCTURES OF BIOPOLYMERS

39.1. *Secondary structures of protein chains (α-helices and β-sheets) are stabilized by hydrogen bonds between the atoms belonging to peptide groups of the main chain.*

One of the most widespread secondary structures of protein molecules, an α helix conformation, is stabilized by hydrogen bonds, which the i-th link of the protein chain forms with the $(i+3)$-th and $(i-3)$-th links (Fig. 7.1a). All amino acid residues are outside the α-helix, thus making its external surface quite irregular.

Another secondary structure typical for proteins is illustrated in Figure 7.1b. It represents a pleated sheet 0.5-nm thick, consisting of a few hydrogen-bonded sections of chain. The length and number of β sections (i.e., the length and width of the β sheet) are not fixed by the secondary structure itself, just as the length of α-helices, but rather are determined by the spatial form of the protein as a whole (*see* Sec. 44).

It was shown (L. Pauling and P. Corey, 1953) that the stereochemistry of polypeptide chains permits the existence of only two short-range forms, α-helices and β-sheets, accompanied by the creation of a regular network of hydrogen bonds. Accordingly, these forms are found in proteins to produce extended sections of secondary structure.

39.2. Some proteins may form secondary structures in the form of triple-stranded helices.

The most abundant (by weight) protein, collagen, exists like other fibrous proteins in living systems in the form of helic braids woven of three strands. It should be noted that the primary structure of these proteins is quite distinctive, (e.g., every third amino acid in a collagen is glycine).

39.3. The secondary structure of DNA has the form of a double helix stabilized by hydrogen bonds linking pairs of complementary bases.

In accordance with their biological function, DNA chains are woven by two, with their primary structures being controlled by a strict complementarity rule: T is always opposite to A, and C is always opposite to G. The complementary bases are linked by hydrogen bonds (AT by two and GC by three bonds), so the base pairs are flat. The chains thus linked via side groups intertwine to form a helix. Apart from a right-hand helix (or a B form, known since J. Watson and F. Crick's discovery), there may sometimes be a left-hand helix (or a Z form). The linking of the bases in the Z form obeys the same rule. The essential parameters of the double helix can be seen in Figure 7.1c. Significantly, the Watson-Crick pairs are located between the chains (i.e., inside the helix), so in distinction to the α helix, the double helix is quite homogeneous in its external structure (irrespective of the primary structure).

39.4. The formation of helical secondary structure drastically increases the persistent length and, consequently, is referred to as a helix–coil transition.

In a non-helic conformation, a polypeptide chain possesses rotational-isomeric flexibility, and the length of its Kuhn segment comprises 5–7 links (i.e., equals ~ 1.8 nm). Regarding the α helix, its flexibility may only be of a persistent nature (provided that the hydrogen bonds are intact), and the corresponding length of the Kuhn segment reaches ~ 200 nm. The difference in Kuhn segments is so large that in a certain range of molecular masses, the following situation may arise. In the initial state, the chain forms a genuine (Gaussian or swollen) coil, because its contour length considerably exceeds the persistent length. After helix formation, the same chain appears as an almost absolutely stiff rod, because its length is less than the persistent length. Of course, even in a helical state, the longer chain as a whole would form a coil.

Quite similarly, a single strand in DNA possesses rotational-isomeric flexibility, and the Kuhn segment length is ~4 nm. The double helix possesses persistent flexibility, and its Kuhn segment is ~100 nm long (equal to approximately 340 base pairs).

The noted difference in the lengths of Kuhn segments makes one regard the decay of helical secondary structure of biopolymers (because of an increase in temperature or concentration of molecules competing for hydrogen bonds) as a kind of melting;[51] it is usually called a *helix–coil transition*, to which the next section is devoted. Here, however, we make two more remarks on the flexibility of biopolymers prone to form a helix.

First, note that the lengths of the Kuhn segments in helices are much greater than the helix pitch. Therefore, their persistent flexibility may be regarded as isotropic in the plane perpendicular to the helix axis. Certainly, the small-scale properties of helices (significant in biologic terms) are anisotropic, but this anisotropy is effectively averaged out over the great length of a segment. Note also that the Kuhn segment depends weakly on the primary structure.[f]

Second, the flexibility mechanism is rather peculiar near the helix–coil transition. Under these conditions, flexibility is determined by chain sections with broken helices, which play the role of free joints because of the small length of a Kuhn segment. Hence, the free-joint mechanism of flexibility is in fact realized.

39.5. *The simplest experimental methods for the observation of the helix–coil transition allow one to observe the variations in the length of the Kuhn segment of the chain and in the fraction of chains located in helical sections (i.e., degree of helicity).*

The helix–coil transition results in changes in a great number of observable polymer characteristics. In particular, the transition is clearly noticeable in hydrodynamic experiments. An abrupt change in the length of the Kuhn segment is observed by the variation of such parameters as the viscosity of the solution, coefficient of sedimentation, electrophoretic mobility, and so on.

Spectroscopic methods provide another approach. On helix formation, the side groups of the links come together considerably. In DNA, additionally, the plates of the bases fold into a parallel stack. This leads to considerable growth in the interaction between the links and, as it appears, diminishes light absorption. This so-called hypochromic effect is easily observed in DNA's, in which the absorption decreases (at the wavelength 260 nm, which is characteristic for the bases) by 25% to 30% as a result of helix formation.

The links of natural biopolymers are not mirror symmetric and, therefore, are optically active. Because a helix itself is not mirror symmetric, helix formation creates a strong increase in optical activity. Most often, this is found by observing circular dichroism effects, that is, the difference between the absorp-

---

[f]In the α helix, this is the case, because the predominant contribution to the flexibility is made by the uniform helical back bone but not the external side groups. For the double helix, it is because the helix is practically homogeneous along the length (because the bases are linked into complementary pairs).

tion of the right- and left-circularly polarized light.

The helix–coil transition is also observed by the calorimetric method, in which heat absorption is measured.

All of these techniques are obviously integral. They only follow evaluation of the helic content (i.e., the fraction of links located in helic portions of an entire macromolecule). Other more sophisticated methods (on which we will not dwell here) make it possible to obtain more detailed information, specifically for a DNA, about the helic content of a particular section of the heteropolymer chain.

40. HELIX-COIL TRANSITION IN SINGLE- AND DOUBLE-STRANDED HOMOPOLYMERS

40.1. *Because of the cooperative nature of conformations, links can exist in two clearly differing discrete states: helices, and coils; the junction between helical and coil sections carries a large positive free energy.*

On transition from a free-coil conformation to a helic one, the system gains energy from hydrogen bond formation and loses entropy because of the rigid fixing of the chain sections between consecutive hydrogen bonds within the helix (i.e., the number of possible realizations of the given state is reduced). This is why the helical state is stable (thermodynamically favorable) at low temperatures[g] and the coil state stable at high temperatures.

The energy gained per link on helix formation is of the order of a hydrogen bond energy (i.e., $\Delta E \sim 0.1$–0.2 eV). The entropy loss ΔS on fixing the link conformation is of the order of a few units. From these estimates, it follows that the free energies of helic and coil conformations should coincide, that is, their difference $\Delta f = \Delta E - T \Delta S$ should become zero at a temperature $T^* = \Delta E / \Delta S$ on the order of room temperature. Obviously, the helix–coil transition must occur near the temperature T^*.

If the turns of a helix appeared and decayed independently of one another, then the probability of a helical state for an arbitrary link would be determined by the Boltzmann law $[\sim \exp(-\Delta f/T)]$, that is, would vary gradually with temperature. The characteristic width of the temperature interval for this variation would be of the order of the temperature T^* itself. In fact, the helix–coil transition occurs in a quite narrow temperature interval because of the cooperative interdependence of the conformations of neighboring links and helix turns.

Let us clarify this circumstance in more detail. The formation of one bond effectively fosters the formation of another close by. For example, the presence of a hydrogen bond between the i-th and $(i+3)$-th links in an α helix requires fixing the conformations of three peptide groups: $i+1$, $i+2$, and $i+3$. The next

---

[g]In the simplest case, the formation of hydrogen bonds results in energy gain. If the solution contains molecules competing for hydrogen bonds, however, then the entropic component of the process can be essential. For the sake of simplicity, we do not dwell on this subject; it should only be noted that in this case, the helix–coil transition may be induced not by a temperature variation but by a change in the solvent composition.

bond between the $(i+1)$-th and $(i+4)$-th links furnishes the same energy gain but requires fixing only one group: $i+4$ (i.e., leads to a much smaller entropy loss). Quite analogous reasoning pertains to the double helix of DNA: the linking of one base pair dramatically facilitates the linking of a neighboring pair, because there is a strong attraction (a so-called stacking interaction) between the planes of neighboring pairs.

The ideas presented here lead to the conclusion that each link of the spiraling chain can be regarded as a system with two clearly separate states: the helical and the "molten" (coil). Certainly, there is a marginal area between the helic and coil sections; however, cooperative effects reduce this area to a mere point junction. Of course, this junction is thermodynamically unfavorable. Hydrogen bonds provide no energy gain near this junction and no entropy gain follows from the freedom of choice of rotational-isomeric conformations.

Hence, helix formation is characterized by two energy parameters. Having chosen the coil state of the link as a reference point, the free energy of the helic link equals Δf and the free energy of the junction between the helix and the coil Δf_s. For the sake of convenience, Δf and Δf_s are usually replaced by the so-called Bragg-Zimm parameters:

$$s \equiv \exp(-\Delta f/T), \quad \sigma \equiv \exp(-2\Delta f_s/T). \tag{40.1}$$

Depending on the temperature, the free energy Δf may be positive or negative (i.e., the value of s may be more or less than unity). At the same time, the temperature dependence of Δf_s is usually insignificant ($\Delta f_s > 0$), and according to what has been said:

$$\sigma \ll 1. \tag{40.2}$$

As a rule, in biopolymers, $\sigma \sim 10^{-3}\text{–}10^{-4}$. The smallness of σ is a quantitative expression of cooperativity. Note the factor 2 in the definition of σ: the statistical weight of one junction is traditionally denoted as $\sigma^{1/2}$.

40.2. The helix–coil transition in a single-stranded homopolymer occurs in a very narrow temperature interval, which becomes more narrow as the cooperativity parameter decreases.

It has already been noted that helix formation does not violate the one-dimensional nature of the chain, because hydrogen bonds are formed between links located near one another along the chain. Consequently, the helix–coil transition can readily be described by the method of the transfer operator discussed in Sec. 6. Because every link has only two states (helic "h" and coil "c"), the transfer operator is reduced to a 2×2 matrix, which can be written explicitly as

$$\hat{Q} = \begin{pmatrix} Q_{cc} & Q_{ch} \\ Q_{hc} & Q_{hh} \end{pmatrix} = \begin{pmatrix} 1 & \sigma^{1/2} \\ s\sigma^{1/2} & s \end{pmatrix}. \tag{40.3}$$

The quantity $\exp(-\varphi_\alpha/T)$ (*see* subsection 6.2) defining the energy of individual links equals 1 at $\alpha=c$ and s at $\alpha=h$. As for the "bond" matrix $g(\alpha',\alpha)$, it must be symmetric (Hermitian) and therefore, up to a constant factor, equal to

$$\hat{g}=\begin{pmatrix} 1 & \sigma^{1/2} \\ \sigma^{1/2} & 1 \end{pmatrix}.$$

According to Eq. (6.5), this should yield the expression (40.3).

Following the general theory (*see* Sec. 6), we must find the largest eigenvalue and corresponding eigenvector of the matrix $\hat{Q}$:

$$\begin{pmatrix} 1 & \sigma^{1/2} \\ s\sigma^{1/2} & s \end{pmatrix}\begin{pmatrix} \psi_c \\ \psi_h \end{pmatrix}=\Lambda\begin{pmatrix} \psi_c \\ \psi_h \end{pmatrix}. \tag{40.4}$$

The calculation is elementary in this case. For Λ, we obtain the quadratic equation $[\mathrm{Det}(\hat{Q}-\Lambda\hat{E})=0]$:

$$(\Lambda-1)(\Lambda-s)=s\sigma.$$

Its maximum root equals

$$\Lambda=(1/2)\{1+s+[(s-1)^2+4s\sigma]^{1/2}\}. \tag{40.5}$$

Recall that $-TN\ln\Lambda$ is the free energy of the chain. Regarding the eigenvector, it is determined by Eq. (40.4) up to a constant normalization factor. It is convenient to make the eigenvector obey the condition $\psi_c+\psi_h=1$; then, ψ_c and ψ_h can be interpreted as the probabilities of the coil and the helic states, respectively, for the terminal links of the chain. Writing Eq. (40.4) as a system of two linear equations, finding ψ_c/ψ_h from any of two and taking into account the normalization condition, we obtain

$$\psi_h=1-\psi_c=(\Lambda-1)/(\Lambda-1+\sigma^{1/2}) \tag{40.6}$$

In experiment, however, one does not observe the degree of helicity of the terminal link but rather the helical content of all of the links. This quantity is simply the "density" n_α introduced in subsection 6.6; according to Eqs. (6.20) and (6.10), we have in this case

$$\begin{pmatrix} n_c \\ n_h \end{pmatrix}=\mathrm{const}\cdot\begin{pmatrix} \psi_c^2 \\ \psi_h^2/s \end{pmatrix},$$

where the factor const is found from the normalization condition $n_c+n_h=N$. Hence, we obtain for the helical content $\vartheta=n_h/(n_c+n_h)$,

$$\vartheta=1/2+(s-1)/2[(s-1)^2+4s\sigma]^{1/2}. \tag{40.7}$$

Let us discuss this result.

For $\sigma=1$ (i.e., in the absence of cooperativity), Eq. (40.7) yields ϑ $=s/(1+s)$, the trivial result of the Boltzmann distribution for independent two-level particles with the ratio of the probabilities of h and c states being equal to s. On the other hand, for $\sigma=0$, Eq. (40.7) yields $\vartheta=\{0$ for $s<1;\ 1$ for $s>1\}$, that is, we have in this case an abrupt transition occurring at $s=1$. This result is also trivial. In fact, the condition $\sigma=0$ corresponds to infinite surface energy Δf_s (i.e., to the strict prohibition against junctions between helic and coil sections). It is clear that in this case, the chain as a whole is induced to transform into a helic state. In the real situation of a small but finite, value of σ one can write

$$\vartheta \cong \begin{cases} 1-\sigma/s, & s>1, \quad s-1\gg\sigma, \\ 1/2+(s-1)/4\sigma^{1/2}, & |s-1|\ll\sigma, \\ s\sigma, & s<1, \quad 1-s\gg\sigma. \end{cases} \tag{40.8}$$

The dependence $\vartheta(s)$ is plotted in Figure 7.2. Thus, the helix–coil transition proceeds in a narrow interval of the variation of s. The middle point of the transition (where $\vartheta=1/2$) corresponds (as expected) to $s=1$ (i.e., to the temperature T^*). The width of the transition region, which is ordinarily determined by the slope of the plot $\vartheta(s)$ at $\vartheta=1/2$ (Fig. 7.2), constitutes $\Delta s=4\sigma^{1/2}$.

To estimate the width of the transition interval on the temperature scale, note that for a small temperature variation, $(s-s^*)/s^* \sim (T-T^*)/T^*$. Hence, $\Delta T \sim T^*\sigma^{1/2}$ (because $s^*=1$). For instance, for a homopolymer DNA chain $(\sigma \sim 10^{-4})$, the melting interval width equals approximately 0.5 K. Also, interestingly, the dependence of s on the helical content of the terminal link $\psi_h(s)$ differs from $\vartheta(s)$, even though being qualitatively similar to it.

40.3. *The mean lengths of helical and coil sections are finite and independent of the total chain length even as $N \to \infty$.*

To comprehend the picture of the helix–coil transition, it is necessary to examine how the helical and the coil sections are distributed along the chain.

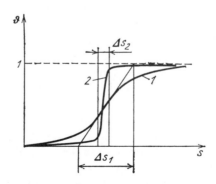

FIGURE 7.2. Degree of helicity as a function of s in the region of the helix–coil transition in the absence of cooperativity $\sigma=1$ (*curve 1*) and the presence of strong cooperativity $\sigma \ll 1$ (*curve 2*).

Clearly, this distribution is controlled by the parameter σ: the less σ is, the more unfavorable become the junctions and the longer the homogeneous chain sections. Specifically, as the appearance of one additional junction leads to the free energy increment Δf_s, it can easily be shown that the number of junctions in the chain is given by the derivative $\partial(N \ln \Lambda)/\partial \ln \sigma$. What is significant is that this number of junctions is proportional to N; therefore, the total number of links in two adjoining helic and coil sections (i.e., the length of the chain containing two junctions) is $k=(2\Lambda/\sigma)(\partial\Lambda/\partial\sigma)^{-1}$. The mean lengths of the helic and the coil sections are obviously equal to $k_h=\vartheta k$ and $k_c=(1-\vartheta)k$, respectively. Simple calculations yield

$$k_h=\frac{1+s+[(s-1)^2+4s\sigma]^{1/2}}{1-s+[(s-1)^2+4s\sigma]^{1/2}},\tag{40.9}$$

$$k_c=\frac{1+s+[(s-1)^2+4s\sigma]^{1/2}}{-1+s+[(s-1)^2+4s\sigma]^{1/2}}.\tag{40.10}$$

As expected, the lengths of helices k_h increase and of coil sections k_c decrease monotonically with the growth of s. The total length $k=k_h+k_c$ is the shortest at $s=1$:

$$k_{\min}=(2k_c)_{s=1}=(2k_h)_{s=1}=1+\sigma^{-1/2}\gg 1.\tag{40.11}$$

It can be seen that because of cooperativity ($\sigma\ll 1$), the length k proves to be fairly large. Significantly, however, this length is independent of the total chain length N (i.e., it remains finite even as $N\to\infty$). To comprehend the profound meaning of this fact, the nature of the helix–coil transition should be discussed in terms of general thermodynamics; this is the subject of the next two subsections.

40.4. Although frequently regarded as a melting of helices, the helix-coil transformation is not a phase transition.

The fact that every link of a helix-forming chain can reside in either of two clearly separated states implies that the free energy of the link has two minima. For $s<1$, the minimum corresponding to the coil state is deeper; conversely, for $s>1$, the free energy minimum of the helic state is deeper. This makes one compare the helix–coil transition to a first-order phase transition. Indeed, there is a certain analogy here. In particular, helic sections existing for $s<1$ (and similarly, coil sections existing for $s>1$) can be treated as heterophase fluctuations, "islets" of the less favorable phase in a "sea" of the more favorable one. The characteristic size of such islets (the critical size of a nucleus), however, is known to grow indefinitely on approach to the point of a genuine phase transition. Accordingly, equilibrium in ordinary systems corresponds to a merging of all of the islets of either phase and a subsequent separation of the whole sample into two macroscopic, immiscible phases. This picture contrasts sharply with that observed in the helix–coil transition, where as we have seen, even at the transition point itself ($s=1$) the lengths of the islets are finite even though great, $\sim \sigma^{-1/2}$ [see Eq. (40.11)]. One can compare the coexistence of helic and coil sections to a fog consisting of liquid droplets suspended in gas. Under normal conditions (e.g., in the atmosphere), however, the fog is metastable, whereas for a helix-forming chain the intermixed state corresponds to statistical equilibrium.

It should also be noted that the infinite growth of the critical nucleus leads to singular behavior of the thermodynamic functions (e.g., to an entropy jump) at the first-order phase transition point. Accordingly, the helix–coil transition (because the size of the islets is finite), is not characterized by genuine singularities in thermodynamic behavior at any point but rather spreads over a region of finite width.

Certainly, the fact that the helix–coil transition has a finite, non-zero width is not yet sufficient by itself to conclude that it cannot be classified as a phase transition. Indeed, the globule–coil transition discussed in Sec. 21, being a phase transition, also has a finite width. The principal difference lies in the fact that the spreading of the globule–coil phase transition is only caused by the finite size of the system. One can say that the transition region appears when the typical size of a single heterophase fluctuation exceeds the size of the entire sample. The probabilities of finding the system in either state become equal in order of magnitude. Clearly, the width of the transition region should diminish with the growth of the number of particles, as in fact is observed in globule–coil transitions, from which follows the formal definition given earlier: a phase transition is one whose width tends to zero as $N \to \infty$. This definition is, of course, totally equivalent to the conventional one, according to which the thermodynamic functions are singular at the phase transition point in the limit $N \to \infty$.

For a long chain ($\sim \sigma^{1/2}$), the width of the helix–coil transition is independent of N; consequently, it is not a phase transition. It is relevant to ask here: why is there no genuine separation in the chain into macroscopic helic and coil phases? The answer is given by the Landau theorem.

40.5. *In a one-dimensional system, as distinguished from systems in higher dimensions, the equilibrium coexistence of macroscopic phases is impossible; this is known as the Landau theorem.*

The actual reason for the non-phase nature of a helix–coil transition is associated with the fact that the ideal chain is one-dimensional. We explain this as follows. Let us consider an island of a new phase of size R in the d-dimensional system of volume V. Its separation into two smaller islands leads to an increase in surface area, a loss in surface energy $\sim R^{d-1}$, and simultaneously to a gain in free energy $\sim T \ln V$ because of the independent translational motions of both halves throughout the volume V. If $d > 1$, then fragmentation of sufficiently large islets is unfavorable *a fortiori*, so these tend to merge, causing separation of a macroscopic phase. In a one-dimensional system, however, the surface energy (i.e., the energy of interphase junctions) is independent of the size of the islets, and the fragmentation of too long sections of a homogeneous phase is thermodynamically favorable. (A more detailed proof of the Landau theorem can be found in Sec. 163 of Ref. 26.)

40.6. *For a short chain, the width of the helix–coil transition region is anomalously large.*

Apart from theoretic interest, the concepts presented earlier permit one to make a practical conclusion. If the whole chain is shorter than the characteristic length $\sigma^{-1/2}$ [*see* Eq. (40.11)], then the transition width (always equal to $1/k_{min}$) proves to be not of order $\sigma^{1/2}$ but of order $1/N$, which is larger (for short chains).

40.7. *A helix–coil transition can be initiated by a change in the pH value of the medium; in this case, the transition is accompanied by a sharp change in the average charge of the molecule.*

Because ionized groups in a helix are closer to one another than in a coil, their ionization makes the helic state less favorable; in other words, the ionization potential in the helic state is reduced: $pK_h < pK_c$. Consequently, on variation of the pH, the constant s also varies. The helix–coil transition occurs when the pH value is such that $s(pH) = 1$. In this situation, the titration curve of the macromolecule has a remarkable shape. On one side of the transition it coincides with the "coil" curve [i.e., is described by Eq. (37.5), with $pK = pK_c$], whereas on the other, it corresponds to $pK = pK_h$. Within the transition region, a crossover occurs (Fig. 7.3).

40.8. *Other rearrangements of secondary structure can be described similar to the helix–coil transition.*

The transition between a right-hand helical B form and a left-hand helical Z form proceeds in DNA's with some primary structures on a change in concentration of the nonpolar component of the solvent. Along with the Z form–coil transition, this is investigated by a procedure similar to that described in this section. The analysis of β folds in proteins, being more complicated, is be given here (*see* Ref. 3).

The end of this section is devoted to the description of the helix–coil transition in a double-stranded homopolymer. A real double-stranded DNA is never homogeneous. However, on the one hand, synthetic homopolynucleotides stim-

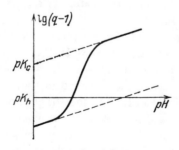

FIGURE 7.3. Mutual influence of helix–coil transition and degree of charge of a polymer chain.

ulate some interest and are investigated; on the other, the model of a homopolymer helps to examine theoretically the effects of a double-stranded structure. A specific subject of the next subsection is the technique used to study a double-stranded polymer. It is more convenient, however, to explain the technique by applying it to the already-solved, simple problem of a single-stranded homopolymer.

*40.9. The application of the method of generating functions to study the helix–coil transition is based on the interpretation of the positions of interphase junctions as internal dynamic variables of the system.

This subsection tackles the helix–coil transition by the method of generating functions presented in subsection 6.10. Calculation of the generating function $\mathscr{L}(p)$ [see Eq. (6.29)] is easy for a single-stranded homopolymer. We begin by writing the ordinary partition function Z_N as a sum of statistical weights $\rho(\Gamma)$ over all of the states of spiralization Γ: $Z_N = \Sigma \rho(\Gamma)$. The helical content can be characterized totally by the lengths of all helical h_i and coil c_i chain sections and their number k ($1 \leqslant i \leqslant k$, $2k \leqslant N$). As long as the chain is ideal, the adjoining sections only interact via the junctions, so the statistical weights of different pairs from coil and helical sections are independent, that is,

$$\rho(\Gamma) = \prod_{i=1}^{k} \rho(h_i, c_i).$$

Also taking into account the evident condition $\sum_{i=1}^{k} (h_i + c_i) = N$, we reduce the calculation of $\mathscr{L}(p)$ to a summation of the geometric progression:

$$\mathscr{L}(p) = \sum_{N} \sum_{k,\{h\},\{c\}} p^N \prod_{i} \rho(h_i, c_i) \delta\left(\sum (h_i + c_i) - N \right)$$

$$= \sum_{k=1}^{\infty} \zeta^k(p)$$

$$= \zeta(p)/(1 - \zeta(p)), \tag{40.12}$$

$$\zeta(p) \equiv \sum_{h,c} \rho(h,c) p^{h+c}. \tag{40.13}$$

Obviously, $\zeta(p)$ is the generating function of a chain segment comprising one helical and one coil section. According to Eqs. (40.1) and (40.3), the statistical weight $\rho(h,c)$ is determined as follows. The factor s is ascribed to each helical link and unity to each coil one. The additional factor σ is ascribed to a coil section as a whole (i.e., to both its ends). Eventually, $\zeta(p)$ is also reduced to a geometric progression:

$$\zeta(p) = \sigma \sum_{h=1}^{\infty} (sp)^h \sum_{c=1}^{\infty} p^c = \sigma[sp/(1-sp)][p/(1-p)]. \tag{40.14}$$

Let us analyze the result obtained. The singularities of $\mathscr{L}(p)$ correspond to the condition $\zeta(p)=1$, that is, to the roots of the equation

$$(sp-1)(p-1)=\sigma sp^2. \tag{40.15}$$

[It can easily be shown that the singular points of $\zeta(p)$ realized for $p=1$ and $p=1/s$ do not correspond to the singularities of $\mathscr{L}(p)$]. All of the singularities are simple poles lying on the real axis, with the pole closest to zero being a smaller root of Eq. (40.15). Comparing Eqs. (40.15) and (40.5), one can see that for $\Lambda=1/p$ [see Eq. (6.31)], we obtain the same result.

40.10. *Coil sections of the double chain are loops; an entropic disadvantage of long loops results in further cooperativity.*

In a single-stranded polymer, the link conformations of a nonhelic section are as free as those of an ordinary coil. This is not the case for a double-stranded polymer: the conformations of the coil section are restricted by the condition that the chains should converge at the ends of the section, thus forming a closed loop (Fig. 7.4). The statistical weight of such a loop state equals the probability that the free-end links of the two chains joined at one end converge within the small volume $v \ll a^3$.

Suppose, that both chains are phantom (i.e., cross one another freely) and have no excluded volume, (i.e., are ideal). Let the point at which one of their joined ends is located be taken to be the origin; then the probability of their other ends converging within the volume v is written as

$$\int P_c(r_1)P_c(r_2)\theta_v(r_1-r_2)d^3r_1d^3r_2, \tag{40.16}$$

where $P_c(r)$ is the Gaussian distribution (4.1) for a chain of c links; r_1, r_2 the radius vectors of the chain ends; and $\theta_v(r)$ a step function differing from zero only within the volume v. The function $P_c(r)$ almost does not vary on motion of the chain end within the small volume $v \ll a^3$. Therefore, the function θ_v in Eq. (40.16) can be replaced by a delta function, that is, Eq. (40.16) is transformed to

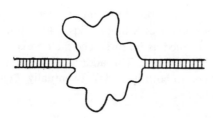

FIGURE 7.4. Loop section in helix–coil transition in a double-stranded macromolecule.

$$v \int P_c^2(r)d^3r = (3/4\pi)^{3/2}(v/a^3)c^{-3/2}. \tag{40.17}$$

Consequently, instead of the constant value of σ (see subsection 40.9), the partition function of the coil section in the double-stranded polymer depends on the section length c:

$$\sigma_{\text{ef}}(c) = \sigma c^{-3/2} \tag{40.18}$$

(the constant factors are included σ by an unessential redefinition of that quantity.)

The formula (40.18) is referred to as the *Stockmayer formula*. Significantly, it leads to a power dependence of the quantity $\sigma_{\text{ef}}(c)$ (the so-called loop factor) on the length c. Such a dependence signifies that the fluctuation motions along the chain of the helix–coil junction points are not free, even at $s=1$; the ends of the coil section attract each other along the chain with some entropic force. From qualitative considerations, it is clear that this leads to a decrease of the fraction of long untwisted sections and an increase in cooperativity of the transition.

Before moving to the quantitative analysis of the role of the loop factor, recall that its expression (40.18) is obtained from the assumption of an ideal nature of both chains. In reality, however, chains may be located in good solvent and, most importantly, be non-phantom, so knots in either of them and entanglements with each other can be formed only by very slow diffusion from the ends. Apparently, the equilibrium knotting does not have enough time to set in. Therefore, it is worthwhile to write the loop factor as

$$\sigma_{\text{ef}}(c) = \sigma c^{-\alpha}. \tag{40.19}$$

At present, the exponent α is not yet calculated exactly, so it should be regarded as a phenomenological parameter.[h]

*40.11. *The loop factor leads to an abrupt sharpening of the helix–coil transition, which turns into a phase transition for $\alpha > 1$.*

The easiest way to account for the loop factor is to use the method of generating functions. The formula (40.12) apparently remains applicable, while the expression (40.14) for $\zeta(p)$ changes. This is because the partition function $\rho(h, c)$ in Eq. (40.13) acquires the additional factor (40.19). Thus, we have

$$\zeta(p) = \sigma \sum_{h=1}^{\infty} (sp)^h \sum_{c=1}^{\infty} p^c c^{-\alpha} = \sigma[sp/(1-sp)]\varphi_\alpha(p),$$

$$\varphi_\alpha(p) \equiv \sum_{c=1}^{\infty} p^c c^{-\alpha}. \tag{40.20}$$

---

[h]Note that the examination of a helix–coil transition in a braid of k ideal chains would yield $\alpha = 3(k-1)/2$.

To investigate the pole $\mathscr{L}(p)$, let us examine the equation $\zeta(p)=1$ [see Eq. (40.12)]. Its root lies on the real axis (otherwise we would obtain an absurd complex expression for the free energy), so we can rewrite the equation $\zeta(p)=1$ in the form

$$1/s-p=\sigma p\varphi_\alpha(p) \tag{40.21}$$

and solve it graphically (Fig. 7.5).

The coefficients of the power series for $\varphi_\alpha(p)$ are positive, and the series obviously converges for $p<1$. Consequently, the function $\varphi_\alpha(p)$ grows monotonically in the interval $0\leqslant p\leqslant 1$. The curve in Figure 7.5 shows the corresponding graph of the function $p\varphi_\alpha(p)$, while the straight lines correspond to the left-hand side of Eq. (40.21) for different values of s.

If $\alpha\leqslant 1$, then $\varphi_\alpha(p)$ diverges at $p=1$. As seen from Figure 7.5, Eq. (40.21) in this case has a solution for any s, and this solution varies smoothly with s. One can easily realize that with a decrease in σ, this dependence acquires the properties of a "switch" from $p\cong 1/s$ for $s\gg 1$ to $p\cong 1$ for $s\ll 1$. In other words, its behavior becomes qualitatively similar to that discussed earlier in terms of the simple theory with $\alpha=0$.

If $\alpha>1$, then on the contrary, the series for $\varphi_\alpha(p)$ converges for $p=1$, that is, $\varphi_\alpha(p=1)$ is some finite number. Then, for finite $s=s_{cr}=(1+\sigma\varphi_\alpha(1))^{-1}$, the root of Eq. (40.21) becomes unity, that is, the free energy (6.31) (and also the helic content) equals zero for $s<s_{cr}$. Consequently, a real phase transition occurs at $s=s_{cr}$. To find the order of the transition, one must determine the asymptotic behavior $\varphi_\alpha(p)$ for $p\to 1$ and, thereby, the behavior of the free energy for $s\to s_{cr}$. This could be found very easily by substituting the integral for the sum over c

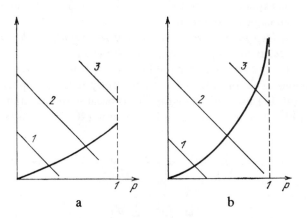

a   b

FIGURE 7.5. Graphic solution of Eq. (40.21) for $\alpha>1$ (a) and $\alpha<1$ (b). The curves represent functions $\sigma p\varphi_\alpha(p)$; straight lines $1/s-p$ correspond to different values of s and are enumerated in the order of descending values of s.

values in the definition of the function $\varphi_\alpha(p)$. Omitting the actual calculations[i] we give the answer here. When $\alpha > 2$, the helix–coil transition, accompanied with a sudden change in helic content, proves to be a first-order phase transition. For $1 < \alpha < 2$, a smooth phase transition is predicted, with the temperature dependence of the helical content showing only a breakpoint but not a jump (as in the previous example).

This may seem to overrule the Landau theorem mentioned in subsection 40.5; however, this is not so. The double helix with loops is not a purely one-dimensional system. The loop factor (40.19) has a primarily three-dimensional origin, and it describes a definite entropic long-range interaction along the chain. The presence of a long-range interaction makes the Landau theorem inapplicable.

41. HELIX-COIL TRANSITION IN HETEROPOLYMERS

41.1. *In heteropolymers, the helicity constants of the links of various types are different, and the character of the helix–coil transition therefore may depend on the primary structure.*

Let us examine the analysis of the role that the heterogeneous primary structure plays in the helix–coil transition. Recall that the general structure of α helices in proteins and of double helices in DNA's depends weakly on the type of links. One may expect, however, that the energies of helix formation and also the constants s to differ perceptibly for different links. In fact, experimental studies of the helix–coil transition in synthetic polynucleotides poly-AT and poly-GC show that the temperatures of the semi-transition, for which $s = 1$, differ by approximately 40 °C. Similar values are also typical for proteins. The difference in the "melting" temperatures often proves to be much larger than the width of the transition $\sim T\sigma^{1/2}$.

To make things clear, we consider the simplest system, a heteropolymer comprising two types of links, for example, A and B. Let T_A and T_B denote the semi-transition temperatures for the poly-A and poly-B. It is assumed that

$$|T_A - T_B| \gg \sigma^{1/2}(T_A + T_B)/2.$$

Now, suppose that the primary structure is as follows: half the length of the molecule is a homopolymer poly-A and the other half poly-B. On the temperature change, the helic content of the molecule initially grows from 0 to 1/2 within an interval of width $\sim \sigma^{1/2}$ near T_A, then grows from 1/2 to 1 within the same interval near the temperature T_B (Fig. 7.6a). The corresponding differential melting curve (DMC), that is, the derivative of the degree of helicity with respect to temperature, has two narrow peaks at temperatures T_A and T_B (Fig. 7.6b).

---

[i]A more detailed investigation of the situation for $\alpha > 1$ requires that allowance should be made for the bifurcation that the function $\varphi_\alpha(p)$ has at $p = 1$. Then, in the corresponding reverse transformation formula (6.30) there appears an integration along the sides of a cut. This integration yields the logarithmic correction to the free energy of the coil state

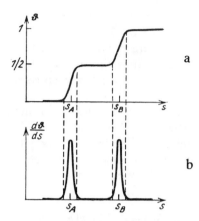

FIGURE 7.6. Degree of helicity (a) and differential melting curve (b) of a hypothetic two-block copolymer.

Suppose now that the primary structure is short periodic, for example, ABABAB... The length of cooperativity now is much greater than the length of the period, and an effective averaging of the link properties therefore is to be expected with the emergence of a single narrow peak on the DMC at some intermediate temperature.

*41.2. In a real heteropolymer, the helix–coil transition proceeds by consecutive meltings of quite definite sections, whose primary structures possess a sufficiently higher concentration of low-melting links.

Experimental research of the helix–coil transition in a real DNA has disclosed primarily that its transition region is about one order of magnitude wider in comparison with that for homopolymers. Next, when a DNA of moderate length is investigated, a non-uniform growth of the degree of helicity in the transition interval becomes perceptible. This shows more clearly after differentiation: a typical DMC of a real DNA represents a sequence of well-defined peaks (Fig. 7.7). More sophisticated experimental methods can reveal in any specific DNA the chain sections whose melting is responsible for each of the peaks.

The situation described becomes quite comprehensible if one recalls the high degree of cooperativity of helix formation ($\sigma \ll 1$), because of which only sufficiently long sections can melt. We clarify this in more detail using an example of a heteropolymer comprising links of two types, A and B.

Let Δf_A and Δf_B denote the differences between the free energies of the helic and coil states for links A and B, respectively. These quantities depend on the temperature and become zero at the points T_A and T_B. For simplicity, suppose that in the whole transition region, the linearization of the temperature dependences of Δf_α

$$\Delta f_\alpha = \Delta S(T - T_\alpha) \quad (\alpha = A, B) \tag{41.1}$$

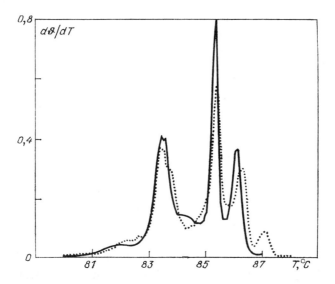

FIGURE 7.7. Experimental (*dotted line*) and theoretical (*solid line*) differential melting curves of DNA (Vologodskii A. V., Amyrykyan B. R., Lyubchenko Y. L., Frank-Kamenetskii M. D.-J. Biomol. Struct. Dyn.-1984-v.2-N1-p.131).

is accurate enough and that the factor ΔS (equal to the entropy loss from transition to the helical state) is independent of the type of link: $\Delta S_A = \Delta S_B = \Delta S$. Suppose also that the surface energy Δf_s (or the cooperativity parameter σ) is also independent of the type of link.

Of particular interest is the temperature interval between T_A and T_B, in which spiralization of one of the components is thermodynamically favorable while that of the other is not. To be specific, let $T_A > T_B$ so that $\Delta f_A < 0$, $\Delta f_B > 0$.

To specify the primary structure, we introduce the quantities $x_A(t)$ and $x_B(t)$ [with $x_A(t) + x_B(t) = 1$]: if the link with number t belongs to type A, then $x_A(t) = 1$ [and $x_B(t) = 0$]; if it is of type B, then $x_A(t) = 0$ [and $x_B(t) = 1$]. In fact, these are the local concentrations of types A and B at the "point" t of the given primary structure.

Let us now define the function $F(t)$:

$$F(t) = \sum_{\tau=1}^{t} [\Delta f_A x_A(\tau) + \Delta f_B x_B(\tau)]. \qquad (41.2)$$

Its meaning is simple: if the chain section between points t_1 and t_2 transforms into the helic state, then the free energy changes by the amount

$$F(t_1, t_2) = F(t_2) - F(t_1) + 2\Delta f_s \qquad (41.3)$$

(the last term corresponds to the appearance of two unfavorable helix–coil junctions). Figure 7.8 shows the function $F(t)$ plotted for an arbitrarily chosen primary structure. The curve descends on the sections where A particles predominate and rises on B-enriched sections. (For a homopolymer, this graph is a straight line with slope Δf.) From the curve $F(t)$, it is easy to find the sections of the heteropolymer whose spiralization under given conditions (i.e., for the given Δf_A, Δf_B, Δf_s) is thermodynamically favorable as well as the sections that are likely to remain in the coil state. It is immediately clear that portions of the graph rising by a value of $2\Delta f_s$ are definitely unfavorable for helix formation [*see* Eq. (41.3)]. Next, the portion of $F(t)$ descending by the same amount $2\Delta f_s$, even separated by small ($<2\Delta f_s$) rises, is favorable for overall helix formation. Finally, the ends of the helic section must correspond to rises of $F(t)$ by $2\Delta f_s$ (possibly separated by small, $<2\Delta f_s$, descents). All of this is illustrated by Figure 7.8.

The seemingly awkward considerations listed are in fact very simple. They indicate that in the considered case, the heteropolymer chain really separates into coil and helic sections not via fluctuations but rather according to its primary structure. As the temperature grows, Δf_A and Δf_B increase, the descents of $F(t)$ become less steep and the rises steeper, and each helic section becomes unfavorable at a certain temperature and is transformed into a coil. The corresponding DMC peak appears at this temperature.

This relates to a section of heteropolymer of moderate length. In a very long chain with complicated (*see* subsection 38.5) primary structure, there are diverse sections, and the number of DMC peaks grows proportional to the chain length producing a spread (relative to the smooth transition) picture. What is the width of the transition region in this case? We examine this problem by example of a heteropolymer with a random primary structure.

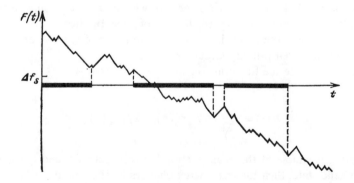

FIGURE 7.8. Dependence $F(t)$ for one realization of the primary structure. The sections indicated would be in a helical state at the given Δf_s.

*41.3. *The helical content of an infinitely long, random heteropolymer has a definite (and not random) value because of the property of self-averaging.*

The degree of helicity of the chain ϑ is a good example of an additive quantity. Denoting the probability that the link of number t is in the helical state by ϑ_t, one can derive the degree of helicity of the entire chain by summing or averaging:

$$\vartheta = \sum_{t=1}^{N} \vartheta_t / N. \tag{41.4}$$

Such additive quantities possess the important property of self-averaging.[44,45] The point is that the value of ϑ_t for the definite link t depends not on the primary structure of the whole chain but only on the nearby section having a length of the order of the cooperativity length. For more remote links, thermal fluctuations of the states and, consequently, the values of ϑ_t are independent. If the total length of the chain is long enough, then its helic content (41.4) results from the averaging over many practically independent blocks. Finally, if the primary structures of these blocks are random and independent, then the value of ϑ as $N \to \infty$ is essentially definite because of the law of large numbers.

The property of self-averaging can also be clarified from the following viewpoint. With overwhelming probability, the degree of helicity will equal or approximate its mean value for a random realization of the primary structure. There are primary structures with substantially different values of ϑ (e.g., the homopolymers poly-A and poly-B); however, the probability of their realization is vanishingly small and the deviation of ϑ from the mean only of order unity. For quantities (discussed later) that do not possess the property of self-averaging, the vanishingly rare realizations yield a large deviation from the average. Accordingly, the picture observed in studies of such characteristics is random, depending substantially on whether any atypical primary structures are present in the chosen ensemble.

*41.4. *The "melting" of a long, random heteropolymer proceeds gradually in a very wide temperature interval; the width of the interval is determined by a variety of random aggregates of links in the primary structure and is proportional to the logarithm of the small parameter σ of cooperativity (cf. the power of σ for a homopolymer).*

Being interested in the behavior of a self-averaging quantity (the helicity content), we may assume that it can be characterized by the composition of the system, with the fine features of the primary structure being of no importance. Therefore, we denote the mean concentration of A links in the primary structure (i.e., the probability of finding A at a randomly chosen point of the chain) by x. We obtain

$$x_A(t) = x + \xi(t), \quad x_B(t) = 1 - x - \xi(t),$$

where $\xi(t)$ are independent random quantities with zero mean.

Because $\langle\xi\rangle=0$, the infinite chain can be expected to "melt" at the mean temperature

$$T_x=xT_A+(1-x)T_B.$$

Indeed, rewriting Eq. (41.2) taking account of the designations (41.1), we obviously obtain

$$F(t)=t\Delta S[(T-T_x)-(T_A-T_B)\eta_t];$$

$$\eta_t\equiv\sum_{\tau=1}^{t}\xi_\tau/t.$$

For long lengths t, fluctuations of the composition of the primary structure, that is, of the quantity η_t, can be disregarded to demonstrate that $F(t)$ becomes zero (i.e., the free energies of the phases coincide at $T=T_x$).

For the terminal sections, however, the transition temperature may differ from T_x because of random aggregates of links of different types. Consider, for instance, the region $T>T_x$, where there are rare islets of helices surrounded by a sea of coils. According to Eq. (41.3), a section of length l turns helic at the temperature T provided that $F(t+l)-F(t)+2\Delta f_s=0$, that is,

$$\eta_l=\frac{T-T_x+2\Delta f_s/(l\Delta S)}{T_A-T_B}.$$

Because of the independence of ξ_τ, the quantity η_l obeys Gaussian statistics:

$$P_l(\eta)=(2\pi\langle\xi^2\rangle/l)^{-1/2}\exp(-l\eta^2/2\langle\xi^2\rangle);\quad\langle\xi^2\rangle=x(1-x).$$

Therefore, the probability that a chain section becoming helic at the temperature T and length l is proportional to

$$P_l\sim\exp\left\{-\frac{[(T-T_x)l^{1/2}+2l^{-1/2}\Delta f_s/\Delta S]^2}{2(T_A-T_B)^2x(1-x)}\right\}.$$

This quantity has a sharp maximum at

$$l(T)=\frac{2\Delta f_s/\Delta S}{T-T_x}. \tag{41.5}$$

Longer aggregates of high-melting A links are improbable, and shorter ones must be too pure, which makes them improbable as well. The quantity $l(T)$ (41.5) is a kind of optimal value. Consequently, for each temperature $T>T_x$, helices of length $l(T)$ provide the dominant contribution to the general helic length, and the helic content can be written as

$$\vartheta(T) \sim P_{l(T)} \sim \exp[-(T-T_x)/\Delta T],$$

$$\Delta T = (T_A - T_B)^2 x(1-x)\Delta S/\Delta f_s. \tag{41.6}$$

The scale of an optimal section (41.5) grows sharply as T approaches T_x. At this point, however, our considerations become invalid, because with an increase in ϑ helix formation in different sections stops being independent. Quite similarly, the case $T < T_x$ can be analyzed, where the helices contain rare melted regions, and for the quantity $1-\vartheta$, an expression symmetric to Eq. (41.6) is obtained. A more detailed analysis (which we omit here) yields the following exact (under the condition $2\Delta f_s \gg (|f_A - f_B|)$ expression:

$$\vartheta(T) = \frac{1}{2} - \frac{\sinh[(T-T_x)/\Delta T] - [(T-T_x)/\Delta T]}{4\sinh^2[(T-T_x)/2\Delta T]}. \tag{41.7}$$

It is easy to demonstrate that the estimate (41.6) is derived from the last expression for $|T - T_x| \gg \Delta T$.

Significantly, Eq. (41.7) shows that the width ΔT of the transition diminishes with the growth of Δf_s according to the power law (41.6), whereas in a homopolymer, the corresponding value decays exponentially [$\sigma^{1/2} = \exp(-\Delta f_s/T)$]. In other words, the dependence of the width of the helix–coil transition on the parameter of cooperativity $\sigma \ll 1$ is logarithmic for a heteropolymer [$\sim (-\ln \sigma)^{-1}$] and follows a power law for a homopolymer ($\sim \sigma^{1/2}$). This effect is stipulated by the difference in spiralization energies for various links, so that ΔT is naturally proportional to $(T_A - T_B)^2$ and the width of the transition has a maximum value at $x = 1/2$ and diminishes for $x \to 0$ and $x \to 1$, (i.e., when the heterogeneity of the chain decreases because of a decrease in the fraction of one of the components).

Hence, the qualitative picture of the helix–coil transition in a heteropolymer can be described using the idea about the consecutive melting (on a temperature increase) of longer and longer sections enriched by the higher-melting component. No doubt, this treatment by no means overrules the role of ordinary thermal fluctuations. The latter can manifest themselves both in the random displacements of boundaries between helic and coil states and in the activation emergence of weakly unfavorable helices or the disappearance of weakly favorable ones. The investigation of these effects calls for a more consistent theory, and it should be noted that such a theory is quite complicated conceptually, some of its problems have not yet been solved, and, generally, is outside of the scope of this book. The essentials of this theory, however, arouse deep interest, associated not only with melting of heteropolymers but with other phenomena observed in disordered polymer systems. Accordingly, a few remarks are due here.

*41.5. *A consistent theory of a heteropolymer involves the analysis of non-commuting transfer matrices; the Green function of a heteropolymer is not a self-averaging quantity.*

While constructing the formal theory, it is natural to resort to the method of the transfer operator (*see* Sec. 6). Because of the specifics of a heteropolymer, we have several different operators corresponding to different links. In the simplest case, these are the matrices $\hat{Q}^{(A)}$ and $\hat{Q}^{(B)}$ of the type (40.3), with parameters s_A and s_B. It is easy to realize that the Green function (6.2) for a heteropolymer is expressed not via the single operator (40.3) [*see* Eq. (6.7)] but rather a product of different operators:

$$G = \prod_{\tau=1}^{N} \hat{Q}^{(\tau)}, \qquad (41.8)$$

where τ are the numbers of links and the operator $\hat{Q}^{(\tau)}$ either $\hat{Q}^{(A)}$ or $\hat{Q}^{(B)}$ depending on the type of the link τ in the given primary structure.

What is important is that the different operators $\hat{Q}$ do not commute with one another:

$$\hat{Q}^{(A)}\hat{Q}^{(B)} \neq \hat{Q}^{(B)}\hat{Q}^{(A)}.$$

This property has a direct and clear physical meaning. It is this property that leads to the dependence of the Green function (41.8), and of all other physical properties of the polymer, on the primary structure, not only on the number of links of different types.

The extreme values of the Green function (corresponding to the pure homopolymers poly-A and poly-B) are proportional to Λ_A^N and Λ_B^N, respectively (i.e., differ very strongly). Accordingly, if we average the Green function over all possible primary structures[j] of the heteropolymer, some exotic sequences (e.g, of homopolymer type), whose probability is vanishingly small, will provide a substantial (and possibly overwhelming) contribution to the average value. Thus, the average Green function fails to give a reasonable characteristic of the heteropolymer. More complicated and detailed characteristics must be investigated.

A special approach is needed to tackle the helix–coil transition in a double-stranded heteropolymer with the presence of the loop factor (40.19), because the method of the transfer operator (even a non-commuting one) cannot be applied directly in this case. We do not describe here the methods for the analysis of such a system because of their complexity, but a final formulation should be given: as in the case of a homopolymer, the transformation becomes a phase transition for

---

[j]If the neighboring links [i.e., the operator cofactors in Eq. (41.8)] are statistically independent, then the averaging of the product (41.8) is reduced to the averaging of the cofactors. Consequently, the averaged Green function has a "homopolymer" structure, being determined by a power of one average transfer operator.

$\alpha > 1$, with the transition behavior being the same as for the homopolymer with averaged s.

This relates to those approximate predictions about the melting of a heteropolymer, which can be made on the basis of insufficient information on the statistical properties of primary structure. According to subsection 38.5, the alternative approach is possible only for a specific, completely known primary structure. In this case, one must use for a single-stranded heteropolymer the obvious recursive relation

$$G^{(t+1)} = \hat{Q}^{(t)} G^{(t)} \tag{41.9}$$

and also to compute the corresponding Green functions one after another. There is a generalization of this recursive algorithm (also intended for computer-aided calculations) that allows the role of the loop factor (40.19) to be taken into account while considering the melting of a double-stranded DNA heteropolymer of moderate length. To illustrate the adequacy of such an approach, Figure. 7.7 compares the experimental DMC data for a DNA with the corresponding numeric calculations.

*41.6. *A homopolymer with mobile ligands adsorbing onto it is described by the averaged Green function of a heteropolymer.*

We mentioned earlier that the average value of the Green function (41.8) for a heteropolymer is determined by the contribution of atypical primary structures. This circumstance has a clear-cut physical meaning: the average Green function describes the other physical object (the so-called mobile heteropolymer) or the homopolymer with ligands. This system is realized when the solution contains molecules (ligands) capable of reversibly attaching to homopolymer links and thereby affecting the free energy of helix formation. In this situation, the chain represents at any moment of time a heteropolymer, because it contains links of two types; free, and with a ligand. The primary structure of the chain, that is, the number and distribution of ligands, however, is not fixed, so all possible sequences are present in the ensemble. Hence, the partition function of the chain with the ligands actually corresponds to the summation and averaging over the primary structures, because in the mobile heteropolymer, the most probable are the thermodynamically favorable structures, which are not at all typical for the genuine heteropolymer with frozen disorder.

It should be noted that the helix–coil transition in the presence of ligands is a real and important phenomenon. For example, many planar molecules are capable of wedging in between neighboring base pairs in the DNA double helix. The corresponding substances are frequently used as a tool in DNA research (e.g., one can monitor the optical activity or circular dichroism within the absorption band of the ligand). Therefore, we dwell briefly on the theory of the helix–coil transition in the presence of ligands.

Theoretically, the system under consideration can be described by several methods. One may introduce a transfer matrix of rank 4 (according to the number of states available for each link, namely, helic, coil, and with or without ligands). It also is possible to use generating functions, with $\zeta(p)$ in Eq. (40.13) being treated as a matrix. In all cases, the investigation is rather simple conceptually, so we suggest that the reader analyze the system, assuming that both spiralization constants s (for chain sections with and without ligands) and both ligand sorption constants (on helic and coil sections) are known. For simplicity, one can assume the parameter σ to be common and, moreover, the adsorption of the ligands to be quite strong, disregarding changes of their concentration in the surrounding solution.

In general terms, the situation is reduced to a redistribution of the ligands among helic and coil sections. This process has no effect on general non-phase properties of the helix–coil transition or on the fluctuating distribution of helic and coil sections along the chain. Clearly, as the number of ligands in the polymer varies, the system is gradually transformed from one extreme homopolymer regime to the other (i.e., from the chain with no ligands to the chain filled completely and uniformly by ligands). Naturally, the temperature of the helix–coil transition varies monotonically as this occurs. Interestingly, the width of the transition region does not vary monotonically: it is most narrow for both extreme homopolymer regimes and broadest for a certain intermediate number of ligands.

42. COIL TERTIARY STRUCTURES OF CLOSED-RING DNA

42.1. *Many biologically important properties of DNA's are of topological origin.*

Relatively short DNA's encountered in some simple biological systems (e.g., plasmids, viruses, and so on) often (and possibly always) have the shape of closed rings. It should be emphasized that in a native circular DNA, either strand is necessarily linked with itself and not with the other, because the chemical structure of the sugar-phosphate backbone of DNA excludes the crosswise connection. It is difficult to judge whether long DNA macromolecules incorporated in cell nuclei form closed rings, but it is known that they sometimes comprise relatively free loops whose ends are firmly attached to special dense protein formations.

The biologic significance of the topological properties of such structures is already clear because of the existence in living cells of topoisomerases, which are special enzymes capable of realizing a mutual crossing of sections in both single- and double-stranded DNA's. There are even some enzymes (DNA-gyrase) that transform a DNA into a thermodynamically unfavorable topological state by twisting the strands of the double helix.

Hence, the development of basic physical ideas about tertiary structures in DNA's calls for a preliminary examination of topological properties. It is remarkable that these properties appear even in isolated DNA molecules extracted from a cell or virus and those diluted in good solvent. Because of the stiffness of DNA, the initial theoretical study of such a system does not demand that we take into account volume interactions. Indeed, in a DNA, the ratio of the effective segment length $l \approx 100$ nm to the thickness $d \approx 2$ nm is very large: $p = l/d \approx 50$. As shown in subsection 26.6, the volume effects for the corresponding linear chain become essential when the contour length of the chain exceeds $p^3 d$, which corresponds to approximately 10^6 base pairs. In fact, in a ring chain, the topological properties already begin to depend substantially on volume interactions at smaller lengths. Nevertheless, there exists even in this case a sufficiently wide range of chain lengths in which the topological properties are independent of volume interactions. This is the situation considered in the following subsection.

42.2. *The state of a closed-ring DNA is characterized by two topological invariants: the type of knot formed by the double helix as a whole, and the linking number of one strand with the other.*

As we already know, the topological state of a ring polymer strongly affects all of its physical properties, even in the absence of volume interactions (*see* Sec. 11): the type of knot determines the size of macromolecules, virial coefficients of their interaction, and so on. Certainly, all of this also relates in full measure to circular DNA's, both closed-ring (i.e., those in which both strands are closed into rings) and open-ring (i.e., those in which only one strand is closed while the other is open). The state of a closed-ring DNA, however, is additionally characterized by another topological invariant, that is, the linking number of strands with each other. This quantity is appropriately denoted by Lk. Recall that the linking number of two contours with each other is defined as an algebraic (i.e., accounting for the direction) number of crossings made by one of the contours through the surface spanning the other [i.e., the Gauss integral (11.6)].

The spatial structure of closed-ring DNA's depends dramatically on the value of Lk. This is one of the most significant characteristics of DNA, because as we have repeatedly emphasized, for the absolute majority of ordinary polymers, the spatial structure is determined only by volume interactions. This characteristic of DNA is believed to play a vital role in the functioning of relevant biologic systems. Admittedly, it usually is not easy to examine in detail how topology affects the properties of complex biologic structures. From a physical viewpoint it is natural to explore initially the topologic effects on the spatial shape of a closed-ring DNA in dilute solution. It should be noted, however, that this problem has not yet been solved in its full formulation. For instance, attempts to calculate exactly the gyration radius of a closed-ring DNA as a function of Lk have failed; still, some fundamental facts and notions established in this area have made it possible to solve a number of important problems. In particular, the torsional stiffness of the DNA double helix was found for the first time from experimental data. The next subsection is devoted just to these facts and notions.

42.3. *The minimum of the energy of a closed-ring DNA corresponds to the superhelical state; the number of twists in a superhelix depends on the order of strand linking, Lk.*

Let γ denote the number of base pairs per turn of a linear, free double helix. We know that usually $\gamma \approx 10$, but the value of this quantity may depend on external conditions (temperature and, generally, concentration in the solution of such substances that can wedge in the double helix between base pairs).

Now imagine a linear DNA consisting of N base pairs. Suppose we locate it on a plane, bend it into a circle, and link the ends without introducing any torsional stress. In the closed-ring DNA thus obtained, $Lk = N/\gamma$ (as expected, $Lk \sim N \gg 1$); However, such a value of Lk is not mandatory. Indeed, just before linking the ends, we can twist one of them an arbitrary number of times τ around the axis (Fig. 7.9a). Then, we obtain

$$Lk = N/\gamma + \tau. \tag{42.1}$$

The quantity τ is called the *number of supertwists* (the origin of this term is commented on later); frequently, it is superseded by the quantity $\sigma = \tau\gamma/N$, called the *density of supertwists*. (This is the number of supertwists per turn of the initial helix.)

A DNA closed with some non-zero number of supertwists is in a torsionally stressed state. Because neighboring segments are not coaxial, the torque makes one segment rotate around the axis of the other, thus displacing the chain from the plane in which the DNA was initially located. Obviously, elastic equilibrium of the system will correspond to a conformation of the type shown in Figure

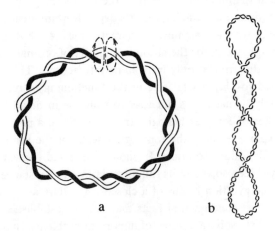

a   b

FIGURE 7.9. Possible twisting of the double helix before closing to form a ring (a) and the resulting superhelix (b).

7.9b: the double helix as a whole forms a superhelix[k] giving the terms for the quantities τ and σ. Of course, the quite symmetric superhelic shape shown in Figure 7.9b corresponds only to the minimum of the elastic energy, just as the rectilinear form of the linear persistent chain. As in the linear chain, the bending fluctuations of the superhelix are very large when no attractive volume interactions are present. Even though the comprehensive theory of these fluctuations has not yet been developed, it is qualitatively clear that the transformation of a molecule into a superhelical form makes it more compact by reducing, for example, its radius of gyration, or more precisely, by making the swelling parameter less than unity.

42.4. *Experimentally, the superhelical state manifests itself in the growth of mobility of the molecule in the process of formation of a superhelix; a real DNA is always negatively superspiralized.*

A decrease in the coil size on formation of a superhelix and an increase in the number of superturns leads to the growth of such characteristics of the macromolecule as mobility in an external field, sedimentation constant, and so on. Specifically, a change in electrophoretic mobility during formation of a superhelix is so conspicuous that it can be successfully used to separate fractions of molecules differing in values of τ or Lk by unity (i.e., to measure these discretely changing quantities accurately). One can also observe changes in the superhelical state resulting from the experimental variation of the value of γ by introducing into the solution small molecules (similar to the ligands described in subsection 41.6), which by wedging in between base pairs in the double helix modify the degree of its equilibrium twisting.

In this way, it was found that a real closed-ring DNA essentially always resides in the superhelical state, with the density of supertwists being negative ($\sigma \approx -0.05$). The negative sign of σ means that the torsional stress in the superhelix tends to untwist the double helix. Clearly, this promotes a partial despiralization (via the helix–coil transition occurring over a portion of the molecule); in fact, this often leads to a change in the secondary structure (*see* subsection 42.10). If the ordinary right-hand, double-stranded secondary structure is stable enough [i.e., the helix formation constant s (40.1) is sufficiently high], however, then the equilibrium and fluctuations of the spatial form of the superhelix are determined by the balance between torsional and bending elastic forces. This is because there is a strict geometrical relationship between torsional twisting and bending of a closed-ring DNA.

42.5. *The axial twisting of strands around each other may differ from the order of their linking by the amount of writhing, which depends on the spatial form of the axis of the double helix.*

The distance between strands in a DNA (i.e., the thickness of the double helix $d \approx 2$ nm) is much less than its intrinsic curvature radius (i.e., the Kuhn segment length $l \approx 100$ nm). Therefore, we can, first, speak about the spatial shape of the

---

[k]We advise the reader to learn this "experimentally" by twisting in his or her hands a thick rope or, better still, a rubber pipe.

double helix as a whole or about its axis C and, second, describe the degree of twisting by the number of turns that one strand makes around the other (i.e., by the twisting parameter Tw). If the axis of the double helix forms a plane contour, then one can see from Figure 7.9a that Tw equals the order of linking of the strands, Lk. It was formally proved by J. White in 1969[23] that

$$Lk = Tw + Wr, \qquad (42.2)$$

where the writhing Wr equals

$$Wr = \frac{1}{4\pi} \oint_C \oint_C [dr_1 \times dr_2] \frac{r_{12}}{|r_{12}|^3}; \quad r_{12} = r_1 - r_2. \qquad (42.3)$$

The linking Lk is the Gauss integral (11.6) for the contours of two strands and is also a topologic invariant, whereas Wr (42.3) is the analogous integral for one contour and is not an invariant. White's theorem (42.2) characterizes the geometry of the closed, smooth, two-sided strip (i.e., having two edges corresponding to two DNA strands and not one edge as in the Moebius loop).

Before proving the relation (42.2), we first cite the exact expressions for the quantities Lk and Tw. If the spatial forms of the contours of both strands (C_1 and C_2) are known, then the order of their linking can be given by the Gauss integral (11.6)

$$Lk = \frac{1}{4\pi} \oint_{C_1} \oint_{C_2} \frac{[dr(s_1) \times dr(s_2)] r_{12}}{|r_{12}|^3}, \qquad (42.4)$$

where s_1 and s_2 are the lengths along the contours C_1 and C_2, respectively (see subsection 11.3). On the other hand, the degree of strand twisting associated with torsional deformation can be written as

$$Tw = \frac{1}{2\pi} \oint_C ds \ \varphi(s), \qquad (42.5)$$

where $\varphi(s)$ is the angle (per unit length of the chain) through which the base pairs are turned relative to one another.

From the previous discussion about the closing of a plane contour, it follows that the quantities Tw and Lk are interrelated, even though Eqs. (42.4) and (42.5) do not show this. It is easy to realize that the relation between Tw and Lk is actually associated with the fact that the strands of the double helix do not form arbitrary contours; the strands follow each other over the entire length at a distance that is very small with respect to the curvature radius, which evidently is of the order of the persistent length of the double helix. This is exactly why a change in the order of linking Lk necessitates a change in the degree of twisting, provided that one of the contours is fixed.

This circumstance can be explained by depicting a DNA in the form of a narrow, bent, closed strip, with its edges running along the sugar-phosphate backbones of the both strands. Let $r(t)$ denote the radius vector of the point t lying on the axis of this strip (i.e., the axis of the double helix). Next, let $a(t)$ denote the unit vector perpendicular to the axis of the strip and lying on the strip surface. Then, the radius vectors of points on opposite edges of the strip equal $r_1 = r(t) + \varepsilon a(t)$ and $r_2 = r(t) - \varepsilon a(t)$, where 2ε is the width of the strip. These are the equations for both contours $r_1(s_1)$ and $r_2(s_2)$, which are to be inserted in the expression (42.4) for Lk. Because Lk is a topological invariant (see subsection 11.3), it does not depend on ε [provided that ε is already small enough, so that mutual topology of $r_1(s_1)$ and $r_2(s_2)$ contours does not change any more with further decrease of ε]. This is why we are interested in the limit $\varepsilon \to 0$. It is impossible, however, to set $\varepsilon = 0$ directly because of the singularity in the integral (42.4) at $s_1 = s_2$.

Let us consider in the integral along the second contour (with respect to s_2) [see Eq. (42.4)], a δ-neighborhood of the point s_1 (the concrete value of δ is chosen later). Within the small δ-neighborhood, the contour $r(t)$ can be regarded as a section of a straight line:

$$r(t) \cong r(s) + \dot{r}(s)(t-s); \quad \dot{r}(t) \cong \dot{r}(s) \equiv \dot{r},$$

provided that $|t-s| < \delta$ is much less than the length of the Kuhn segment in the double helix, $\delta \ll l$. Here, $\dot{r}$ is obviously the unit vector of the corresponding tangent. Then,

$$dr_1 = \dot{r} ds_1 + \varepsilon da(s_1); \quad dr_2 = \dot{r} ds_2 - \varepsilon da(s_2);$$

$$[dr_1 \times dr_2] = \varepsilon ds_1 ds_2 [(\dot{a}(s_1) + \dot{a}(s_2)) \times \dot{r}];$$

$$r_{12} = (s_1 - s_2)\dot{r} + \varepsilon(a(s_1) + a(s_2));$$

$$(r_{12})^2 = (s_1 - s_2)^2 + 2\varepsilon^2(1 + a(s_1)a(s_2)). \qquad (42.6)$$

(Here, we took into account that the unit vectors $\dot{r}(s)$ and $a(s)$ are mutually perpendicular.) Inserting the relations (42.6) in Eq. (42.6), we find that the contribution of the δ-neighborhood to Lk equals

$$\int_0^L ds_1 \int_{s_1-\delta}^{s_1+\delta} ds_2 \; \varepsilon^2 \frac{(a(s_1)+a(s_2))[(\dot{a}(s_1)+\dot{a}(s_2)) \times \dot{r}]}{\{(s_1-s_2)^2 + 4\varepsilon^2(1+a(s_1)\dot{a}(s_2))\}^{3/2}}$$

$$\cong \frac{1}{2\pi} \int_0^L ds \; \dot{a}(s)[\dot{r}(s) \times a(s)] 2\varepsilon^2 \int_{s-\delta}^{s+\delta} \frac{ds'}{\{s'^2 + 4\varepsilon^2\}^{3/2}}$$

$$\cong \frac{1}{2\pi} \int_0^L ds \; \dot{a}(s)[\dot{r}(s) \times a(s)] 2\varepsilon^2 \int_{-\infty}^{+\infty} \frac{ds'}{\{s'^2 + 4\varepsilon^2\}^{3/2}},$$

where the value of the last integral with respect to s' is easily evaluated to give $1/2\varepsilon^2$. We choose δ to be smaller than the characteristic length of $a(s)$ rotation, so $a(s_1) \cong a(s_2)$ in the range of integration (first simplification); on the other hand, if $\delta \gg \varepsilon$, then the integration can be extended to infinity. Such a choice for δ is possible in the $\varepsilon \to 0$ limit.

The rest of the integral (42.4) has no singularities, so we can directly set $\varepsilon = 0$, that is, to regard the contours C_1 and C_2 as coinciding. It is remarkable that after this operation, the double integral has no singularity at $s_1 = s_2$, because for a smooth curve (with curvature radius $\sim l$), the vector product of unit tangents $[\dot{r}(s_1) \times \dot{r}(s_2)]$ tends to zero with the points converging as $(s_1 - s_2)^2/l$. Therefore, the condition $\delta \ll 1$ allows one to set $\delta \to 0$ in this integral and to remove the non-physical auxiliary parameter δ. Eventually, we obtain $Lk = Tw + Wr$ (42.2), where

$$Tw = \frac{1}{2\pi} \int ds \ \dot{a}(s) [\dot{r}(s) \times a(s)], \qquad (42.7)$$

$$Wr = \frac{1}{4\pi} \oint_C \oint_C ds_1 ds_2 [r(s_1) \times r(s_2)] \frac{r_{12}}{|r_{12}|^3} \qquad (42.8)$$

The expression (42.8) obtained for the writhing obviously coincides with the one given earlier, (42.3). The expressions for the twisting, (42.7) and (42.5), are also identical. This can easily be understood if we recall that the direction of the axis of the double helix, $\dot{r}(s)$, changes over lengths $\sim l$ but remains practically invariant over the length of one turn of the double helix, where the vector $a(s)$ makes a complete turn around the vector $\dot{r}(s)$. Consequently, $\dot{a}(s) = \varphi(s)[a(s) \times \dot{r}(s)]$, and the substitution of this expression into Eq. (42.7) immediately transforms it into Eq. (42.5).

42.6. *The writhing is a geometrical characteristic of the spatial form of the contour of a closed polymer.*

Moving to the discussion of Eq. (42.2), it should first be noted that its left-hand side contains the topological invariant Lk, a quantity taking only integer values. The right-hand side, however, comprises two geometrical characteristics, Tw and Wr, each of which separately (in contrast to their sum) can vary continuously.

Of principal importance is that the writhing (42.3) depends only on the spatial form of the axis of the double helix (or the strip), but it is independent of the way in which the DNA strands (or the strip) are wound on this axis. Consequently, the writhing can also be defined for a single-strand polymer, including (in formal terms) a non-closed polymer.

The writhing is a dimensionless number that remains constant with a similarity transformation of the contour (i.e., without a change in its form). The writhing of the spatial curve equals zero when the curve has a plane or a center of symmetry. Clearly, this is because the definition of writhing (42.3) features a vector product. Consequently, in a certain sense, the writhing of the curve may be regarded as a measure of its mirror asymmetry or chirality.

The writhing may seem to be a continuous function of the form of a curve. This is not exactly the case. The relationship of the writhing with topologic properties shows in the fact that it changes continuously with a change in the shape of the curve until the segments of the curve cross one another; the crossings make the writhing jump from the value 2 to −2 (depending on the direction of the crossing). We suggest that the reader practice by proving this.

For a polymer chain, the writhing fluctuates together with the conformation. Next, we discuss the statistical properties of the writhing.

42.7. The conformational entropy of a ring polymer with fixed topology is supposed to be a quadratic function of the writhing.

Recall that the Gaussian distribution of the end-to-end distance for an ordinary, linear ideal chain (4.1) is associated with the conformational entropy of the macroscopic state (with a fixed end-to-end distance), which is proportional to that distance. A broad generalization of this is provided by the Lifshitz formula for the conformational entropy of a linear chain in a state specified more definitely (i.e., in terms of a smoothed density distribution [see Sec. 9]).

Turning to the physics of ring polymers, it is also natural to explore their conformational entropy. In doing this, we should not take into account (just as for the linear chains) volume interactions and torsional stiffness. It should be noted, however, that even though no comprehensive theory has been developed in this field so far, the tentative evaluations and computer-simulation data[25] given here nevertheless provide a sufficiently clear qualitative picture.

First, it is obvious that the conformational entropy of a ring polymer strongly depends on the topologic type of the polymer knot (i.e., for a DNA, on the first of the topological invariants mentioned in subsection 42.2). Next, we must agree on how to define the macroscopic state whose conformational entropy is of interest to us. Keeping in mind the subsequent analysis of torsional elasticity effects, we consider macroscopic states with a given value of the writhing. Then, by analogy with a linear chain and its end-to-end distance, the problem is reduced to a statistical analysis of the probability $P(Wr)$ of writhing values for an immaterial ring of given topology.

Of course, the average value of the writhing, $\langle Wr \rangle$, depends on the type of knot. It is immediately obvious that for mirror-symmetric knots, $\langle Wr \rangle = 0$. In particular, this is valid for the most important case of an unknotted ring (or strictly speaking, of a trivial knot). For a mirror-asymmetric knot, $\langle Wr \rangle \neq 0$, with the values of $\langle Wr \rangle$ for right- and left-handed versions coinciding in modulus and having opposite signs.

Computer-simulation experiments show that the probability distribution of the writhing in the region of its small fluctuations is very close to a Gaussian:

$$P(Wr) = \text{const} \cdot \exp[-(Wr - \langle Wr \rangle)^2/2\langle (\Delta Wr)^2 \rangle],$$

$$\Delta Wr \equiv Wr - \langle Wr \rangle, \tag{42.9}$$

According to the same data, the writing variance depends linearly on the chain length:

$$\langle (\Delta \mathrm{Wr})^2 \rangle = cL/l. \tag{42.10}$$

In this formula, the factor l^{-1} is introduced for reasons of dimensionality, L/l is the number of effective segments in the chain, and c is the dimensionless constant. According to computer-simulation experiments, $c \approx 0.1$.

In essence, Eqs. (42.9) and (42.10) signify that the writing is a thermodynamically additive quantity. The definition (42.3), however, does not show this explicitly. To understand the situation qualitatively, it is useful to consider the following example. Suppose that an unknotted contour has the form of a M-turn helix (or superhelix in the case of a DNA) with a very small (infinitesimal) pitch (Fig. 7.10) and a short (infinitesimal) connector linking its ends. There are 2^M such contours, because two directions are feasible for each turn of the helix; in other words, the connector can pass either inside or outside each coil.

Let us now find the distribution of the writing in this set of 2^M contours. The writing in this situation obviously equals the sum of independent contributions of all M turns. Each contribution equals either 1 or -1 with an equal probability of $1/2$, because an alteration of the position of the connector with respect to any turn changes the writing by ± 2. [The reader can confirm this by direct evaluation of the integral (42.3)]. Thus, we see that the writing is additive in this particular model; for large M, its distribution is indeed reduced to the normal one (42.9), with $\langle \mathrm{Wr}^2 \rangle = M$, that is, $\langle \mathrm{Wr}^2 \rangle$ is proportional to the chain length in exact correspondence with Eq. (42.10).

No doubt, a real superhelix is not shaped like the model shown in Figure 7.10 (cf. Fig. 7.9b), and its fluctuations involve primarily violations of the spatial arrangement of supertwists. The calculation performed for the model of Figure 7.10 must be regarded merely as an illustration. Still, the expressions (42.9) and (42.10) are believed to be true, because the corollaries following from them have been confirmed experimentally (*see* the next subsection).

For appreciable deviations from the average exceeding $\sim 5 \langle (\Delta \mathrm{Wr})^2 \rangle^{1/2}$, the writing probability distribution deviates substantially from normal. This determines the structures of strongly twisted superhelical DNA's. To investigate the non-Gaussian part of the distribution, special approaches should be taken, gener-

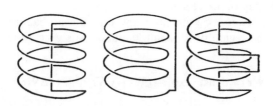

FIGURE 7.10. Conformations of M-turn helix with different values of writing.

ating in reality or via statistical sampling in a computer simulation such strongly twisted structures. This book does not examine these problems.

As mentioned, the writhing probability distribution can be interpreted in terms of conformational entropy:

$$S(\mathrm{Wr}) = -(\mathrm{Wr} - \langle \mathrm{Wr} \rangle)^2/2\langle(\Delta \mathrm{Wr})^2\rangle. \tag{42.11}$$

This relation is quite analogous to Eq. (8.1). As in Eq. (8.1), the entropy (42.11) for the chain with the persistent stiffness mechanism (and, generally, with any stiffness mechanism prohibiting free internal rotation) involves elastic bending energy.

*42.8. *The free energy of superhelical stress is proportional to the square of the density of superturns (i.e., the Hooke law is valid); the corresponding effective modulus of elasticity depends on the bending and torsional stiffness of a polymer.*

Knowing the conformational entropy (42.11), we then can find the free energy of the double helix, taking into account its torsional stiffness (but neglecting volume interactions as before). In fact, the elastic energy of torsional deformation is determined by the twisting and evidently equals

$$E_{\mathrm{tors}} = (g/2N)[2\pi(\mathrm{Tw} - N/\gamma)]^2, \tag{42.12}$$

because according to Eq. (42.1), the torsional stress is absent at $\mathrm{Tw} = N/\gamma$. Here, N is the number of base pairs and g the modulus of torsional elasticity per pair.[1] In addition, we took into account that the modulus of elasticity diminishes N-fold on joining of N "springs" in series. [The factor 2π in Eq. (42.12) is associated with the definition of the twisting.]

Consequently, the partition function of the chain in a state with given topology (e.g., for a given value of Lk) equals[m]

$$Z = \int \exp(-E_{\mathrm{tors}}(\mathrm{Tw})/T) \, \exp(S(\mathrm{Wr})) \, \delta(\mathrm{Tw} + \mathrm{Wr} - \mathrm{Lk})d\mathrm{Tw} \, d\mathrm{Wr}.$$

The calculation of this partition function is elementary, because the integral is Gaussian. To avoid unwieldy formulas, we only cite the result for the case $\langle \mathrm{Wr} \rangle_0 = 0$, which corresponds to the unknotted state of the double helix or the state in which the double helix forms a mirror-symmetric knot. The symbol $\langle ... \rangle_0$ here denotes averaging without allowing for torsional stiffness, that is, when $g = 0$ [which was in fact supposed in the derivation of Eq. (42.11) and even

---

[1]Here is the definition of g: the turn of two neighboring pairs through an angle φ with respect to one another requires the energy $(1/2)g(\varphi - 2\pi/\gamma)^2$, where $2\pi/\gamma$ is the angle between neighboring base pairs in a nonstressed double helix. In fact, in a DNA, $g \approx 110$ kcal/(mol rad), that is, $g/T \approx 200$ 1/rad at room temperature. The corresponding rms fluctuation of the angle of helic rotation of neighboring base pairs equals $\Delta\varphi = (T/g)^{1/2} \approx 0.07 \approx 4°$.

[m]This is the integral over conformations. Of all conformational variables, the torsional energy E_{tors} depends only on the writhing. That is why the writhing is so convenient to take as one of the generalized coordinates. From this viewpoint, $P(\mathrm{Wr})$ is the Jacobian of transition to these coordinates.

earlier]. This example corresponds to a DNA with one broken strand, in which one strand is closed to form a ring and thus fixes the topology while the other is broken to permit free torsional relaxation. The result takes the form

$$\mathscr{F} = -T \ln Z = \frac{\tau^2}{2L} \left[\frac{T \cdot l_{\text{tors}}}{1 + c l_{\text{tors}}/l} \right]. \tag{42.13}$$

Here, we used l_{tors} to designate the quantity that is naturally called the *effective torsional segment*:

$$l_{\text{tors}} = 4\pi^2 (g/T)(L/N), \tag{42.14}$$

where L/N is the double-helix length per pair of bases. It is interesting to compare the expression (42.14) with Eq. (2.9), giving the ordinary effective segment in terms of the bending modulus of a persistent chain. It is evident that l_{tors} is the length over which the angle of fluctuation twisting becomes of the order unity. In a real DNA, $L/N \approx 0.34$ nm, $l_{\text{tors}} \approx 2600$ nm, and $l \approx 100$ nm.

The most significant feature of the result (42.13)[n] is that $\mathscr{F} \sim \tau^2$. We have already mentioned that the superhelic state is stressed; then, the quantity τ can be treated as a strain. The expression $\mathscr{F} \sim \tau^2$ signifies that the emerging stress grows linearly with superhelic strain (i.e., the Hooke law is valid). The corresponding proportionality coefficient represents the effective modulus of elasticity.

42.9. *The relationship between bending and torsional elasticity defines what fraction of supertwists is realized via axial twisting of the strands and what fraction via an increase in the writhing (i.e., via twisting of the double helix as a whole in space).*

Were the torsional stiffness very high, rotational fluctuations would be impossible (i.e., the relation $\text{Tw} = N/\gamma$ would hold). In addition, the expression $\text{Wr} = \text{Lk} - N/\gamma = \tau$ would necessarily be valid, meaning that chain fluctuations would be restricted to conformations with fixed writhing (i.e., to superhelices with τ twists).

On the other hand, if the bending stiffness was very high, then the chain would take on the conformation with the least possible curvature, that is, for the unknotted system, the shape of a plane circle. In this case, there would be $\text{Wr} = 0$, and consequently, $\text{Tw} - N/\gamma = \tau$, that is, the superturns would act so as to completely untwist the double helix itself (or in a hypothetic case $\tau > 0$, to twist it still more tightly).

In the general case of the arbitrary torsional stiffness g, the following relation is obvious for the total number of superturns (42.1):

$$\langle Wr \rangle_g + \langle Tw - N/\gamma \rangle_g = \tau,$$

---

[n]Historically, the expression (42.13) was not derived but rather found experimentally: in the presence of sufficient amount of topo-isomerases in a DNA, the equilibrium Boltzmann distribution $\exp(-\mathscr{F}/T)$ over topo-isomerases (i.e., over τ) sets in, which can be observed. This is exactly how the value of l_{tors} or g was found (M. D. Frank-Kamenetskii *et al.*, 1979)

Let us examine the distribution between the torsionally and conformationally induced superturns. From Eq. (42.13), it can easily be seen that

$$\langle \mathrm{Wr} \rangle_g = (1/Z) \int \exp(-E_{\mathrm{tors}}/T) \exp S \cdot \delta(\mathrm{Tw} + \mathrm{Wr} - \mathrm{Lk}) \mathrm{Wr} \, d\mathrm{Tw} \, d\mathrm{Wr}.$$

Omitting simple calculations, for $\langle \mathrm{Wr} \rangle_0 = 0$, we obtain

$$\frac{\langle \mathrm{Wr} \rangle_g}{\langle \mathrm{Tw} - N/\gamma \rangle_g} = \frac{c l_{\mathrm{tors}}}{l} \approx 2.7,$$

in complete agreement with the qualitative considerations given above.

42.10. *The stresses caused by negative superspiralization may bring about a transformation of the secondary structure of some DNA sections with special sequences.*

We have already mentioned that negative superspiralization tending to untwist the strands of the double helix could result not only in torsional stresses and spatial bendings but also in modifications in the secondary structure. As a rule, the superhelic stress is insufficient to destroy the secondary structure, even though the helix–coil transition over a small length of the molecule could of course release the stresses totally. In reality, the superhelic stress can be released not only by breakdown but also by a transformation of the secondary structure.

For example, it is obvious that a transition of a portion of the macromolecule from the right-hand B helic state to the left-hand Z helic state (*see* subsection 39.3) favors a decrease in superhelicity. The properties of a Z form in DNA are such that the indicated transition is thermodynamically favorable and actually proceeds only on sections where the purine and pyrimidine nucleotides alternate.

Another special sequence in the primary structure, a palindrome (*see* subsection 38.1), can also favor stress release. The untwisting of such a section proves to be favorable, because its untwisted strands can then form separate self-helices with a cross-like structure. Certainly, the palindrome must be quite long for this to happen. There are some other non-canonic structures that can be formed at certain sections of DNA.

43. GLOBULAR TERTIARY STRUCTURES OF DNA AND RNA

43.1. *Giant DNA molecules exist in biological systems in a very complex globular state.*

Even for polymers, DNA molecules are extremely large, often reaching 10^9 base pairs, which corresponds to a contour length $\sim 10^8$ nm. Because the statistical segment of the double helix ~ 100 nm, the size of a Gaussian coil in such a DNA might be of the order 10^5 nm. Meanwhile, in real biological systems, a DNA is known to be packed within a volume with linear dimensions ~ 100 nm. The characteristic volume of the double helix 10^8 nm long corresponds to a sphere of radius ~ 100 nm. Consequently, it is clear that real DNA is in a packed globular state; however, this globule is very complicated. Its structure has many intermediate (between the double helix and the globule as a whole) hierarchic levels. For example, the simplest is represented by a so-called nucleosome, a protein particle a few nanometers long on which the DNA double helix is wound as a thread on a spool (the double helix makes two or three turns around the spool as a common wire). The nucleosomes in turn combine to form the structures of the next level, and so on. It is significant that, first, the proteins (being macromolecules themselves) act in chromatin as small-scale ligands and, second, the chromatin itself is a dynamic functioning system ensuring DNA replication, transcription, and so on. It is evident that a simpler model must be found for an initial physical investigation.

43.2. *Although a DNA double helix is stiff and charged, there are some solvents favoring its being in a globular state; a DNA is efficiently compressed by adding another flexible polymer to the solution; and usually, DNA globules have a toroid shape and a liquid-crystalline internal structure.*

At first glance, an elementary model of the condensed state could be a globule that a double helix should form as a whole in poor solvent. Choosing a solvent for this purpose is not easy, however, not only because the DNA links are negatively charged and repel one another and the double helix has a high bending stiffness,° but also because of the need to avoid effects of the solvent on the internal microscopic structure of the globule. Both of these problems can be solved most efficiently by using a dilute or semidilute solution of another polymer that is flexible enough and not attracted to the DNA. Under these conditions, the immiscibility of stiff and flexible polymers provokes globulization of the double helix. In the process, the emerging DNA globule expels almost all of the flexible chains, because otherwise, each link of these chains might form unfavorable contacts. The thermodynamic gain resulting from expelling a flexible chain from the DNA globule is thus proportional to the number of chain links, whereas the entropy loss from this process does not exceed a value of order unity for the whole chain. It can be said that DNA compression is caused by the osmotic pressure the coils of a flexible polymer exert on the "walls" of the globule (Fig. 7.11). The absence of coils inside the globule guarantees their noninterference in its internal structure.

The DNA globulization in polymer solution just described is sometimes referred to as ψ *condensation*. The Greek letter ψ was chosen to serve as an abbreviation for polymer-solvent-induced (PSI) condensation. The globule itself can be pictured correctly by assuming the DNA double helix to be a persistent (worm-like) polymer. The analysis shows that a DNA of moderate length forms a toroidal globule, and Figure 7.12 illustrates how the molecules are located

---

°A growth of bending stiffness removes the helix–coil transition point away from the θ point, that is, inhibits a globule formation even for a long chain [*see* Eq. (21.8)], especially as it pertains to chains of real moderate length, for which the size of the globule is compared with the effective segment and globule formation involves a very substantial bending of the chain.

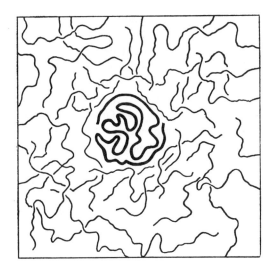

FIGURE 7.11. Globular state of a stiff polymer chain in a solution of flexible chains.

within the toroid. From a local viewpoint, the elements of the globule are seen to be regions with orientational ordering (i.e., regions of a liquid-crystalline phase).

A DNA can also be globulized using moderately long chains, each containing several positively charged links (e.g., these may be proteins with an appropriate primary structure). By forming salt bonds with negatively charged DNA links, such chains serve as cross-links between remote sections of the DNA. The internal structure of the globule, however, strongly depends in this case on the properties of the cross-links themselves.

In conclusion, it should be noted that the role of topological constraints and torsional stiffness in the collapse of a closed-ring DNA and its globular structure still needs to be investigated thoroughly.

43.3. *The spatial structure of a single-stranded RNA is formed by the twisting of complementary double-helix sections located far from one another in the chain; this is called a cloverleaf structure.*

Unlike DNA, RNA resides in a living cell, usually as a single strand without its complementary counterpart. If, however, conditions are such that the spiralization constant is large ($s > 1$), then the chain sections tend strongly to interlace

FIGURE 7.12. Toroid globule.

in pairs. Because RNA is a heteropolymer, however, only those pair sections whose primary structures turn out randomly to be complementary can interlace. Moreover, two interlacing strands should be oppositely directed. Hence, the helic regions are formed in a quite definite fashion. The helix formation of the first section produces a structure with one loop and two tails, the next helix formation is possible either for two loop sections or for tail sections, and so on. What is significant is that the interlacing of two sections from different loops (even though these sections are complementary) is strictly prohibited because of topological constraints. A theoretically possible interlacing of a tail and a loop is also essentially forbidden. The resulting structure is called a cloverleaf structure. The number of helices in a real RNA is very large, reaching many hundreds. Of course, the cloverleaf subsequently forms a certain spatial structure of globular type, whose properties and distribution of RNA sections depend on the volume interactions of both helical and linear portions of the chain.

The following important physical question arises here: does the cloverleaf structure that is observed experimentally correspond to thermodynamic equilibrium for the given RNA macromolecules, or does the emergence of some helices fix the structure kinetically? This question has not yet been solved, but the comments that we make in Sec. 44 in connection with the analogous problem for proteins are quite applicable here as well.

A homopolymer RNA, had it existed in nature, would not have formed cloverleafs, because there would have been no complementary sections in it. Even so, we may consider a periodic heteropolymer with a short repeating interval [e.g., poly(AU), poly(GC), poly(AGCU), and so on], in which the complementary connection is feasible on any (with a possible shift by a fraction of the period) section. Thus, such a polymer forms cloverleafs with an annealed structure (or "floating" loop structure). Such a structure was reported by P. G. de Gennes [Rep. on Progress in Physics, v. 32, n. 2, p. 187–205 (1969)].

44. TERTIARY STRUCTURES OF PROTEINS

Most proteins functioning within a cell in a globular state are appropriately called *globular proteins*, and it appears that the globular proteins are the most complex molecules known. Typically, protein chains comprise from 100 to 10,000 links, with corresponding dimensions of protein globules reaching tens of nanometers. It should be noted immediately that the current physical concepts about globular proteins are far from comprehensive. Researchers come across some sophisticated problems associated with the limits of applicability of mechanical and statistical treatments, because the behavior of globular proteins has both mechanical and statistical aspects. The almost-complete set of modern methods of molecular physics and biochemistry is used for the experimental investigation of globular proteins.

In this book, we touch on the properties of globular proteins only briefly, paying primary attention to the polymer aspect. More extensive information is given in Refs. 9, 46, and 47.

44.1. *The native structure of each globular protein is characterized by a quite definite layout of the chain and possesses the property of self-organization.*

The most extraordinary and novel feature of globular proteins is the uniqueness of their spatial structure in the native (i.e., found in nature) state. This signifies that all globules of a given protein (i.e., all polypeptide chains with a given primary structure) are quite identical in their tertiary structure. Of course, this property by no means reduces to the smallness of fluctuations in the ordinary globule (*see* subsections 7.2 and 20.1), where only such rough characteristics of the spatial structure like density, size, and so on fluctuate weakly. In the native state of a protein globule, the shape of the entire polypeptide chain is specified in a unique way; in other words, the coordinates of most atoms ($\sim N$, where N is the polymerization degree) are accurately fixed to within thermal vibration amplitudes.

It has been established experimentally for many proteins that a polypeptide chain placed in appropriate external conditions (temperature, pH, and so on) spontaneously begins forming the "correct" native spatial structure. This phenomenon is called *self-organization*. It plays a key role in the current physical conceptions about the biologic functions of proteins, because the determinacy in the function of the protein is associated with the unique determinacy of its spatial structure.

Placing a protein molecule in "bad" external conditions, one can, of course, destroy the native structure; this phenomenon is called *denaturation*. It can be induced by heating, changing the pH value of the solution (i.e., charging the links of the protein), or adding to the solution molecules (a denaturant) reducing the hydrophobic effect. On denaturation, a protein loses its ability to function (e.g., an enzyme protein loses its catalytic activity). Denaturation is frequently reversible, however. Taking a denatured protein back to the appropriate medium, one can repeat the process of self-organization, which is called (in this case) a *renaturation*.

In a certain sense, the uniqueness of the spatial structure makes a protein globule and a crystal similar. Speaking about this analogy, however, we should stress only the fact of the unique determinacy of crystalline structure and never its spatial periodicity. The spatial structure of any protein globule is extremely irregular and nonuniform, because it consists of amino acid residues of different types. By this criterion, a protein globule can in some respect be compared to well-known disordered systems, for example, glasses.

Recall, however, that in physics, a glass is a substance frozen in a nonequilibrium state. Possessing a tremendously large relaxation time (greatly exceeding the time for any reasonable physical experiment or observation), the glass remembers its random fluctuation structure that exists at the moment of preparation. Melting and subsequent glass formation lead to a total change of its microscopic structure (i.e., to a memory loss). It is because of these non-equilibrium properties that the structure of the glass is disordered (i.e., can be formed in a multitude of different arrangements). From a thermodynamic viewpoint, the entropy of the glass continues to differ from zero down to zero-temperature.

We have already mentioned the analogy between the primary structure of biopolymers and the structure of a glass. This analogy is close indeed[p] (even though of little interest and not very productive), because the relaxation time is incomparably long with respect to the duration of primary structure rearrangements and the breakdown of the primary structure leads to a total loss of linear memory. As to the similarity between the tertiary structure and the structure of the glass, it is restricted to the fact of irregularity, because of the capability for self-organization signifies that the memory of the tertiary structure is not erased on denaturation.

At the same time, the property of self-organization does not mean that the native tertiary structure necessarily corresponds to total thermodynamic equilibrium with only the linear memory fixed. Indeed, a statistical system, all of whose N particles are different (even the identical amino acids almost always have different neighbors in the chain), has many structures ($\sim \exp N$), and their sorting takes a prohibitively long time. It should thus be assumed that the native state corresponds to such a minimum in the free energy that does not necessarily coincide with the absolute one but is thermodynamically stable enough (i.e., surrounded by sufficiently high energy barriers) and always easily accessible from a kinetic point of view.

44.2. *The fluctuating thermal motion of atoms in a native protein globule does not violate the uniqueness of tertiary structure; relaxation times of some perturbations in proteins are very long.*

Of course, the strict definiteness of the native structure does not mean that it has no fluctuations; on the contrary, fluctuation motions and motions associated with strain (e.g., caused by adsorption by a globule of some smaller molecule) are important for many biologic processes. For example, the catalytic center of some enzymes is located at the bottom of a sufficiently narrow "pocket," so the input of substrate and the yield of product proceed via diffusion (i.e., because of the mobility of various elements of the tertiary structure).

In comparison with ordinary condensed media (e.g., crystals, amorphous substances, and so on), however, protein globules are characterized by peculiar properties of the fluctuations and motions. For example, some defects involving atom transpositions are strictly prohibited because of topologic constraints, excluding mutual intersection of chains. There are some other types of perturbations that practically do not occur in proteins or occur much more seldom than might be expected to be found in equilibrium according to their activation energy. The complex nature of motions in a protein globule leads to immense relaxation times, up to hundreds of milliseconds observed in some globules and

---

[p]Here, we deal with the already synthesized chain. The formation of the primary structure is, of course, far from being accidental; it is determined by the biosynthesis.

quite uncommon for objects of such a moderate size.

It should be noted that fluctuations do not violate the unique determinacy of the tertiary structure. Even if the amplitudes of some motions are relatively large, they still do not change the essential characteristics of the tertiary structure associated, for example, with chain topology, relative arrangement of secondary structure blocks, and so on. In other words, the conformational entropy of a protein globule in the native state is very small.

44.3. *The compactness of a protein globule is maintained primarily by the hydrophobic effect; hydrophobic links are located mainly inside the globule and hydrophilic ones on the surface.*

As a rule, approximately half of all amino acids of a globular protein belong to the class of hydrophobic acids (*see* subsection 37.5). For the coil state of the chain, it is mostly the aversion of these acids to water that makes the globular state thermodynamically favorable.

This is why the simplest model of a protein chain uses the rough conception of two sorts of links; hydrophobic, and hydrophilic. Within the framework of this model, it can be easily realized that there should be a kind of interglobular separation inside the protein molecule. Imagining that the hydrophobic links are strongly prohibited from contacting water while the hydrophilic (on the contrary) from being isolated from it, it becomes immediately clear that the globule must consist of a solid hydrophobic nucleus and a hydrophilic shell enclosing it on all sides.

It is useful to note that the nature of the stability of the hydrophobic nucleus of the globule is similar to that of an oil drop suspended in water. In fact the division of the globule into a hydrophobic nucleus and a hydrophilic shell is far from being credible, but this is still quite reasonable as a first approximation. Moreover, it allows an initial primitive evaluation of the relationship between rough characteristics of the primary and the tertiary protein structures; we show this here. Suppose that the chain comprises N links, of which $(1-\vartheta)N$ are hydrophobic and ϑN hydrophilic. Next, let each link (irrespective of type) occupy the volume $(4/3)\pi r^3$, in the globule so that the total volume of the globule equals $V=(4/3)\pi r^3 N$. If the surface area of the globule is denoted by A, then the volume of the hydrophilic shell equals $2Ar$. The same value should be obtained on multiplication of the number of hydrophilic links by the volume of a single link. Thus, we immediately obtain

$$\vartheta = 2Ar/V. \qquad (44.1)$$

The fraction ϑ of hydrophilic links characterizes the primary structure, and the surface-to-volume ratio characterizes the shape of the globule (i.e., the tertiary structure). Equation (44.1) provides an evaluation of the relationship between the structures.

Suppose, for example, that the globule is spheric. Then, $A=4\pi R^2$, $V=(4/3)\pi R^3$, and Eq. (44.1) yields $\vartheta=6N^{-1/3}$. If in a real chain the value of ϑ somewhat exceeds $6N^{-1/3}$, then the globule is oblate, its surface area being larger than that of a sphere of the same volume. If the value of ϑ is slightly less than $6N^{-1/3}$, then the globule has a prolate shape (with a smaller surface area). If ϑ differs strongly from $6N^{-1/3}$, however, then the discussed structure becomes impossible: for $\vartheta \gg 6N^{-1/3}$, the globulization is quite unfavorable, because almost all of the chain is hydrophilic; for $\vartheta \ll 6N^{-1/3}$, the hydrophobic links would be found on the surface, and the chain would form a string of several small globules (a so-called quaternary or domain structure).

Note that in this subsection, we use the essential fact that N is not very large for protein molecules. Indeed, if, for example, $\vartheta = 0.6$ and $\vartheta = 6N^{-1/3}$, then $N = 1000$.

44.4. *The globule is a system of stiff blocks of secondary structure with their surfaces bristling with side groups of amino acids; van der Waals interactions between the side groups of neighboring blocks fix the details of the tertiary structure.*

Inside the hydrophobic nucleus of a protein molecule, there are virtually no water molecules. Therefore, water molecules cannot compete for hydrogen bonds, and it is very favorable thermodynamically for the chain sections located in the nucleus to form secondary structure. The energy of stabilization of the secondary structure is such that the corresponding sections of α helices and β sheets behave as essentially indestructible stiff blocks. The form of the secondary structure (α or β) emerging on the given chain section depends on both the primary structure of this section and the tertiary structure of the whole protein molecule. The point is that the elements of the secondary structure in globular proteins are small, actually comprising less than 20 links (a few turns of α helix or sheets of β structure). Physically, these lengths are determined not only by the primary structure (i.e., not only by the factors discussed in subsection 39.1) but also by the tertiary structure. Clearly, exceedingly long elements of the secondary structure will not fit inside the hydrophobic nucleus and therefore prove to be unfavorable.

Among amino acids, there are some (e.g., proline) in whose vicinity a polypeptide chain bends readily, because the volume of the side group is small. As a rule, the proline-containing sections are not built in the secondary structure and form bridges between the blocks, and the other sections form secondary structure of some kind. In specific calculations of the secondary structure of a real protein, the influence of the tertiary structure can roughly be taken into account by introducing an effective external field (well) of appropriate configuration acting on the links (cf. subsection 6.2). The solution algorithm for this problem is basically clear, even though the corresponding transfer matrices are complicated because of the presence of various types of secondary structure and inhomogeneous external fields (*see* subsection 41.5). Nevertheless, the multiplication of these matrices can be performed successfully using modern computers. At present, the problem of the calculation of secondary structures for real proteins is basically solved (O. B. Ptitsyn, and A. V. Finkelstein, 1980).

The number of blocks of secondary structure in a globule is small (usually approximately 10). It therefore is clear that many of them (and sometimes all) come out to the surface of the globule, at least with one of their sides. Correspondingly, the hydrophobic and hydrophilic links should alternate in the primary structure so that the elements of secondary structure that come out to the surface consist mainly of hydrophobic links while the elements inside the globule consist mainly of hydrophilic ones. Hence, the microscopic separation in a protein globule proceeds with the participation of the secondary structure (Fig. 7.13).

Small displacements of stiff blocks in the nucleus of the globule hardly have any effect on the hydrophobic interaction energy (i.e., on the gain in the free energy in comparison with the placing of all links in water) provided that no water gets into the nucleus (i.e., the slits on the surface are narrow enough [~0.3 nm] to stop penetration of water molecules).

As mentioned in subsection 39.1, the surfaces of the blocks of secondary structure are formed by side groups of amino acids. Consequently, for a real protein with complex primary structure, these surfaces are also complex and irregular. Therefore, the small displacements of the blocks described earlier, being non-essential for the hydrophobic effect, are quite significant for the determination of the energy of short-range van der Waals interactions between the side groups. Thus, it is clear that the interactions are capable of providing a fine adjustment of all of the details of the native tertiary structure.

It should also be pointed out that the nuclei of protein globules are similar to ordinary molecular crystals in their packing density and the absence of void space.

44.5. *Electrostatic interactions play an essential role only far from the isoelectric point of a given protein; charged links are located only on the surface of the globule.*

If the pH value of the solution is far from the isoelectric point (*see* subsection 37.4) of a protein, then many links carry charges of the same sign, and their repulsion can be expected to result in a weakening of globule stability or even in its transition to a coil. The electrostatic energy of the globule is easy to evaluate.

First, note that the charged links cannot be located inside the protein globule,

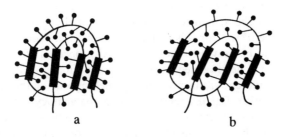

a   b

FIGURE 7.13. Sketch of a protein globule as a system of stiff blocks of secondary structure with attached side groups in native (a) and molten (b) states

because the dielectric constant of the nonpolar nucleus of the globule (equal to a few units) is much less than that of water ($\varepsilon \approx 80$) and the repulsion of the charges inside the globule would be too strong. Assuming the charges to be distributed uniformly over the surface, we obtain the required evaluation of the electrostatic energy of the globule:

$$E_{el} = (Q^2/\varepsilon R)[1 + (R/r_D)]^{-1},$$

where Q is the charge of the globule, R its radius, and r_D the Debye radius of the solvent (typically ~ 1 nm). The charge Q depends on the pH value of the medium, $Q(pH)$; the charge increase on withdrawal from the isoelectric point in any direction leads to a growth in E_{el} and, in the final analysis, to the denaturation of the globule.

44.6. *Together with coil and native globular states, the diagram of states of a protein molecule also includes a molten globular state.*

Experimental investigations have shown that the denaturation of globular proteins proceeds very sharply, and it is accompanied by a large entropy jump of approximately one entropic unit per amino acid residue (i.e., the heat of transition $\sim T$ per residue). We know that such characteristics are typical for a globule–coil transition (*see* Sec. 22) or a globule–globule transition associated with transformation of the internal microscopic structure of the nucleus (*see* Sec. 22). It is known from experimental data that both possibilities are realized.

Specifically, the thermal denaturation occurring as the temperature rises must proceed as a globule–globule transition, because the hydrophobic effect does not weaken on heating (*see* subsection 37.5) and the compactness of the globule thus is not violated. In fact, experiments show that the volume of the molecule increases only slightly in this process and that the secondary structure usually does not change strongly (i.e., the hydrophobic nucleus remains almost inaccessible to water). Such a denatured state of the globule is referred to as a *molten globule state*. In the molten state, a slight increase in the volume and distance between blocks of the secondary structure leads to a loss in the energy of van der Waals interactions of side groups of amino acids belonging to different blocks, but it also simultaneously provides an entropy gain. This is because both the rotation and vibration of the same side groups become less restrained. The competition of these two factors brings about the melting (i.e., the first-order phase transition in the structure of the nucleus of the protein globule). The typical dependence $\mu^*(n)$ for this situation is plotted in Figure 3.11a.

Attention should be drawn to the exceptional properties of this unified transition occurring in a heterogeneous system. For example, ordinary solid solutions melt quite differently, via preliminary melting out of the lowest-melting composition (eutectic). Equally relevant here is the analogy with the helix–coil transition (*see* Sec. 41), where heterogeneity and step-by-step melting broaden the transition region by one order of magnitude. Cooperativity of melting of the globule stems from the indestructibility of blocks of the secondary structure. A decrease in density is possible only because of disengagement of whole blocks, which is inevitably accompanied with melting out of all sorts of residues; for the entropy of motion of various blocks, it is negligibly small as their number is small (~ 10).

Certainly, a globule–coil transition can also occur in a protein molecule. For example, it is inevitable when the charge of the globule rises sufficiently because of withdrawal from the isoelectric point. The globular phase also must disintegrate, because the concentration of some denaturant (a substance weakening the hydrophobic effect) is increased in the solution. These considerations are summarized in Figure 7.14, where a typical diagram of states for a protein molecule is shown.

44.7. *Self-organization of the tertiary structure of globular protein proceeds in two stages: a rapid globule–coil transition followed by the slow formation of native structure in the globule.*

Moving to the dynamics of a self-organization process, it should immediately be noted that it is not completely understood at present. Nevertheless, some qualitative circumstances are assumed to be clear enough.

First, the reliability of self-organization is assured when the process, having started in a coil chain, proceeds simultaneously through all of its sections. In particular, this means that the secondary structure must form at the earliest phase of the process. Of course, it may happen that the initially formed secondary structure does not coincide with the native one; in this case, it must transform itself at a further phase of the process.

The next spontaneous step is the general contraction of the chain accompanying formation of a hydrophobic nucleus. Thus, the intermediate point on the way to self-organization is a globular state, which is not native. It is assumed that this is the state of a melted globule.

Finally, the concluding stage of self-organization is the fixing of all of the details of tertiary structure in the compact globular state. Apparently, this stage takes the most time. The theory of self-organization of a globular protein must provide the answers to the following questions: how do we describe in terms of statistical physics the dynamics of the globule–coil transition producing the roughly correct parameters of the tertiary structure (e.g., the topology of the chain), and how do we predict all of the details of the tertiary structure (i.e., the

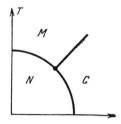

FIGURE 7.14. Diagram of states of protein: *N*—native globule, *M*—molten globule, *C*—coil state. Concentration of denaturant in solution is laid off on the abscissa.

coordinates of nearly all atoms) in a chain with a given primary structure and the already-emerged rough structure of the globule? The development of such a theory is a task for the future.

References

1. P. Flory, *Principles of Polymer Chemistry* (Cornell University Press, Ithaca, NY, 1953).
2. M. V. Volkenstein, *Configurational Statistics of Polymeric Chains* (Intescience, New York, 1963).
3. T. M. Birshtein and O. B. Ptitsyn, *Conformations of Macromolecules* (Interscience, New York, 1966).
4. P. J. Flory, *Statistical Mechanics of Chain Molecules* (InterScience, New York, 1969).
5. H. Yamakawa, *Modern Theory of Polymer Solutions* (Harper and Row, New York, 1971).
6. M. V. Volkenstein, *Molecular Biophysics* (Academic Press, New York, 1977).
7. M. V. Volkenstein, *Biophysics* (Nauka Publishers, Moscow, 1988). (in Russian).
8. P. G. de Gennes, *Scaling Concepts in Polymer Physics* (Cornell University Press, Ithaca, NY, 1979).
9. C. R. Cantor and T. R. Schimmel, *Biophysical Chemistry* (Freeman and Co., San Francisco, 1980).
10. Yu. Ya. Gotlib, A. A. Darinsky; and Yu. E. Svetlov, *Physical Kinetics of Macromolecules* (Khimiya Publishers, Leningrad, 1986) [in Russian].
11. M. Doi and S. F. Edwards, *Theory of Polymer Dynamics* (Academic Press, New York, 1986).
12. V. G. Dashevsky, *Conformational Analysis of Macromolecules*, (Nauka Publishers, Moscow, 1987) [in Russian].
13. V. G. Rostiashvili, V. I. Irzhak, and B. A. Rosenberg, *Glass Transition in Polymers* (Khimiya Publishers, Leningrad, 1987) [in Russian].
14. J. des Cloiseaux and G. Jannink, *Polymeres en solution* (Les Editions de Physique, Paris, 1987) in French.
15. A. V. Vologodskii, *Topology and Physics of Circular DNA* (CRC Press, ???, FL, 1992).
16. J. C. Kendrew, *A Thread of Life* (G. Bell and Sons, London, 1965).
17. L. R. G. Treloar, *Introduction to Polymer Science* (Wykeham Publications, London, 1970).
18. M. D. Frank-Kamenetskii, *Unraveling DNA* (VCH Publishers, NY, 1993).
19. A. Yu. Grosberg and A. R. Khokhlov, *Physics of Chain Molecules* (Znaniye, Moscow, 1984) [in Russian].
20. A. Yu. Grosberg and A. R. Khokhlov, *Physics in the World of Polymers* (Nauka Publishers, Moscow, 1989) [in Russian].
21. V. E. Eskin, *Light Scattering by Polymer Solutions* (Nauka Publishers, Moscow, 1973) [in Russian].
22. R. P. Feynman and A. R. Hibbs, *Quantum Mechanics and Path Integral* (McGraw-Hill, New York, 1965).
23. R. H. Crowell and R. H. Fox, *Introduction to Knot Theory* (Ginn and Company, Boston-New York, 1963).
24. V. G. Boltyansky and V. A. Yefremovich, *Graphic Topology* (Nauka Publishers, Moscow, 1982) [in Russian].
25. M. D. Frank-Kamenetskii and A. V. Vologodskii, Topological Aspects of the Physics of Polymers: the Theory and its Biophysical Applications. Usp. Fiz. Nauk **134**, 641 (1981) [Sov. Phys. Usp. **24**, 679, (1981)]
26. L. D. Landau and E. M. Lifshitz, *Statistical Physics* (Pergamon Press, Oxford, 1988).
27. R. D. Mattuck, *A Guide to Feynman Diagrams in the Many-Body Problem* (McGraw-Hill, New York, 1967).
28. Shang-Keng Ma, *Modern Theory of Critical Phenomena* (W. A. Benjamin, London, 1976).
29. I. M. Sokolov, Dimensionalities and Other Geometric Critical Exponents in Percolation Theory, Usp. Fiz. Nauk **150**, 221 (1986) [Sov. Phys. Usp. **29**, 924 (1986)].
30. I. M. Lifshitz, Some Problems of the Statistical Theory of Biopolymers, Zh. Eksp. Teor. Fiz.

Nauk **55**, 2408 (1968) [Sov. Phys. JETP **28**, 1280 (1969)].

31. P. G. de Gennes, *The Physics of Liquid Crystals* (Clarendon Press, Oxford, 1974).
32. S. Chandrasekhar, *Liquid Crystals* (Cambridge University Press, Cambridge, 1977).
33. A. S. Sonin, *Introduction into Physics of Liquid Crystals* (Nauka Publishers, Moscow, 1983) [in Russian].
34. A. N. Semenov and A. R. Khokhlov, Statistical Physics of Liquid–Crystalline Polymers, Usp. Fiz. Nauk **156**, 427 (1988) [Sov. Phys. Usp. **31**, 988 (1988)].
35. J. R. Mayer and M. Goeppert Mayer, *Statistical Mechanics* (John Wiley and Sons, New York, 1977).
36. L. R. G. Treloar, *The Physics of Rubber Elasticity*, 3rd ed. (Clarendon, Oxford, 1975).
37. A. L. Efros, *Physics and Geometry of Disorder* (Nauka Publishers, Moscow, 1982) [in Russian].
38. Yu. L. Klimontovich, *Statistical Physics* (Nauka Publishers, Moscow, 1982) [in Russian].
39. L. D. Landau and E. M. Lifshitz, *Fluid Mechanics* (Pergamon Press, Oxford, 1986).
40. L. D. Landau and E. M. Lifshitz, *Electrodynamics of Continuous Media* (Pergamon Press, Oxford, 1960).
41. A. N. Tikhonov and A. A. Samarsky, *Equations of Mathematical Physics* (Nauka Publishers, Moscow, 1972) [in Russian].
42. Yu. M. Romanovsky, N. V. Stepanova, and D. S. Chernavsky, *Mathematical Biophysics* (Nauka Publishers, Moscow, 1984) [in Russian].
43. A. L. Leninger, *Biochemistry* (Worth Publishers, New York, 1972).
44. I. M. Lifshitz, S. A. Gredeskul, and L. A. Pastur, *Introduction to the Theory of Disordered Systems* (John Wiley and Sons, New York, 1988).
45. J. M. Ziman, *Models of Disorder* (Cambridge University Press, Cambridge, 1979).
46. G. E. Shulz and R. H. Schirmer, *Principles of Protein Structure*, (Springer-Verlag, Berlin, 1979).
47. T. E. Creighton, *Proteins: Structure and Molecular Properties* (W. H. Freeman Co., San Francisco, 1983).
48. A. Z. Patashinsky and V. L. Pokrovsky, *Fluctuation Theory of Phase Transitions* (Pergamon Press, New York, 1979).
49. F. Osawa, *Polyelectrolytes* (Marcel-Dekker, New York, 1971).
50. M. Eigen and P. Schuster, *The Hypercycle* (Springer-Verlag, New York, 1979).
51. A. A. Vedenov, A. M. Dykhne, and M. D. Frank-Kamenetskii, The Helix–Coil Transition in DNA, Usp. Fiz. Nauk **105**, 479 (1971) [Sov. Phys. Usp. **14**, 715 (1972)].
52. C. Tanford, *Physical Chemistry of Polymers* (John Wiley and Sons, New York, 1965).

Subject Index